MEMS/NEMS Devices and Applications, 3rd Edition

MEMS/NEMS Devices and Applications, 3rd Edition

Guest Editors

Ching-Liang Dai
Zhi-Xuan Dai

Basel • Beijing • Wuhan • Barcelona • Belgrade • Novi Sad • Cluj • Manchester

Guest Editors
Ching-Liang Dai
Department of Mechanical Engineering
National Chung Hsing University
Taichung
Taiwan

Zhi-Xuan Dai
Department of Bio-Industrial Mechatronics Engineering
National Chung Hsing University
Taichung
Taiwan

Editorial Office
MDPI AG
Grosspeteranlage 5
4052 Basel, Switzerland

This is a reprint of the Special Issue, published open access by the journal *Micromachines* (ISSN 2072-666X), freely accessible at: https://www.mdpi.com/journal/micromachines/special_issues/E6H71230WR.

For citation purposes, cite each article independently as indicated on the article page online and as indicated below:

Lastname, A.A.; Lastname, B.B. Article Title. *Journal Name* **Year**, *Volume Number*, Page Range.

ISBN 978-3-7258-8081-2 (Hbk)
ISBN 978-3-7258-8082-9 (PDF)
https://doi.org/10.3390/books978-3-7258-8082-9

Contents

About the Editors

Ching-Liang Dai

Ching-Liang Dai is a Lifetime Distinguished Professor in the Department of Mechanical Engineering at National Chung Hsing University, Taiwan. His research interests include MEMS, CMOS-MEMS, microsensors, microactuators, micro energy harvesting devices, and nanotechnology. His research focuses on CMOS-compatible MEMS fabrication technologies and integrated microsystems for sensing and energy harvesting applications. He was elevated to IEEE Senior Member and elected as a Fellow of the Institution of Engineering and Technology (FIET). He has also been listed among the World's Top 2% Scientists.

Zhi-Xuan Dai

Zhi-Xuan Dai is an Assistant Professor in the Department of Bio-Industrial Mechatronics Engineering at National Chung Hsing University, Taiwan. His research interests include MEMS, mechatronics integration, mechanical design, embedded systems, Internet of Things (IoT), control engineering, and biomechanics. His research focuses on the development and integration of MEMS devices, intelligent mechatronic systems, embedded control technologies, and IoT-based applications. He has participated in interdisciplinary research projects related to smart systems, microsystems integration, and advanced engineering technologies. His recent studies emphasize the integration of MEMS technologies, embedded systems, and intelligent control methods for industrial, biomedical, and smart sensing applications.

Preface

Recently, nanoelectromechanical system (NEMS) and microelectromechanical system (MEMS) technologies have been employed to develop various microdevices and microstructures. Many sensors and actuators have been manufactured and commercialized using technologies such as pressure sensors, accelerometers, gyroscopes, tactile sensors, thermal sensors, flow sensors, optical sensors, image sensors, microphones, magnetic sensors, chemical sensors, gas sensors, biosensors, microchannels, ink jet heads, optical switches, RF switches, micromirror, motors, relays, resonators, filters, and energy harvesters. NEMS/MEMS devices have been widely applied in various fields.

Ching-Liang Dai and Zhi-Xuan Dai
Guest Editors

micromachines

Editorial

Editorial for the Special Issue on MEMS/NEMS Devices and Applications, 3rd Edition

Zhi-Xuan Dai [1], Takahito Ono [2] and Ching-Liang Dai [3,*]

1 Department of Bio-Industrial Mechatronics Engineering, National Chung Hsing University, Taichung 402, Taiwan
2 Department of Mechanical Systems Engineering, Tohoku University, Sendai 980-8579, Japan
3 Department of Mechanical Engineering, National Chung Hsing University, Taichung 402, Taiwan
* Correspondence: cldai@dragon.nchu.edu.tw; Tel.: +886-4-2284-0433

1. Introduction

Microelectromechanical systems (MEMS) and nanoelectromechanical systems (NEMS) have experienced rapid and sustained development in recent years and have become key enabling technologies for intelligent sensing [1–4], energy harvesting [5–10], biomedical devices [11], and Internet of Things (IoT) [12] applications. A wide variety of MEMS/NEMS devices have already been developed and successfully deployed in practical systems. For example, capacitive accelerometers are extensively used for motion sensing in smartphones, wearable electronics, automotive safety systems, and structural or vibration monitoring, as their proof mass–spring architectures convert acceleration into measurable capacitance variations with excellent scalability [13]. Vibrating gyroscopes complement accelerometers in inertial measurement units and play a critical role in camera stabilization, navigation, robotics, and vehicle stability control by providing accurate angular-rate information in a compact, low-power form [14]. Pressure sensors, including capacitive, piezoresistive, and emerging resonant types, are widely applied in tire-pressure monitoring, medical catheters, respiratory equipment, and industrial process control, where advances in materials and packaging continue to enhance sensitivity and long-term reliability [15]. MEMS microphones are now the standard audio interface in smartphones, smart speakers, and wearable devices, supporting voice recognition, noise cancellation, and beamforming with high sensitivity and a small form factor [16]. Infrared sensors are extensively applied in thermal imaging for night vision, firefighting, industrial inspection, and human presence detection by converting infrared radiation into electrical signals [17]. For reconfigurable radio-frequency systems, RF MEMS switches offer low insertion loss, high isolation, and near-zero static power consumption, supporting tunable filters, phase shifters, and reconfigurable antennas, particularly in satellite and space-constrained communication platforms [18]. In three-dimensional perception, scanning micromirrors enable compact LiDAR beam steering for autonomous vehicles, drones, and robotic systems, offering reduced size, high scanning speed, and cost advantages over conventional mechanical scanners [19]. In minimally invasive healthcare, microneedle arrays create micron-scale pathways through the stratum corneum, enabling painless transdermal drug and vaccine delivery as well as biofluid sampling for point-of-care diagnostics and theranostic patches [20]. Gas sensors, especially metal–oxide-based sensors integrated with microheaters, are used for environmental monitoring, industrial safety, and indoor air-quality control, enabling rapid and selective detection of hazardous gases at low power consumption [21]. Surface acoustic wave (SAW) sensors serve as highly sensitive platforms for chemical and biosensing, where functionalized piezoelectric substrates and interdigitated transducers convert mass loading

 https://doi.org/10.3390/mi17020205

or viscoelastic changes into measurable frequency or phase shifts [22]. In lab-on-a-chip automation, microfluidic membrane valves enable reliable on-chip flow routing, metering, and isolation, facilitating portable analytical systems and integrated sample preparation without bulky external equipment [23]. Finally, MEMS microgrippers support precise micromanipulation in microassembly and biomedical handling of cells, microtissues, and delicate micro-objects, where compliant mechanisms and compact actuation allow accurate pick-and-place operations and gentle gripping at the microscale [24].

In the previous 2nd Edition Special Issue, a broad range of NEMS/MEMS devices was reported [25], covering diverse functional categories such as actuators [26–29], antennas [30], sensors [31–35], and energy harvesters [36], reflecting the rapid expansion of micro- and nano-scale system technologies. Building upon this foundation, the present Special Issue further consolidates recent advances in MEMS/NEMS device design, fabrication innovation, sensors, actuators, and microscale components, highlighting both fundamental developments and application-oriented progress. As illustrated in Figure 1, the contributions in this Special Issue can be systematically classified according to device functionality. In the actuator category, reported works include a light-controlled ion drag pump [Contribution 1], electrothermal actuators [Contribution 2], a high-frequency resonator [Contribution 3], and electrothermal co-actuation devices [Contribution 4], demonstrating diverse actuation mechanisms and structural strategies. The sensor-related studies encompass a capacitive micromachined ultrasonic transducer (CMUT) [Contribution 5], an impedance microsensor [Contribution 6], and a DC electric field meter [Contribution 7], addressing challenges in precision sensing, robustness, and environmental adaptability. In the area of energy harvesting, this Special Issue features a CMOS-compatible photovoltaic generator [Contribution 8] and a rotary piezoelectric–electromagnetic hybrid energy harvester [Contribution 9], emphasizing self-powered operation and efficient energy conversion. Additionally, several key microcomponents are presented, including stacked metal–insulator–metal capacitors [Contribution 10], an electro-optic liquid crystal device [Contribution 11], a P-channel vertical double-diffused metal–oxide–semiconductor (VDMOS) transistor [Contribution 12], and a tunnel field-effect transistor [Contribution 13], which collectively underscore ongoing progress in microelectronic technology.

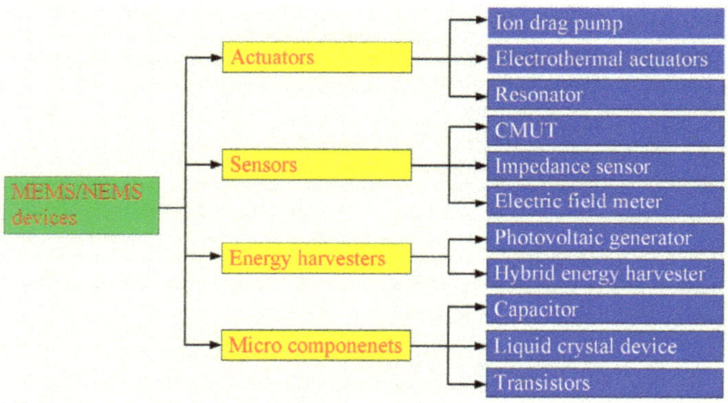

Figure 1. Classification of MEMS/NEMS devices presented in this Special Issue.

2. MEMS/NEMS Actuators in Special Issue

Tang et al. [Contribution 1] report an experimental investigation of a light-controlled ion drag pump that eliminates conventional wired high-voltage supplies by exploiting the photoelectric properties of lanthanum-modified lead zirconate titanate (PLZT) ceramics. The central research theme is to demonstrate how optically generated high photovoltage

 https://doi.org/10.3390/mi17020205

can be directly coupled with electrohydrodynamic (EHD) charge injection to realize wireless, non-contact fluid transport. The main innovation lies in integrating the anomalous photovoltaic effect of PLZT with an ion drag pumping mechanism, enabling kilovolt-level driving voltages under ultraviolet illumination without external power conditioning, which fundamentally differs from traditional EHD pumps relying on bulky, noise-prone high-voltage sources. Methodologically, the study combines theoretical modeling, numerical analysis, and systematic experiments to clarify the relationships among photovoltage generation, electric field formation, charge injection, and fluid motion. The device consists of a PLZT ceramic patch serving as a light-activated voltage source and a multilayer pump chip composed of a dielectric substrate, patterned serrated electrodes, a microfluidic channel, and a cover layer. Under UV irradiation, the PLZT generates a rapidly rising photovoltage that drives field emission at the emitter electrodes, forming ions within a dielectric fluid; the resulting Coulomb force transports the fluid toward the collector electrodes. The fabrication approach emphasizes relatively simple planar electrode patterning and dielectric channel assembly, ensuring electrical insulation and preventing breakdown while maintaining compatibility with microfabrication techniques. Experimentally, key structural parameters, which include channel height, electrode spacing, number of electrode pairs, and inter-pair gaps, are varied to quantify their effects on response time, flow rate, and output pressure. The results show that higher light intensity accelerates activation and enhances performance, while reduced electrode spacing and increased electrode pair numbers significantly improve pressure and flow. An optimal electrode gap is identified that balances opposing electric fields and hydrodynamic losses. Importantly, the pump achieves flow rates on the order of hundreds of microliters per minute and stall pressures approaching 1 kPa under purely optical control. Overall, the work establishes a new light-driven pumping paradigm with low electromagnetic interference and strong spatial controllability, offering valuable design guidelines for future applications in microfluidics, soft robotics, and optically addressable microsystems.

Tang et al. [Contribution 2] explore the comprehensive electrothermal–thermomechanical analysis of a V-shaped micro electrothermal actuator, aiming to establish an accurate and efficient numerical framework for predicting its driving behavior under electrical excitation. The primary innovation of the work lies in the introduction of the polynomial point interpolation collocation method (PPCM) as a strong-form, meshfree computational approach to solve the highly coupled, nonlinear multiphysics problem inherent in electrothermal actuation. Unlike conventional finite-element methods that rely on meshing and may suffer from remeshing or numerical instability, the proposed PPCM directly constructs local polynomial-based interpolation functions at discrete nodes, enabling stable and efficient analysis for microscale structures. The research methodology integrates electrothermal modeling with thermomechanical deformation analysis in a sequential yet fully coupled manner. First, the temperature field generated by Joule heating is obtained by solving nonlinear heat transfer equations that account for temperature-dependent electrical resistivity and thermal conductivity. Subsequently, the resulting temperature distribution is incorporated into a thermoelastic model to predict actuator displacement. The actuator is fabricated from polysilicon and adopts a planar V-shaped beam configuration, a geometry well known for amplifying lateral displacement through asymmetric thermal expansion. The structural dimensions, material properties, and boundary conditions are defined to reflect realistic MEMS fabrication constraints, while the numerical formulation avoids excessive geometric simplifications. An incremental load strategy is embedded into the PPCM to enhance convergence when handling strong material nonlinearities. To validate the proposed approach, numerical predictions of temperature and displacement are systematically compared with both finite-element simulations and experimental measurements

 https://doi.org/10.3390/mi17020205

obtained from infrared thermal microscopy and prior mechanical characterization. The results demonstrate that the PPCM achieves temperature predictions within approximately 10% of experimental values and displacement errors as low as 2 μm at higher driving voltages, closely matching FEM results. Overall, this study confirms that the PPCM provides a robust, accurate, and computationally efficient tool for analyzing electrothermal microactuators, offering significant potential for future MEMS design optimization and multiphysics simulation without the limitations of traditional mesh-based methods.

Lin et al. [Contribution 3] investigate a high-frequency film bulk acoustic resonator (FBAR) operating near 6.5 GHz, with the primary objective of mitigating lateral acoustic wave leakage that severely limits the quality factor of conventional high-frequency resonators. The core innovation is the introduction of a Mo/SiC composite microstructure strategically embedded along the boundary of the active resonant region to create multiple acoustic impedance discontinuities, thereby reflecting laterally propagating acoustic waves back into the resonator. Unlike prior approaches that rely mainly on longitudinal Bragg reflectors or electrode-based frames, this work exploits the strong acoustic impedance contrast between molybdenum and silicon carbide to enhance in-plane energy confinement without compromising structural integrity. The research methodology integrates acoustic impedance theory, finite-element multiphysics simulations, and experimental validation to systematically optimize microstructure thickness and width while suppressing spurious resonance modes. Structurally, the device adopts a sandwich configuration consisting of Mo top and bottom electrodes, a $Sc_{0.2}Al_{0.8}N$ piezoelectric film, and an air cavity, fabricated on a SiCOI substrate to ensure high crystalline quality of the piezoelectric layer. The Mo/SiC composite microstructures are laterally distributed near the resonator edge, with carefully optimized sectional widths to balance acoustic reflection and mode purity. Fabrication is realized through an advanced thin-film bonding and transfer process, combined with sputtering, dry etching, wafer bonding, chemical–mechanical polishing, and XeF_2 cavity release, enabling precise control of multilayer thickness and microstructure geometry. Experimental radio-frequency (RF) measurements confirm that the optimized resonator achieves a series resonance frequency of 6.488 GHz with a parallel-mode quality factor of 310, representing a 51.2% improvement over the baseline FBAR while maintaining acceptable electromechanical coupling. Statistical measurements across multiple devices further demonstrate good repeatability and stability. Overall, this work provides both a practical fabrication route and a robust structural design strategy for suppressing lateral acoustic losses, offering important guidance for the development of next-generation low-loss RF resonators for advanced wireless communication systems.

Tang et al. [Contribution 4] investigate a multi-electrothermal co-actuation device for safety-and-arming mechanisms in loitering munition fuzes, targeting reliable large displacement under extreme overload. The main contribution is a co-actuation architecture that integrates multiple V-shaped electrothermal actuators, where primary units generate displacement and auxiliary units provide position holding and bidirectional control for improved robustness. An electrothermal–mechanical coupled model is established and validated through finite-element simulations and extensive experiments. The actuator beams use stainless steel 304 to enhance strength and overload tolerance, and the structure is fabricated by laser processing to precisely pattern thick metal components. Key parameters, including V-beam number, air-gap thickness, contact type, load, driving voltage waveform, heat transfer, and gas damping, are systematically evaluated. Results show that optimized point–contact interfaces and multi-beam designs increase displacement efficiency, while appropriate thermal management reduces performance degradation. High-overload impact tests further demonstrate stable operation beyond 12,000 g without fracture, with only minor recoverable deformation. Overall, the proposed strategy delivers both enhanced

https://doi.org/10.3390/mi17020205

electrothermal actuation and exceptional mechanical resilience for high-reliability micro-actuation applications.

3. MEMS/NEMS Sensors in Special Issue

He et al. [Contribution 5] report the development of a CMUT-based linear array system for non-contact thickness measurement of marine engineering structures under varying environmental conditions. The work overcomes the limitations of conventional piezoelectric ultrasonic transducers in underwater environments by exploiting MEMS-based CMUT technology, which provides wide bandwidth, low power consumption, and improved acoustic impedance matching with water. The key contribution is the system-level integration of a high-density 16-element CMUT linear array with polyurethane encapsulation and customized transmit–receive electronics, enabling stable and accurate thickness measurements without physical contact. The study combines analytical modeling of CMUT electromechanical behavior, membrane-based structural design, MEMS fabrication, and comprehensive experimental validation. Each CMUT element comprises numerous micro-capacitive cells formed by a single-crystal silicon membrane, a vacuum cavity, and insulated electrodes, allowing efficient bidirectional acoustic–electrical transduction. The array is fabricated using silicon–silicon wafer bonding, followed by cavity etching, wafer-level bonding, substrate thinning, membrane release, dielectric deposition, and metal sputtering to ensure uniform membranes and high yield. Polyurethane encapsulation is applied to provide electrical insulation and environmental protection while minimizing acoustic reflection losses due to its water-like impedance. The complete system integrates high-voltage pulse excitation and low-noise signal reception, with thickness determined using a time-of-flight algorithm enhanced by signal processing techniques. Experimental results show a transmitting sensitivity of 146.82 dB and a receiving sensitivity of −229.55 dB at 1 MHz. Importantly, tests across different water temperatures and salinity levels demonstrate thickness errors within ±0.1 mm, confirming the robustness and reliability of the proposed system for marine applications.

Shi et al. [Contribution 6] propose a novel debris material identification approach based on impedance microsensing, targeting the long-standing challenge of accurately distinguishing metal wear particles in lubricating oil for machinery condition monitoring. Instead of relying solely on optical appearance or inductive amplitude, the study introduces an impedance-based strategy that jointly exploits variations in inductance and alternating-current resistance induced by metal debris passing through a micro-coil sensor. The key innovation is recognizing that different metals exhibit unique combinations of magnetic permeability and electrical conductivity, which lead to distinct, nonlinear inductance–resistance amplitude trends as particle size increases. To validate this concept, the authors combine electromagnetic theory, finite-element simulations, and controlled experiments to establish a database of characteristic amplitude–size curves for multiple metals. The sensing device consists of a coil-based impedance microsensor integrated with a microfluidic channel of 900 μm diameter, where a solenoid coil with precisely defined geometry surrounds the flow path to ensure stable and repeatable electromagnetic excitation. The sensor is fabricated using a casting-based process, including coil winding, mold formation, substrate curing, and channel construction, enabling a compact and robust structure suitable for inline oil monitoring. During operation, metal particles driven by a syringe pump traverse the sensing zone under a high-frequency excitation, and the resulting impedance variations are captured using precision instrumentation and signal processing techniques such as filtering and baseline correction. Experimental results demonstrate that ferromagnetic particles generate positive inductance pulses, while non-ferromagnetic particles produce negative inductance responses, whereas all metals increase AC resis-

 https://doi.org/10.3390/mi17020205

tance to different extents. By jointly analyzing inductance and resistance signatures and comparing them with the pre-established database, the system successfully distinguishes iron, stainless steel, aluminum, copper, and brass particles, even when colors are similar or surfaces are contaminated. Importantly, the method is insensitive to oil transparency, impurities, and surface dirt, overcoming key limitations of optical techniques. Overall, the study delivers a robust, physics-driven solution that enriches debris information beyond size and count, offering improved reliability for early fault diagnosis and advancing the practical capability of intelligent oil condition monitoring systems.

Yang et al. [Contribution 7] report the design, fabrication, and validation of a high-sensitivity three-dimensional DC electric field meter (3D EFM) based on MEMS technology, aiming to overcome the limited sensitivity, humidity susceptibility, and structural instability of previously reported MEMS EFMs. The core research focus is to realize accurate vector electric field measurement under real-world environmental conditions while maintaining compactness and low power consumption. The main innovation lies in a highly sensitive packaging strategy employing an inward-convex metallic cover, which significantly amplifies the electric field intensity within the sensor cavity compared with conventional flat-cover packages, as confirmed through analytical modeling and finite-element simulations. In addition, the study introduces a mechanically robust 1D electric field sensing chip (EFSC) architecture that replaces long flexible beams with short strip-type electrodes anchored on rigid frames, thereby improving vibration stability during resonant operation. The research methodology integrates theoretical electrostatic analysis, COMSOL-based multiphysics simulations, MEMS fabrication, package-level optimization, and full-system experimental calibration. Each EFSC is fabricated using a silicon-on-insulator process and operates in a lateral resonant vibration mode at approximately 3 kHz, driven by low-voltage electrostatic actuation. Three identically packaged EFSCs are incorporated into a cubic 3D EFM structure, where orthogonally arranged detecting electrodes couple the external electric field into the sensor packages. A moistureproof system-level enclosure, consisting of grounded metal chambers, Teflon insulators, tinplate shielding, and semi-rigid coaxial cables, is implemented to suppress humidity-induced polarization and ground potential interference. Wireless data transmission is further adopted to isolate the sensing unit from installation-dependent grounding effects. Experimental results demonstrate that the inward-convex package improves single-axis sensitivity from 1.95 to 7.19 mV/(kV/m), while the complete 3D EFM achieves a sensitivity at least 4.64 times higher than previously reported MEMS-based 3D EFMs. The maximum relative error under arbitrary orientation is limited to 2.2%, linearity remains within 1% even at high humidity, and the resolution of each axis is better than 10 V/m. These results verify that the proposed device offers a robust and practical solution for high-precision DC electric field vector measurement in demanding environments.

4. MEMS/NEMS Energy Harvesters in Special Issue

Chen et al. [Contribution 8] present the design and experimental verification of a CMOS-compatible micro photovoltaic generator (MPG) for compact, self-powered microsystems and IoT applications. The study tackles the limited efficiency of silicon-based MPGs fabricated using standard CMOS processes, where optical shading, shallow junction depths, and dielectric passivation layers reduce photon absorption and carrier collection. The main contribution is a mesh-patterned p–n junction design combined with a simple post-CMOS oxide removal process, which enhances photocurrent generation without modifying the foundry CMOS flow. Compared with conventional interdigitated layouts, the grid-like p-type regions embedded in a lightly doped n-well increase junction perimeter density and shorten lateral carrier transport paths, thereby reducing recombination losses.

https://doi.org/10.3390/mi17020205

TCAD simulations are employed to optimize device geometry, followed by fabrication in a commercial 0.18 μm CMOS process and systematic electrical characterization under controlled illumination. The MPG is realized entirely using standard CMOS materials, including a p-type substrate, n-wells, doped junctions, and aluminum interconnects, ensuring full process compatibility and scalability. After fabrication, a wet etching step removes the oxide layer above the active regions, allowing more efficient light penetration into the depletion zones. The device is wire-bonded and mounted on a printed circuit board for testing. Experimental results indicate that oxide removal increases the short-circuit current by about 20.1%. Under 1000 W/m^2 illumination, the MPG achieves 0.49 V open-circuit voltage, 239 μA short-circuit current, and 90 μW maximum output power, corresponding to 12.9% efficiency. These results demonstrate a practical approach to high-performance CMOS-integrated photovoltaic power sources for autonomous microsystems.

Yao et al. [Contribution 9] present the design and experimental demonstration of a rotary piezoelectric–electromagnetic hybrid energy harvester aimed at efficiently capturing low-frequency rotational mechanical energy commonly found in natural and industrial environments. The central research theme is to overcome the low power density and narrow operating bandwidth of single-mechanism harvesters by integrating complementary transduction principles within a compact mechanical architecture. The key innovation is the incorporation of a planetary gear transmission system with a 2:1 ratio, which effectively amplifies the external excitation frequency before it is delivered to the energy harvesting modules, thereby enhancing electrical output under low-speed rotation. Methodologically, the study combines theoretical modeling, multiphysics simulation, prototype fabrication, and systematic experimental testing. Analytical models are established to describe the kinematics of the slider–crank mechanism, magnetic coupling forces, piezoelectric response, and electromagnetic induction, while COMSOL and ANSYS simulations are used to predict magnetic fields, cantilever deformation, voltage generation, and power output. Structurally, the device integrates four piezoelectric energy harvesters (PEHs) and four electromagnetic energy harvesters (EMHs) arranged symmetrically around the gear system. The PEHs consist of PZT-5H ceramic layers bonded to stainless-steel cantilever beams, each carrying a rectangular NdFeB magnet, whereas the EMHs employ copper coils and cylindrical NdFeB magnets undergoing linear reciprocating motion. Magnetic coupling between moving magnets excites both piezoelectric bending and electromagnetic flux variation without direct mechanical contact, improving durability and reliability. The prototype is fabricated using a combination of 3D printing for the mechanical frame, precision assembly of gears and bearings, epoxy bonding of piezoelectric elements, and coil winding, enabling rapid realization of a robust mesoscale device. Experimental results show that, under an excitation frequency of 7 Hz with eight magnets, the hybrid harvester delivers a combined output voltage of 10.4 V and an average power of 250 mW, with an overall energy conversion efficiency of approximately 67%. Practical demonstrations further confirm its capability to light 60 LEDs and continuously power a temperature–humidity sensor. These results highlight the effectiveness of frequency amplification and hybrid transduction in harvesting low-frequency rotational energy, offering a promising solution for self-powered sensing and distributed IoT applications.

5. MEMS/NEMS Micro Components in Special Issue

Choi et al. [Contribution 10] examine the radio-frequency characteristics of stacked metal–insulator–metal (MIM) capacitors fabricated using RF CMOS technology, aiming to establish an accurate modeling approach for multi-GHz circuit applications. The study focuses on understanding how vertically stacked MIM structures, commonly used to enhance capacitance density, introduce additional parasitic effects that are inadequately

https://doi.org/10.3390/mi17020205

described by conventional single-layer capacitor models. The key contribution is the development of an improved equivalent RF circuit that separately accounts for electrode-dependent parasitic resistance and inductance in each stacked layer, as well as the extra series impedance from the bottom electrode. To extract intrinsic device parameters, the authors compare one-step and two-step de-embedding methods and demonstrate that combined open–short de-embedding is necessary for reliable RF characterization. The methodology integrates wideband S-parameter measurements from 500 MHz to 20 GHz with systematic de-embedding and circuit modeling validated through Y-parameter analysis. The stacked MIM capacitors are fabricated in a standard 0.18 μm CMOS process using two vertically connected TiN/SiNx/TiN capacitor units to double capacitance per unit area. Multiple capacitor sizes and dedicated open and short test structures are implemented to evaluate scaling and parasitic effects. Experimental results show that single-step de-embedding causes under- or overestimation of capacitance and resistance, whereas the proposed two-step approach yields consistent and physically meaningful values. The resulting modified model accurately reproduces capacitance, quality factor, and S- and Y-parameters up to 20 GHz, significantly outperforming conventional models. These findings provide practical guidance for designing compact, high-density capacitors in high-frequency RF CMOS-integrated circuits.

Basu et al. [Contribution 11] investigate a material-based approach to accelerate electro-optic switching in nematic liquid crystal (LC) devices by suppressing mobile ionic impurities using helical carbon nanotubes (hCNTs). The study addresses a long-standing issue in LC technology, where residual ions from materials and interfaces degrade dynamic performance by increasing rotational viscosity and causing flicker and image sticking. The main contribution is the introduction of an ultralow concentration of non-functionalized hCNTs as physical ion traps, exploiting their spring-like geometry to capture both positive and negative ions without altering LC chemistry or requiring complex purification. The research combines material preparation, transient ion current analysis, dielectric spectroscopy, rotational viscosity extraction, and optical switching measurements. Nematic LC E7 serves as the host medium, while hCNTs are uniformly dispersed using ethanol-assisted sonication. The devices employ a planar LC cell structure with indium tin oxide electrodes and rubbed polyimide alignment layers, enabling reproducible electro-optic characterization using standard LC assembly processes. Experimental results show that hCNT doping at 3.1×10^{-3} wt.% reduces free ion concentration by 60–70% across a wide temperature range, leading to a 17% reduction in rotational viscosity and enhanced dielectric anisotropy. Consequently, optical relaxation time is shortened by approximately 25% while maintaining stable textures. The study also identifies an optimal concentration window, as excessive hCNT loading causes aggregation and electrical shorting. Overall, the work demonstrates a simple and effective strategy for improving LC switching speed and reliability.

Danković et al. [Contribution 12] present a comprehensive review of negative bias temperature instability (NBTI) in commercial power p-channel VDMOS transistors, focusing on how device-level degradation affects practical circuit performance. The review aims to bridge fundamental NBTI reliability studies and real application behavior by jointly considering threshold voltage shifts, self-heating, and circuit-level degradation. Instead of concentrating only on defect physics, the authors synthesize extensive long-term experimental results to demonstrate how NBTI-induced parameter changes degrade load-driving capability and logic operation. Key experimental methodologies are reviewed, including accelerated negative gate-bias stress at elevated temperatures, relaxation and annealing processes, and repeated electrical characterization of transfer characteristics. Special emphasis is placed on infrared thermography as an effective tool for visualizing and quantifying self-heating in VDMOS devices under realistic operating conditions. Circuit-level evalua-

 https://doi.org/10.3390/mi17020205

tion methods are also discussed, particularly CMOS inverter configurations incorporating stressed p-channel VDMOS transistors, which reveal changes in transfer curves, rise and fall times, and pulse narrowing effects. From a structural perspective, the review examines commercial p-channel VDMOS devices with polysilicon gates, thick gate oxides, and vertical double-diffused architectures packaged in TO-220 housings, highlighting their inherent susceptibility to NBTI and thermal stress due to high current capability. The consolidated findings show that NBTI leads to a progressive increase in absolute threshold voltage, higher on-resistance, and increased power dissipation, which further intensifies self-heating. This thermal feedback is identified as a critical reliability concern, especially at elevated temperatures. The review also notes that proper thermal management, such as heatsink use, can partially mitigate degradation, while challenges remain for long-term operation and complex circuit conditions.

Chen et al. [Contribution 13] survey the evolution and current state of tunnel field-effect transistors (TFETs) as a promising "Beyond Moore" solution for ultra-low-power electronics, focusing on how quantum-mechanical band-to-band tunneling can overcome the fundamental 60 mV/dec subthreshold swing limit of conventional MOSFETs. The central emphasis of the review is to systematically compare how material selection, device architectures, and fabrication strategies influence TFET switching steepness, on-state current, and leakage behavior, which are the key performance bottlenecks hindering large-scale adoption. Rather than presenting new device demonstrations, the article synthesizes a wide body of experimental and simulation-based studies, organizing them into coherent categories that reflect the historical and technological progression of TFET research. Experimentally reported devices based on Si, Si/Ge heterojunctions, III–V compound semiconductors, two-dimensional materials, and carbon nanotubes are critically examined, highlighting how bandgap size, band alignment, and effective mass determine tunneling efficiency. The review places particular weight on advanced device geometries—such as double-gate, gate-all-around nanowire, L-shaped, U-shaped, symmetric, and line-tunneling TFETs—that expand the tunneling area or enhance electrostatic control to simultaneously improve subthreshold swing and drive current. Complementing experimental results, a substantial portion of the review discusses TCAD-based simulation studies, which explore idealized architectures including dopingless, junctionless, negative-capacitance, heterogeneous-gate, and quasi-broken-gap TFETs, revealing the theoretical performance limits achievable through aggressive band and field engineering. Across these studies, the review identifies recurring methodological tools, including band-diagram analysis, tunneling probability modeling, and benchmarking using metrics such as minimum/average subthreshold swing, Ion/Ioff ratio, and operating voltage. The major conclusions drawn are that TFETs not only exhibit sub-60 mV/dec switching but, in optimized forms, can also achieve steep slopes below 10 mV/dec and competitive on-currents at supply voltages around 0.5 V, particularly when line tunneling or broken-gap heterojunctions are employed. Finally, the review outlines future challenges related to variability, fabrication complexity, and integration while emphasizing emerging application opportunities in ultra-low-power logic, biosensors, and energy-efficient integrated circuits, thereby providing a consolidated roadmap for both device engineers and circuit designers working toward next-generation low-power systems.

Conflicts of Interest: The authors declare no conflict of interest.

List of Contributions:

1. Tang, Y.; Shi, Y.; Lv, Z.; Guo, Z.; Wang, X. Experimental study on the performance of light-controlled ion drag pump based on PLZT ceramic. *Micromachines* **2026**, *17*, 45.

https://doi.org/10.3390/mi17020205

2. Tang, Y.; Qi, A.; Gu, Y.; Zhu, Y.; Li, H.; Gu, D.; Chen, H. Behavior of electrothermal actuator analyzed by polynomial point interpolation collocation method. *Micromachines* **2025**, *16*, 1415.
3. Lin, B.; Zheng, Y.; Li, H.; Ren, Y.; Yang, T.; Wang, Z.; Cai, Y.; Xu, Q.; Sun, C. Enhancement of quality factors in a 6.5 GHz resonator using Mo/SiC composite microstructures. *Micromachines* **2025**, *16*, 529.
4. Tang, Y.; Guo, Z.; Ding, Y.; Wang, X. Study on the driving performance and influencing factors of multi-electrothermal co-actuation devices considering application environments. *Micromachines* **2025**, *16*, 603.
5. He, C.; Luo, M.; Chai, H.; Wang, H.; Zhang, G.; Wang, R.; Cui, J.; Yang, Y.; Zhang, W.; Jia, L. Design and evaluation of a capacitive micromachined ultrasonic transducer(CMUT) linear array system for thickness measurement of marine structures under varying environmental conditions. *Micromachines* **2025**, *16*, 898.
6. Shi, H.; Xie, Y.; Zhang, H. Novel debris material identification method based on impedance microsensor. *Microachines* **2025**, *16*, 812.
7. Yang, P.; Wen, X.; Li, X.; Chu, Z.; Peng, C.; Wu, S. A 3D DC electric field meter based on sensor chips packaged using a highly sensitive scheme. *Micromachines* **2025**, *16*, 484.
8. Chen, H.W.; Lee, C.Y.; Dai, C.L. CMOS-compatible micro photovoltaic generator with post-processing enhanced optical absorption. *Micromachines* **2026**, *17*, 48.
9. Yao, Z.; Li, C. A rotary piezoelectric electromagnetic hybrid energy harvester. *Micromachines* **2025**, *16*, 807.
10. Choi, T.M.; Lee, H.R.; Pyo, S.G. Radio-frequency characteristics of stacked metal–insulator–metal capacitors in radio-frequency CMOS devices. *Micromachines* **2026**, *17*, 54.
11. Basu, R.; Kehr, C.C. Accelerated electro-optic switching in liquid crystal devices via ion trapping by dispersed helical carbon nanotubes. *Micromachines* **2025**, *16*, 457.
12. Danković, D.; Živanović, E.; Veselinović, N.; Đorđević, D.; Petrović, M.; Tasić, L.; Marjanović, M.; Veljković, S.; Mitrović, N.; Davidović, V.; Ristić, G. Examination of impact of NBTIs on commercial power P-channel VDMOS transistors in practical applications. *Micromachines* **2026**, *17*, 52.
13. Chen, S.; An, Y.; Wang, S.; Liu, H. A review of tunnel field-effect transistors: materials, structures, and applications. *Micromachines* **2025**, *16*, 881.

References

1. Aderibigbe, A.O.; Ohenhen, P.E.; Nwaobia, N.K.; Gidiagba, J.O.; Ani, E.C. Advanced sensing techniques in electro-mechanical systems: Surveying the rise of smart sensors and their implications for system robustness. *Eng. Sci. Technol. J.* **2023**, *4*, 323–340. [CrossRef]
2. Lee, C.Y.; Huang, Z.Y.; Hsu, J.Y.; Kuo, T.F.; Dai, Z.X. Measurement circuit design of 3-in-1 microsensor embedded in proton exchange membrane water electrolyzer. *Sensors Mater.* **2025**, *37*, 1537–1550. [CrossRef]
3. Varadan, V.K. Nanotechnology: MEMS and NEMS and their applications to smart systems and devices. *SPIE Smart Mater. Struct. Syst.* **2003**, *5062*, 20–43.
4. Dai, Z.X.; Tsai, R.W.; Shih, Q.H. High-sensitivity magnetic field sensors manufactured utilizing CMOS–MEMS technology. *IEEE Sensors J.* **2025**, *25*, 9540–9549. [CrossRef]
5. Sun, H.; Yin, M.; Wei, W.; Li, J.; Wang, H.; Jin, X. MEMS based energy harvesting for the Internet of Things: A survey. *Microsyst. Technol.* **2018**, *24*, 2853–2869. [CrossRef]
6. Dai, Z.X.; Chen, C.Y.; Chiu, B.C.; Lee, C.Y. Manufacturing and measurement of CMOS-MEMS-based micro thermoelectric generators with long-length thermocouples. *J. Micromech. Microeng.* **2024**, *34*, 075001. [CrossRef]
7. Dai, C.L.; Li, W.H.; Ono, T.; Hsu, C.C. Design and fabrication of CMOS-based thermoelectric microgenerators with enhanced power factor. *IEEE Sensors J.* **2025**, *25*, 41164–41173. [CrossRef]
8. Dai, Z.X.; Chen, C.Y. Fabrication and evaluation of a CMOS-based energy harvesting chip integrating photovoltaic and thermoelectric energy harvesters. *Energy Convers. Manag. X* **2024**, *24*, 100748. [CrossRef]
9. Chen, Y.W.; Wu, C.C.; Hsu, C.C.; Dai, C.L. Fabrication and testing of thermoelectric CMOS-MEMS microgenerators with CNCs film. *Appl. Sci.* **2018**, *8*, 1047. [CrossRef]

 https://doi.org/10.3390/mi17020205

10. Dai, Z.X.; Li, W.H.; Liu, Y.N. Effect of post-processing techniques on the performance of CMOS-based photovoltaic energy harvesters with grid-like architectures. *J. Micromech. Microeng.* **2025**, *35*, 065005. [CrossRef]
11. Mahalakshmi, M.; Diana, D.C.; Ramachandran, R.; Ravva, S.K.; Illuri, B.; Johannah, J.J.; Manikandan, T.T.; Anand, A.J. Review on carbon-based micro and nano electro-mechanical systems for biotechnological application. *Recent Pat. Nanotechnol.* **2025**, *19*, 468–482. [CrossRef]
12. Luo, B.; Will-Cole, A.R.; Dong, C.; He, Y.; Liu, X.; Lin, H.; Huang, R.; Shi, X.; McConney, M.; Page, M.; et al. Magnetoelectric microelectromechanical and nanoelectromechanical systems for the IoT. *Nat. Rev. Electr. Eng.* **2024**, *1*, 317–334. [CrossRef]
13. D'Alessandro, A.; Scudero, S.; Vitale, G. A Review of the Capacitive MEMS for Seismology. *Sensors* **2019**, *19*, 3093. [CrossRef] [PubMed]
14. Gill, W.A.; Howard, I.; Mazhar, I.; McKee, K. A Review of MEMS vibrating gyroscopes and their reliability issues in harsh environments. *Sensors* **2022**, *22*, 7405. [CrossRef] [PubMed]
15. Han, X.; Huang, M.; Wu, Z.; Gao, Y.; Xia, Y.; Yang, P.; Fan, S.; Lu, X.; Yang, X.; Liang, L.; et al. Advances in high-performance MEMS pressure sensors: Design, fabrication, and packaging. *Microsyst. Nanoeng.* **2023**, *9*, 156. [CrossRef]
16. Shah, M.A.; Shah, I.A.; Lee, D.G.; Hur, S. Design approaches of MEMS microphones for enhanced performance. *J. Sensor* **2019**, *2019*, 9294528. [CrossRef]
17. Xu, D.; Wang, Y.; Xiong, B.; Li, T. MEMS-based thermoelectric infrared sensors: A review. *Front. Mech. Eng.* **2017**, *2*, 557–566. [CrossRef]
18. Shao, B.; Lu, C.; Xiang, Y.; Li, F.; Song, M. Comprehensive review of RF MEMS switches in satellite communications. *Sensors* **2024**, *24*, 3135. [CrossRef]
19. Wang, D.; Watkins, C.; Xie, H. MEMS mirrors for LiDAR: A review. *Micromachines* **2020**, *11*, 456. [CrossRef]
20. Rad, Z.F.; Prewett, P.D.; Davies, G.J. An overview of microneedle applications, materials, and fabrication methods. *Beilstein J. Nanotechnol.* **2021**, *12*, 1034–1046. [CrossRef]
21. Asri, M.I.A.; Hasan, M.N.; Fuaad, M.R.A.; Yunos, Y.M.; Ali, M.S.M. MEMS gas sensors: A review. *IEEE Sensors J.* **2021**, *21*, 18381–18397. [CrossRef]
22. Mandal, D.; Banerjee, S. Surface acoustic wave (SAW) sensors: Physics, materials, and applications. *Sensors* **2022**, *22*, 820. [CrossRef] [PubMed]
23. Sesen, M.; Rowlands, C.J. Thermally-actuated microfluidic membrane valve for point-of-care applications. *Microsyst. Nanoeng.* **2021**, *7*, 48.
24. Zhao, Y.; Wu, H.; Zheng, L.; Wang, Y.; Zhu, Y.; Chen, L. Advanced pneumatic microgripper for versatile biomedical micromanipulation. *Precision Eng.* **2024**, *88*, 223–234. [CrossRef]
25. Tsai, Y.C.; Huang, P.C.; Dai, C.L. Editorial for the Special Issue on MEMS/NEMS devices and applications, 2nd Edition. *Micromachines* **2025**, *16*, 189. [CrossRef]
26. Shuaibu, A.H.; Rabih, A.A.S.; Blaquière, Y.; Nabki, F. Laterally actuated Si-to-Si DC MEMS switch for power switching applications. *Micromachines* **2024**, *15*, 1295.
27. Wang, L.; Zhang, S.; Yuan, Q. Strain-induced frequency splitting in PT symmetric coupled silicon resonators. *Micromachines* **2024**, *15*, 1278. [CrossRef]
28. Chimerad, M.; Borjian, P.; Pathak, P.; Fasano, J.; Cho, H.J. A miniaturized, fuel-free, self-propelled, bio-Inspired soft actuator for copper ion removal. *Micromachines* **2024**, *15*, 1208.
29. Shan, Y.; Qian, L.; Wang, J.; Wang, K.; Zhou, P.; Li, W.; Shen, W. Driving principle and stability analysis of vertical comb-drive actuator for scanning micromirrors. *Micromachines* **2024**, *15*, 226. [CrossRef]
30. Li, X.; Zhao, P.; Wang, G.; Li, N.; Zhang, Y. Structural optimization design of magnetoelectric thin-film antenna for bandwidth and radiation enhancement. *Micromachines* **2024**, *15*, 810. [CrossRef]
31. Wang, B.; Corsi, C.; Weiland, T.; Wang, Z.; Grund, T.; Pohl, O.; Bienia, J.M.; Weiss, J.; Ngo, H.D. Screen-printed PVDF piezoelectric pressure transducer for unsteadiness study of oblique shock wave boundary layer interaction. *Micromachines* **2024**, *15*, 1423. [CrossRef]
32. Lv, M.; Li, P.; Miao, J.; Qiao, Q.; Liang, R.; Li, G.; Zhuang, X. Design and optimization of MEMS resonant pressure sensors with wide range and high sensitivity based on BP and NSGA-II. *Micromachines* **2024**, *15*, 509. [CrossRef]
33. Zhou, J.; Xu, D.; Li, X.; Chen, F. Design of a micro-electro mechanical system quad mass gyroscope with compliant mechanical amplification. *Micromachines* **2024**, *15*, 124. [CrossRef]
34. Xu, Q.; Zhao, X.; Qin, Y.; Gianchandani, Y.B. Control software design for a multisensing multicellular microscale gas chromatography system. *Micromachines* **2024**, *15*, 95. [CrossRef]

https://doi.org/10.3390/mi17020205

35. Song, X.; Wu, W.; Zhao, Y.; Cheng, Y.; Liu, L. Structural design and optimization of the milling force measurement tool system embedded with thin-film strain sensors. *Micromachines* **2023**, *14*, 2133. [CrossRef]
36. Chen, C.Y.; Dai, Z.X. Fabrication and characterization of photovoltaic microgenerators using the complementary metal oxide semiconductor process. *Micromachines* **2023**, *14*, 2038. [CrossRef]

 https://doi.org/10.3390/mi17020205

micromachines

MDPI

Article

Radio-Frequency Characteristics of Stacked Metal–Insulator–Metal Capacitors in Radio-Frequency CMOS Devices

Tae Min Choi, Hwa Rim Lee and Sung Gyu Pyo *

School of Integrative Engineering, Chung-Ang University, Seoul 156-756, Republic of Korea
* Correspondence: sgpyo@cau.ac.kr; Tel.: +82-2-820-5781

Abstract

This paper describes the radio-frequency (RF) characteristics of stacked metal–insulator–metal (MIM) capacitors used in RF CMOS technology. To ensure accurate analysis, various de-embedding methods for stacked MIM capacitors were verified, and an improved RF model was constructed accordingly. To develop an equivalent circuit for the improved RF model by analyzing the RF characteristics of stacked MIM capacitors, we compared de-embedding methods for measured stacked MIM capacitors: one-step (open-pattern or short-pattern) de-embedding and two-step (combined open-pattern and short-pattern) de-embedding. For the analysis of stacked MIM capacitors, at least two-step de-embedding was used, while for precise de-embedding, three-step de-embedding using a thru pattern was employed. Based on the measured values obtained using these two-step de-embedding methods, a modified equivalent circuit was constructed. This circuit was analyzed based on various parameters, including MIM capacitance, quality factor, S-parameter, and Y-parameter, and the results were comparatively examined. The findings highlight outstanding accuracy of the modified model, which is maintained even in high frequency bands.

Keywords: stacked MIM; RFIC; CMOS; capacitance

1. Introduction

With the rapid development of wireless communication technology, radio-frequency (RF) resources have become indispensable, with materials, components, and circuits operating in the RF band being in high demand. These RF components and circuits are continuously being researched as the basis for integrated circuits (ICs) that can achieve not only high performance but also miniaturization, weight reduction, and cost-effectiveness [1–6]. Specifically, Si-based RFICs are circuits that integrate active and passive components and utilize CMOS-based active components and passive components, such as resistors, inductors, and capacitors, according to the characteristics of the circuit. The characteristics of each component directly impact the overall performance of such circuits; hence, improving the characteristics of both active and passive components is crucial [7–12]. Capacitors, a representative passive component, are crucial in RFICs, serving various purposes, including impedance matching, DC blocking in bias circuits, bypass circuits for preventing low-frequency oscillation and improving noise characteristics, and degeneration circuits for increased stability and linearity [13–17].

However, metal–insulator–metal (MIM) capacitors, along with inductors, have the disadvantage of occupying a large area. Research is being conducted in two main areas to overcome this limitation. First, increasing the capacitance density involves changing the type or thickness of the insulator [18,19]. Second, various structural design methods for MIM capacitors can be employed to increase the capacitance density per unit area [20–22]. The MIM capacitor used in this study is a stacked-type unit, the structural design of which involves vertically connecting two MIM capacitors in parallel to increase the capacitance density per unit area.

Based on their RF characteristics, typical capacitors do not function as pure capacitors. This is because as the frequency increases, the capacitor is affected by parasitic components, such as parasitic inductance and parasitic resistance, in addition to the original capacitance. Ideally, the impedance of a capacitor should decrease linearly with increasing frequency. However, as the frequency increases, the impedance increases significantly due to the influence of parasitic inductance and parasitic resistance [23–25]. Recent studies have extensively explored RF modeling techniques for conventional MIM capacitors, with efforts focused on accurately capturing frequency-dependent degradation caused by parasitic inductance, resistance, and substrate-related loss mechanisms [17,26–29]. Advanced modeling approaches have incorporated multi-layer dielectric coupling, extended substrate networks, and frequency-dependent parasitic extraction to improve prediction accuracy in the multi-GHz regime [30–32]. However, despite these developments, research on the RF modeling of stacked MIM capacitors remains limited. Stacked architectures introduce additional parasitic paths associated with vertically interconnected electrodes and inter-metal dielectrics, which are not sufficiently addressed in traditional single-layer MIM capacitor models. Due to the increasing demand for high-density capacitors in RF CMOS applications, establishing an accurate RF model that reflects the unique parasitic behavior of stacked MIM structures has become a critical requirement. This study aims to introduce an improved RF model. In contrast to earlier stacked MIM capacitor models, which primarily represented two MIM units connected in parallel with simplified parasitic branches [11], the present work adopted a refined topology that independently accounts for electrode-dependent parasitic paths and the additional series impedance introduced by the bottom electrode. Furthermore, the proposed model employs a rigorous parameter-extraction procedure based on two-step open–short de-embedded S-parameters and validates the extracted elements against the full frequency dependence of the Y-parameter, yielding substantially improved agreement compared to conventional approaches. Unlike previous models typically verified up to approximately 10 GHz, the proposed model demonstrates consistent accuracy from 500 MHz to 20 GHz, highlighting its suitability for high-frequency RF CMOS applications requiring reliable stacked-capacitor behavior across an extended GHz range.

2. Experimental Procedure

2.1. Fabrication of Stacked MIM Capacitors

The MIM capacitors used in the study were manufactured using a 0.18-μm standard CMOS process. The MIM stacked structure with two vertically parallel connected structures corresponded to the stacked structure of $TiN/SiN_x/TiN$ capacitor single units previously implemented in research results [11]. The stacked MIM capacitor module used in the analysis was structured to achieve twice the capacitance per unit area by connecting the first MIM capacitor implemented on Metal 3 and the second MIM capacitor formed on Metal 5 in parallel. The test pattern of the stacked MIM capacitor was laid out to evaluate electrical characteristics such as capacitance density, leakage current, VCC, and TCC. The unit sizes of the first and second MIM capacitors included in each module were 10 μm × 10 μm,

https://doi.org/10.3390/mi17010054

15 μm × 15 μm, 20 μm × 20 μm, 25 μm × 25 μm, 30 μm × 30 μm, and 50 μm × 50 μm. To determine the RF characteristics of the stacked MIM capacitor alone, without any effect of the degree of mismatch for the capacitor, the test pattern of the capacitor in the unit form (excluding the array form) was used in the analysis. Figure 1 briefly illustrates the cross-sectional structure and design rule of the stacked MIM capacitor.

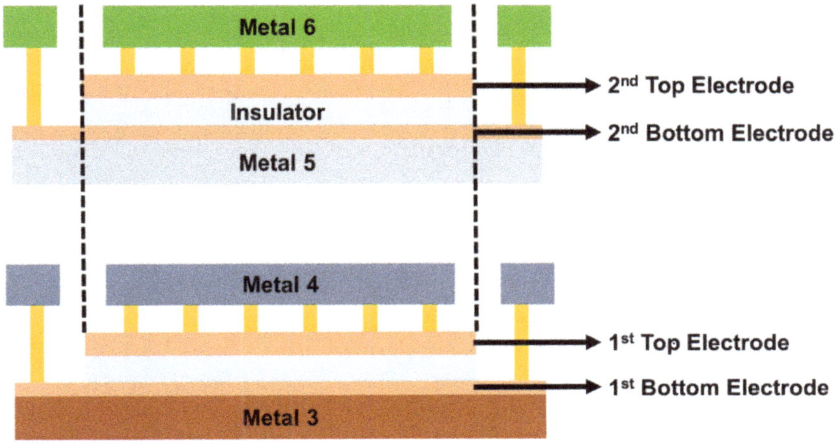

Figure 1. Plan view and cross-sectional view of stacked MIM capacitor.

2.2. RF Measurement Setup and Calibration

The structure of the ground signal from the ground pad was utilized to assess the wafer of the test pattern using a microwave vector network analyzer (8510C, Agilent Technologies, Santa Clara, CA, USA).

Measurements for RF analysis were performed at high frequencies, ranging from several gigahertz to several tens of gigahertz. In such cases, conventional I–V and C–V methods cannot be used, and S-parameter measurements are required instead. S-parameter values measured using a network analyzer can vary significantly depending on the level of calibration, which necessitates accurate calibration. The RF measurement environment for stacked MIM capacitors is summarized below:

1. Measurement range: 500 MHz to 20 GHz;
2. Measurement equipment: HP8510C Vector Network Analyzer (Agilent Technology);
3. RF probe: Cascade Infinity GSG Probe;
4. Calibration method: Two-port short, open, load, and thru calibration.

2.3. De-Embedding Procedure

Accurate RF characterization of stacked MIM capacitors requires the removal of parasitic contributions originating from probe pads, interconnects, and measurement fixtures. To extract the intrinsic RF response of the device under test (DUT), de-embedding was performed using open and short calibration structures fabricated alongside the DUT. The measurement test pattern and the associated parasitic elements introduced by the pad and interconnect structures are illustrated in Figure 2. In Figure 2a, Ports 1 and 2 denote the signal terminals of the two-port VNA measurement and are applied to the signal pads of the DUT. The adjacent pads correspond to the ground (G) terminals of the GSG probe and provide the RF return path. The underlying metal interconnect routing is not left floating; it electrically connects the signal pads to the DUT electrodes and is included in the signal path. The two metal lines beneath the DUT correspond to the interconnect routing between

 https://doi.org/10.3390/mi17010054

the probe pads and the DUT, which introduce additional parasitic elements that must be removed through the de-embedding procedure. Although the simplified illustration may visually suggest a connection to the ground pads, the actual electrical excitation follows the conventional G–S–G contact scheme.

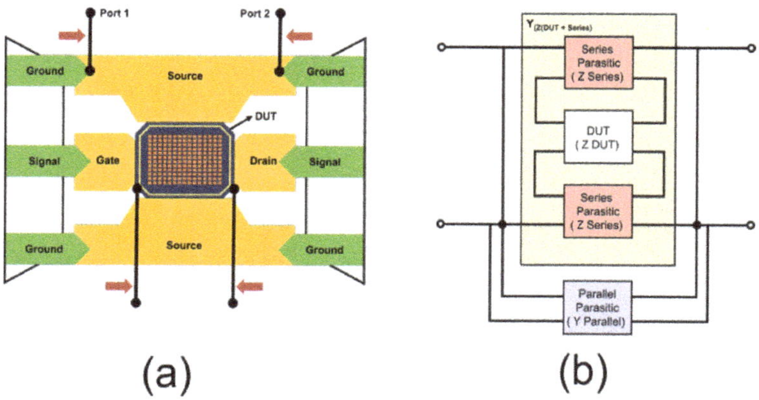

Figure 2. Test patterns with (**a**) extra pattern and (**b**) parasitic elements.

Open de-embedding was first applied to eliminate parallel parasitic components associated with pad and interconnect capacitances. As illustrated in Figure 3, the open de-embedding structure represents these parasitic elements in the admittance domain. At this stage, the physical configuration remains identical to that shown in Figure 3a,b, where the same GSG probe connection and access interconnect routing are assumed, and only the termination condition at the DUT location differs between the short and open structures. The measured S-parameters were converted into Y-parameters, and the parasitic admittance extracted from the open pattern was subtracted from the total admittance, as expressed in Equation (1).

$$Y_{DUT(Open)} = Y_{Total} - Y_{Open}. \tag{1}$$

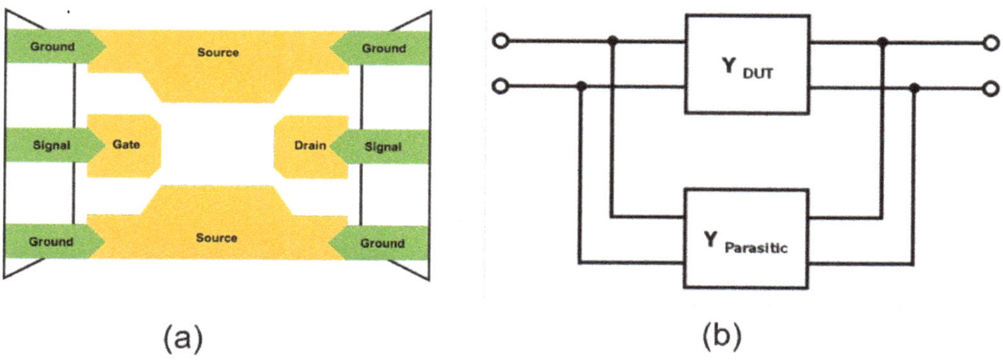

Figure 3. Open de-embedding patterns with (**a**) extra pattern and (**b**) parasitic elements.

Subsequently, short de-embedding was employed to remove series parasitic components introduced by metal interconnects and probe access paths. The short de-embedding structure, shown in Figure 4, expresses these parasitic elements in the impedance domain. In Figure 4a, the metal structure represents the interconnect path that electrically links the probe pad to the DUT. This metal path provides the current flow into the stacked

https://doi.org/10.3390/mi17010054

MIM capacitor and introduces series parasitic resistance and inductance associated with the access routing. Therefore, the short de-embedding pattern is used to characterize and remove these series parasitic components, enabling the accurate extraction of the intrinsic RF response of the DUT. The open-de-embedded Y-parameters were converted into Z-parameters, and the impedance contribution obtained from the short pattern was subtracted, as described by Equations (2)–(7). Accordingly, $Y_{(short\text{-}open)}$ represents the combined admittance of the pad capacitance and access interconnect parasitics defined in the equivalent circuit of Figure 3a, with no additional structural elements introduced.

$$Z_{DUT(Short)} = Z_{Total} - Z_{Short}. \tag{2}$$

$$Y_{DUT(Open)} = Y_{Total} - Y_{Open}, \tag{3}$$

$$Y_{(Short)(Open)} = Y_{Short} - Y_{Open}, \tag{4}$$

$$Z_{DUT(Open)} = Z_{(YDUT(Open))} \text{ (Y-parameter Z-parameter conversion)}, \tag{5}$$

$$Z_{Short(Open)} = Z_{(YShort(Open))}, \tag{6}$$

$$Z_{DUT} = Z_{DUT(Open)} - Z_{Short(Open)}, \tag{7}$$

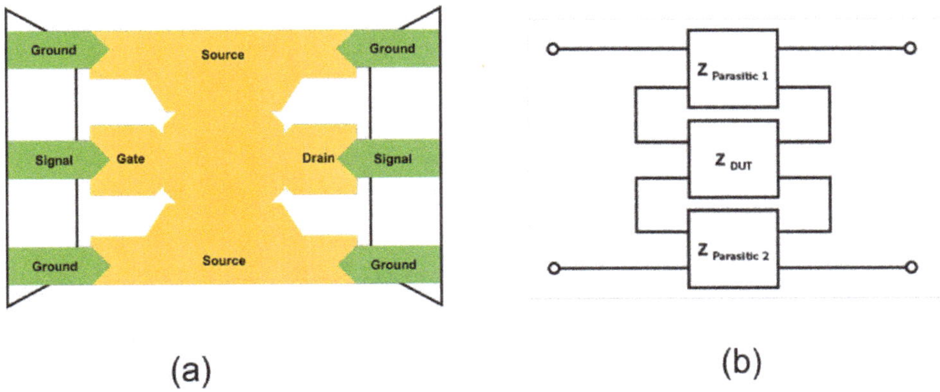

Figure 4. Short de-embedding patterns with (**a**) extra pattern and (**b**) parasitic elements.

Equation (7) is derived under the same physical and electrical configuration as Equations (4)–(6), and therefore shares the identical GSG probe connection and interconnect topology illustrated in Figure 4a,b. This equation does not correspond to a new schematic or connection diagram; rather, it represents a subsequent algebraic de-embedding step that removes the previously identified parasitic admittance to extract the intrinsic DUT response.

Finally, the de-embedded Z-parameters were converted back into S-parameters to obtain the intrinsic RF characteristics of the stacked MIM capacitors, as summarized in Equation (8). This two-step open–short de-embedding procedure enables reliable extraction of capacitance, quality factor, and Y-parameter characteristics over a wide frequency range.

$$S_{DUT} = S_{(ZDUT)} \text{ (Z-parameter S-parameter conversion)}. \tag{8}$$

https://doi.org/10.3390/mi17010054

In this study, the two-step de-embedding approach was adopted because the pad dimensions and interconnect lengths of the test structures are electrically short within the measurement frequency range. Under these conditions, transmission-line effects are negligible, and the additional complexity associated with three-step de-embedding using a thru structure provides limited benefit.

3. Results and Discussion

The de-embedding method removes parallel and serial parasitic components by using both open and short de-embedding patterns. Open de-embedding is performed using the Y-parameter of the test pattern, while short de-embedding is performed using the Z-parameter of the open de-embedded short pattern; the de-embedded patterns are then converted into S-parameters to obtain the true S-parameters of the DUT. RF analysis is performed using the S-parameters thus extracted. Equations (3)–(8) summarize the process of open and short pattern de-embedding.

3.1. RF Characteristics Analysis Based on De-Embedding Method

Precise RF measurements are essential for developing the equivalent circuit for an accurate RF model. To address discrepancies in the effective capacitance values extracted from measured stacked MIM capacitors depending on the de-embedding method, one-step (open-pattern or short-pattern) de-embedding and two-step (combined open-pattern and short-pattern) de-embedding were compared and analyzed. The open de-embedding pattern resulted in under-de-embedding due to residual parasitic components in the area where the DUT was formed. In contrast, the short de-embedding pattern resulted in over-de-embedding as it de-embedded the entire area where the DUT was formed. Using both open and short de-embedding yielded reliable data. Figure 5a displays the effective capacitance of the unit stacked MIM capacitor of the stacked module among the Al-based stacked MIM capacitors. In the case of under de-embedding open de-embedding, the total effective capacitance was significantly larger than that of fine de-embedding open, short de-embedding, and in the case of over de-embedding short de-embedding, the total effective capacitance was smaller than that of fine de-embedding open, short de-embedding. Figure 5b shows the parasitic total effective resistance of the module. In the future, at least two-step de-embedding, i.e., open + short de-embedding, should be used for analyzing stacked MIM capacitors, while three-step de-embedding using thru patterns should be employed for precise de-embedding. The three-step de-embedding method is required when the interconnect routing between the probe pads and the DUT becomes electrically long at high frequencies or when substantial pad-to-pad coupling occurs, making the simple two-step open–short removal insufficient to isolate the intrinsic device response. In such cases, the thru structure is used to extract and remove the transmission-line characteristics of the interconnects, thereby enabling accurate modeling of phase delay and frequency-dependent series impedance. In contrast, the present work employed a two-step method because the pad pitch, metal routing length, and operating frequency range place the interconnects well within the electrically short regime, where transmission-line effects are negligible. Under these conditions, the accuracy improvement gained by switching to three-step de-embedding is typically modest—on the order of 1–2% in extracted capacitance and below 0.5 dB in S-parameter magnitude at the upper frequency limit—while the added complexity does not yield proportional benefit. Therefore, the two-step open–short de-embedding scheme was appropriately adopted here, while emphasizing that the three-step method becomes necessary only when interconnect-induced phase errors or line impedance are no longer negligible.

https://doi.org/10.3390/mi17010054

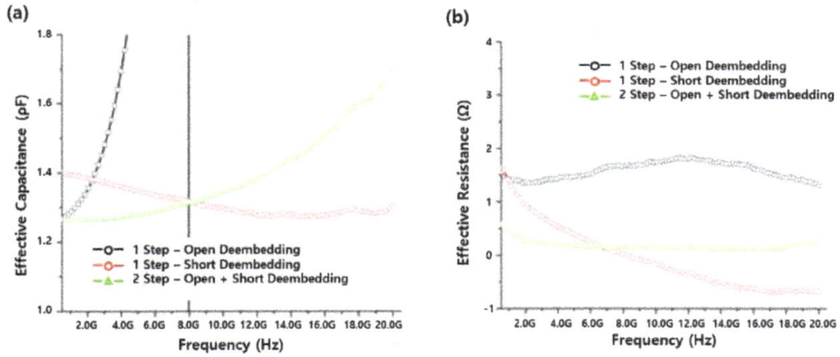

Figure 5. Effective (**a**) capacitance and (**b**) resistance with frequencies for stacked MIM module, obtained using different de-embedding methods.

3.2. RF Modeling Based on Analysis of RF Characteristics of Stacked MIM Capacitors

For a conventional MIM capacitor, RF behavior is typically represented by a model in which the intrinsic capacitance is connected in series with parasitic inductance and resistance associated with the electrode structure, while parasitic capacitance and substrate-related resistance are connected in parallel to account for inter-metal dielectric coupling and substrate loss. This widely adopted model has been shown to effectively describe the frequency-dependent degradation induced by parasitic components [11,33]. Parasitic components extracted using the general equivalent circuit are summarized in Table 1.

Table 1. RF model parameters of typical RF model MIM capacitors.

MIM (μm^2)	10×10	15×15	20×20	25×25	30×30
L_S (pH)	19.04	11.74	6.26	2.61	0.78
R_S (mΩ)	2.28	0.76	0.35	0.22	0.18
C_{MIM} (fF)	216.00	466.40	825.90	1277.00	1782.00
C_{ox} (fF)	4.71	7.59	11.11	1.25	20.02
C_{Sub} (fF)	2902.00	1370.88	836.10	589.08	455.17
R_{Sub} (Ω)	7.16	18.77	35.32	56.83	83.30

However, the stacked MIM capacitors examined in this study differs in that two dielectric capacitor units are vertically connected in parallel. As demonstrated in our previous work [11], each capacitor unit contributes an individual intrinsic capacitance and possesses its own parasitic inductance and resistance components due to the corresponding electrode path. Therefore, when the two MIM capacitor layers are stacked, the parasitic components associated with the top and middle conductive layers must be considered independently, while the bottom electrode introduces an additional series parasitic path. Furthermore, the parasitic coupling between the stacked structure and the inter-metal dielectric/substrate was previously modeled using parallel parasitic capacitance and resistance. To evaluate the validity of the proposed RF model, the measured RF characteristics of the stacked MIM capacitor reported in previous work [11] were used as the reference dataset.

https://doi.org/10.3390/mi17010054

Figure 6 presents the results of analyzing the RF model accuracy of a stacked MIM capacitor for each parameter using the equivalent circuit of a general RF model. The equivalent circuit of a typical MIM capacitor RF model includes the MIM capacitance (C_{MIM}), which acts as a direct capacitor, along with a parasitic inductance (L_s) and resistance (R_s) component occurring in series with this capacitor. Additionally, it was configured with a combination of parasitic capacitance (C_{ox}, C_{Sub}) and parasitic resistance (R_{Sub}) components occurring in parallel with these series components. Figure 7 shows the corresponding results based on the equivalent circuit of the improved RF model. Figure 7A highlights the degree of agreement with the actual total effective capacitance (Cmim1 + Cmim2) as the sum of the formed MIM Capacitances 1 and 2. The measured value and the extracted simulated value were almost identical. Thus, the accuracy was higher when the equivalent circuit of the general RF model (Figure 7A) was used. The accuracy remained high up to high frequencies. Figure 7B shows high accuracy as the quality factor, which is an important parameter for evaluating the characteristics of stacked MIM capacitors. Figure 7C–F compares the magnitude and phase values of the S-parameter, which can be used to evaluate the accuracy of the model, based on the real and imaginary parts of the Y-parameter, respectively. Excellent accuracy of the model is underscored by the fact that the measured value and the extracted simulated value were almost identical up to a high frequency band. For the Y-parameter, which is more important than the S-parameter in the analysis of capacitor characteristics, a significantly more accurate result was obtained when the equivalent circuit of the improved RF model was used with respect to RF model was used. Based on the analysis of the extracted RF model of the stacked module in terms of the MIM capacitance, quality factor, S-parameter, and Y-parameter, excellent accuracy could be observed, given that the measured values and the extracted simulated values were almost identical up to a high frequency band. Considering these results, stacked MIM capacitors can perform stably in higher frequency bands if a ground shield is included to reduce degradation at high frequencies or if the RF isolation of each stacked MIM capacitor is increased to reduce parasitic capacitance.

https://doi.org/10.3390/mi17010054

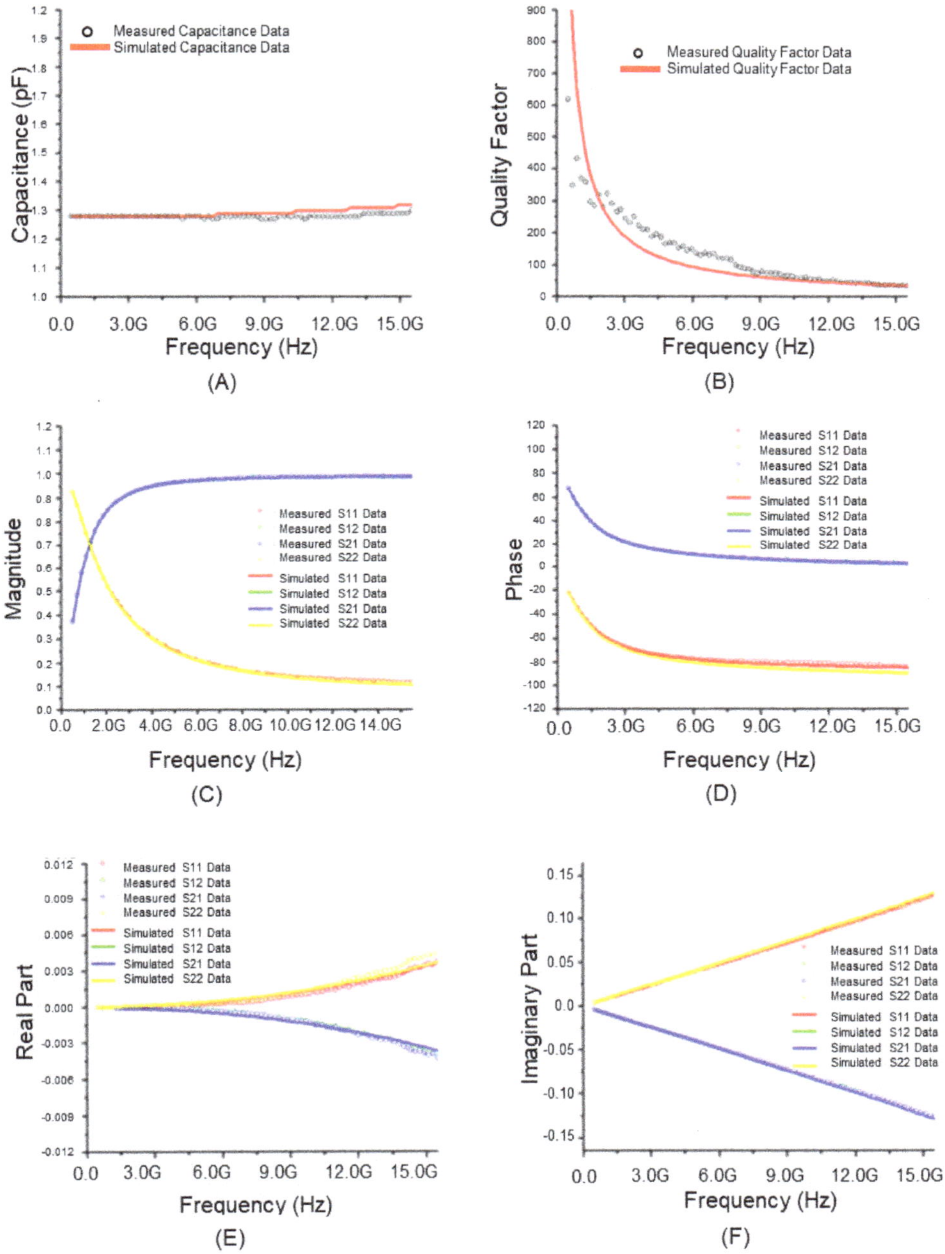

Figure 6. Characteristics of conventional RF model of stacked module (25 μm × 25 μm): (**A**) Total effective capacitance; (**B**) Quality factor; (**C**) Magnitude of S-parameter; (**D**) Phase of S-parameter; (**E**) Real part of Y-parameter; (**F**) Imaginary part of Y-parameter.

https://doi.org/10.3390/mi17010054

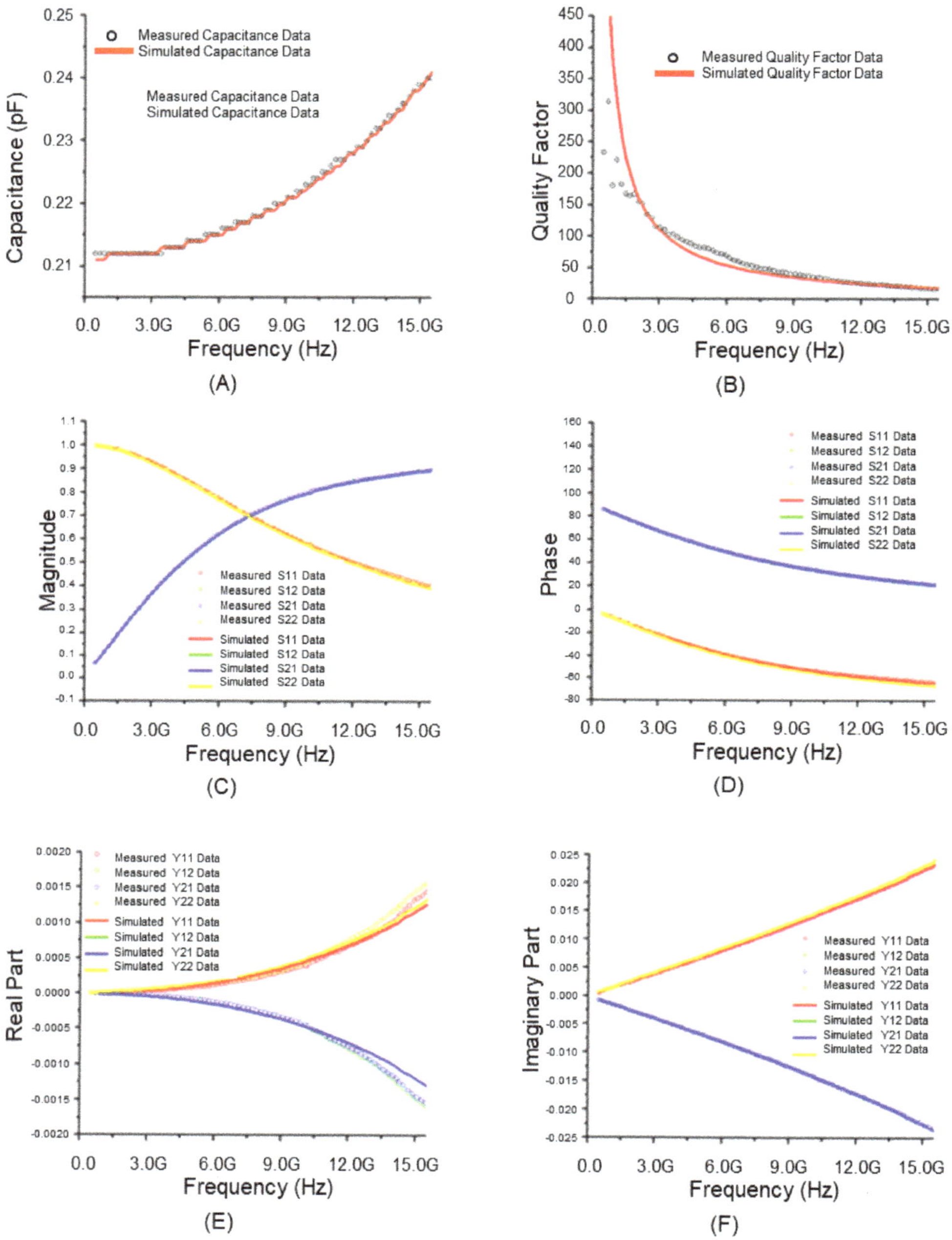

Figure 7. Characteristics of modified RF model of stacked module (10 μm × 10 μm): (**A**) Total effective capacitance; (**B**) Quality factor; (**C**) Magnitude of S-parameter; (**D**) Phase of S-parameter; (**E**) Real part of Y-parameter; (**F**) Imaginary part of Y-parameter.

4. Conclusions

In this study, we sought to identify the problems associated with stacked MIM capacitors and suggest potential solutions by analyzing the RF characteristics of such capacitors. Moreover, we constructed an improved RF model for a stacked MIM capacitor and obtained

https://doi.org/10.3390/mi17010054

accurate simulation data accordingly. The RF analysis was performed using integration processing and an Al-based stacked MIM capacitor. To create an equivalent circuit of an accurate RF model, one-step (open-pattern or short-pattern) de-embedding and two-step (combined open-pattern and short-pattern) de-embedding were compared when analyzing the measured stacked MIM capacitor. The findings suggest that at least two-step de-embedding should be used for the analysis of stacked MIM capacitors and that three-step de-embedding using thru patterns should be employed for precise de-embedding. A modified equivalent circuit was constructed using the measured values obtained via two-step de-embedding. The modified equivalent circuit constructed in this study was analyzed in terms of various parameters, such as MIM capacitance, quality factor, S-parameter, and Y-parameter, and the results were comparatively examined. The findings underscore the excellent accuracy of the modified model, which was sustained up to a high frequency band. The results for the modified equivalent circuit indicate that additional research is needed to manufacture stacked MIM capacitors to reduce degradation at high frequencies, which can be achieved by inserting a ground shield, or to reduce parasitic capacitance, which is possible by increasing the RF isolation of each stacked MIM capacitor. These techniques could allow stacked MIM capacitors to perform stably in higher frequency bands.

Author Contributions: Conceptualization, T.M.C. and S.G.P.; Investigation T.M.C.; Methodology, H.R.L. and S.G.P.; Supervision, S.G.P.; Validation, T.M.C., H.R.L. and S.G.P.; Writing—original draft preparation, T.M.C. and H.R.L.; Writing—review and editing, S.G.P. All authors have read and agreed to the published version of the manuscript.

Funding: This work was supported by the Technology Innovation Program (or Industrial Strategic Technology Development Program-Public-private joint investment semiconductor R&D program (K-CHIPS) to foster high-quality human resources) ("RS-2023-00237003", High selectivity etching technology using cryoetch) funded by the Ministry of Trade, Industry & Energy (MOTIE, KOREA) (1415187674). This work was also supported by the Technology Innovation Program (No. 20022472), funded by the Ministry of Trade, Industry, and Energy (MOTIE, KOREA).

Data Availability Statement: The original contributions presented in this study are included in the article. Further inquiries can be directed to the corresponding author.

Conflicts of Interest: The authors declare no conflicts of interest.

References

1. Levinger, R.; Shumaker, E.; Banin, R.; Ravi, A.; Degani, O. The rise of the digital rfic era: An overview of past and present digital rfic advancements. *IEEE Microw. Mag.* **2022**, *23*, 71–85. [CrossRef]
2. Robertson, I.D.; Lucyszyn, S. *RFIC and MMIC Design and Technology*; IET: London, UK, 2001.
3. Ellinger, F.; Claus, M.; Schröter, M.; Carta, C. Review of advanced and beyond cmos fet technologies for radio frequency circuit design. In Proceedings of the 2011 SBMO/IEEE MTT-S International Microwave and Optoelectronics Conference (IMOC 2011), Natal, Brazil, 29 October–1 November 2011; pp. 347–351.
4. Malm, B.G.; Haralson, E.; Johansson, T.; Ostling, M. Self-heating effects in a bicmos on soi technology for rfic applications. *IEEE Trans. Electron Devices* **2005**, *52*, 1423–1428. [CrossRef]
5. Sia, C.B.; Ong, B.H.; Yeo, K.S.; Ma, J.-G.; Do, M.A. Accurate and scalable rf interconnect model for silicon-based rfic applications. *IEEE Trans. Microw. Theory Tech.* **2005**, *53*, 3035–3044. [CrossRef]
6. Yang, R.; Qian, H.; Li, J.; Xu, Q.; Hai, C.; Han, Z. Soi technology for radio-frequency integrated-circuit applications. *IEEE Trans. Electron Devices* **2006**, *53*, 1310–1316. [CrossRef]
7. Mukhopadhyay, R.; Park, Y.; Sen, P.; Srirattana, N.; Lee, J.; Lee, C.-H.; Nuttinck, S.; Joseph, A.; Cressler, J.D.; Laskar, J. Reconfigurable rfics in si-based technologies for a compact intelligent rf front-end. *IEEE Trans. Microw. Theory Tech.* **2005**, *53*, 81–93. [CrossRef]
8. Bennett, H.S.; Brederlow, R.; Costa, J.C.; Cottrell, P.E.; Huang, W.M.; Immorlica, A.A.; Mueller, J.-E.; Racanelli, M.; Shichijo, H.; Weitzel, C.E. Device and technology evolution for si-based rf integrated circuits. *IEEE Trans. Electron Devices* **2005**, *52*, 1235–1258. [CrossRef]

https://doi.org/10.3390/mi17010054

9. Watson, A.; Mayevskiy, Y.; Francis, P.; Hwang, K.; Srinivasan, G.; Weisshaar, A. Compact modeling of differential spiral inductors in si-based rfics. In Proceedings of the 2004 IEEE MTT-S International Microwave Symposium Digest (IEEE Cat. No. 04CH37535), Fort Worth, TX, USA, 6–11 June 2004; Volume 2, pp. 1053–1056.
10. Sia, C.B.; Ong, B.H.; Lim, W.M.; Yeo, K.S.; Alam, T. Modeling and layout optimization of differential inductors for silicon-based rfic applications. *IEEE Trans. Electron Devices* **2008**, *55*, 1058–1066. [CrossRef]
11. Sul, W.S.; Pyo, S.G. Rf characteristic analysis model extraction on the stacked metal–insulator–metal capacitors for radio frequency applications. *IEEE Trans. Electron Devices* **2014**, *61*, 3011–3013. [CrossRef]
12. Sul, W.S.; Kwon, S.H.; Choi, E.; Cui, Y.; Lee, K.W.; Shim, H.J.; Gao, Y.; Hahn, S.J.; Pyo, S.G. Radiofrequency characteristics of ionized sputtered tantalum nitride thin-film resistor in cmos device. *Electron. Mater. Lett.* **2017**, *13*, 230–234. [CrossRef]
13. Mindan, B.; Hong, L. The analysis of impedance matching problem in rf circuit design. In Proceedings of the 2010 International Forum on Information Technology and Applications, Kunming, China, 16–18 July 2010; Volume 1, pp. 350–353.
14. Chongcheawchamnan, M.; Karacaoglu, U.; Robertson, I.D. Radio-frequency integrated circuits. In *Encyclopedia of RF and Microwave Engineering*; Wiley: Hoboken, NJ, USA, 2005.
15. Chen, C.; Huang, D.; Zhao, Y.; Jin, Y.; Yang, J. An ultra-low-voltage 2.4-ghz flicker-noise-free rf receiver front end based on switched-capacitor hybrid tia with 4.5-db nf and 11.5-dbm oip3. *IEEE J. Solid-State Circuits* **2023**, *58*, 1825–1837. [CrossRef]
16. Molinero, D.; Aghaei, S.; Morris, A.S.; Cunningham, S. Linearity and rf power handling on capacitive rf mems switches. *IEEE Trans. Microw. Theory Tech.* **2019**, *67*, 4905–4913. [CrossRef]
17. Kannadassan, D.; Sivasankaran, K.; Kumaravel, S.; Cheng, C.-H.; Baghini, M.S.; Mallick, P. High-*k* metal–insulator–metal capacitors for rf and mixed-signal vlsi circuits: Challenges and opportunities. *Proc. IEEE* **2024**, *112*, 1610–1631. [CrossRef]
18. Choi, T.-M.; Jung, E.-S.; Yoo, J.-U.; Lee, H.-R.; Yoon, S.; Pyo, S.-G. I–v characteristics and electrical reliability of metal–sixny–metal capacitors as a function of nitrogen bonding composition. *Micromachines* **2025**, *16*, 615. [CrossRef]
19. Choi, T.M.; Jung, E.S.; Yoo, J.U.; Lee, H.R.; Pyo, S.G. Capacitance-voltage fluctuation of si(x)n(y)-based metal-insulator-metal capacitor due to silane surface treatment. *Micromachines* **2024**, *15*, 1204. [CrossRef]
20. Yu, X.; Zhu, C.; Hu, H.; Chin, A.; Li, M.F.; Cho, B.J.; Kwong, D.-L.; Foo, P.D.; Yu, M.B. A high-density mim capacitor (13 ff/μm2) using ald hfo2 dielectrics. *IEEE Electron Device Lett.* **2003**, *24*, 63–65. [CrossRef]
21. Klootwijk, J.H.; Jinesh, K.B.; Dekkers, W.; Verhoeven, J.F.; Heuvel, F.C.v.d.; Kim, H.D.; Blin, D.; Verheijen, M.A.; Weemaes, R.G.R.; Kaiser, M.; et al. Ultrahigh capacitance density for multiple ald-grown mim capacitor stacks in 3-d silicon. *IEEE Electron Device Lett.* **2008**, *29*, 740–742. [CrossRef]
22. Guo, Y.; Wang, S.; Du, X.; Liang, S.; Huang, S.; Peng, S.; Xie, Y.; Ma, M.; Xiong, L. Construction of ultrahigh capacity density carbon nanotube based mim capacitor. *Energy Storage Mater.* **2023**, *63*, 103064. [CrossRef]
23. Gutiérrez-Vicente, V.; Torres-Torres, J.A.; Torres-Torres, R. Broadband s-parameter-based characterization of multilayer ceramic capacitors submitted to mechanical stress through bending tests on a pcb. *Micromachines* **2024**, *15*, 1386. [CrossRef]
24. Ye, Y.; Cheng, K.W.E. Analysis and optimization of switched capacitor power conversion circuits with parasitic resistances and inductances. *IEEE Trans. Power Electron.* **2017**, *32*, 2018–2028. [CrossRef]
25. Gurov, E.V.; Uvaysov, S.U.; Uvaysova, A.S.; Ivanov, I.A. Analysis of the parasitic parameters influence on the analog filters frequency response. In Proceedings of the 2019 International Seminar on Electron Devices Design and Production (SED), Prague, Czech Republic, 23–24 April 2019; pp. 1–7.
26. Aniktar, H.; Savcı, H.Ş. Numerical and measurement based modeling of a mim capacitor in a 0.25 μm sige-c bicmos process. *Prog. Electromagn. Res. C* **2023**, *129*, 173–186. [CrossRef]
27. Wang, L.; Xu, R.-M.; Yan, B. Mim capacitor simple scalable model determination for mmic application on gaas. *Prog. Electromagn. Res.* **2006**, *66*, 173–178. [CrossRef]
28. Mu, J.; Chou, X.; Ma, Z.; He, J.; Xiong, J. High-performance mim capacitors for a secondary power supply application. *Micromachines* **2018**, *9*, 69. [CrossRef] [PubMed]
29. Moon, H.; Yu, S.; Song, S.S.; Nam, I. Characterization and modeling of stacked mim on-chip capacitors with high-capacitance density up to 20 ghz frequency region. *Microw. Opt. Technol. Lett.* **2009**, *51*, 1235–1238. [CrossRef]
30. Cai, W.Z.; Shastri, S.C.; Azam, M.; Hoggatt, C.; Loechelt, G.H.; Grivna, G.M.; Wen, Y.; Dow, S. Development and extraction of high-frequency spice models for metal-insulator-metal capacitors. In Proceedings of the 2004 International Conference on Microelectronic Test Structures (IEEE Cat. No.04CH37516), Awaji, Japan, 22–25 March 2004; pp. 231–234. [CrossRef]
31. Tseng, V.F.G.; Xie, H. Increased multilayer fabrication and rf characterization of a high-density stacked mim capacitor based on selective etching. *IEEE Trans. Electron Devices* **2014**, *61*, 2302–2308. [CrossRef]

 https://doi.org/10.3390/mi17010054

32. Piquet, J.; Cueto, O.; Charlet, F.; Thomas, M.; Bermond, C.; Farcy, A.; Torres, J.; Fléchet, B. Simulation and characterization of high-frequency performances of advanced mim capacitors. In Proceedings of the 35th European Solid-State Device Research Conference, 2005. ESSDERC 2005, Grenoble, France, 16 September 2005; pp. 497–500.
33. Liao, E.B.; Li, H.; Guo, L.H.; Lo, G.Q.; Kumar, R.; Balasubramanian, N.; Kwong, D.L. Rf, dc, and reliability performance of mim capacitors embedded in organic substrates by wafer-transfer technology (wtt) for system-on-package applications. *IEEE Trans. Electron Devices* **2007**, *54*, 425–432. [CrossRef]

https://doi.org/10.3390/mi17010054

micromachines

Article

CMOS-Compatible Micro Photovoltaic Generator with Post-Processing Enhanced Optical Absorption

Hung-Wei Chen [1], Chi-Yuan Lee [2] and Ching-Liang Dai [3,*]

1 Powerchip Semiconductor Manufacturing Corporation, Miaoli 366, Taiwan
2 Department of Mechanical Engineering, Yuan Ze Fuel Cell Center, Yuan Ze University, Taoyuan 320, Taiwan; cylee@saturn.yzu.edu.tw
3 Department of Mechanical Engineering, National Chung Hsing University, Taichung 402, Taiwan
* Correspondence: cldai@dragon.nchu.edu.tw; Tel.: +886-4-2284-0433

Abstract

This work reports the design and realization of a silicon-based micro photovoltaic generator (MPG) fabricated using a standard 0.18 μm complementary metal oxide semiconductor (CMOS) technology. The device harvests optical energy and converts it into electrical power through the photovoltaic effect, leveraging a network of engineered p–n junctions formed within the semiconductor. A grid-structured architecture is adopted, in which patterned p-type regions are embedded inside an n-well platform. This configuration expands the effective junction area, increases carrier-collection paths, and strengthens the internal electric field, thereby enhancing photocurrent generation. To further improve optical coupling, a specialized post-CMOS treatment is introduced. A wet etching is used to selectively remove the silicon dioxide layer that normally covers the junction regions in CMOS processes. Eliminating this dielectric layer enables direct photon penetration into the depletion region minimizes reflection-related losses, resulting in a significant improvement in device performance. Under an illumination intensity of 1000 W/m^2, the fabricated microgenerator delivers an open-circuit voltage of 0.49 V, a short-circuit current of 239 μA, and a maximum output power of 90 μW. The device exhibits an overall energy conversion efficiency of 12.9%, confirming the effectiveness of the grid-like junction design and the post-processing oxide removal.

Keywords: micro photovoltaic generator; output power; conversion efficiency; MEMS; CMOS

1. Introduction

A photovoltaic generator converts incoming light into electrical power by utilizing the photovoltaic effect. Because it directly transforms optical energy into electricity without mechanical motion, a photovoltaic generator is highly reliable, silent, and suitable for long-term operation. Photovoltaic generators are widely used across various fields [1–3]. In consumer electronics, they power small portable devices, sensors, and calculators. In the Internet of Things (IoT), compact photovoltaic units enable self-powered sensor nodes, eliminating the need for frequent battery replacement. In biomedical systems, miniature photovoltaic generators supply energy for wearable or implantable devices. They are also deployed in environmental monitoring stations, smart agriculture, wireless networks, and low-power autonomous microsystems [4,5].

Micro-Electro-Mechanical Systems (MEMS) technology integrates miniature mechanical structures [6–10], sensors [11–15], actuators [16–20], and electronic components on a

https://doi.org/10.3390/mi17010048

single chip using semiconductor fabrication processes. This technology enables the development of highly compact, low-power, and high-performance systems. MEMS technology is also well suited for fabricating micro photovoltaic generators (MPGs). By leveraging complementary metal oxide semiconductor (CMOS)-compatible micromachining, MEMS processes can form finely patterned p–n junctions and optimize light-absorbing structure. MEMS-based MPGs can achieve high power density in a small footprint, making them ideal for self-powered sensors and autonomous microsystems. Hung et al. [21] introduced an MPG consisting of patterned silicon p–n junction cells fabricated using a standard 0.18 μm CMOS process. A localized substrate-removal technique enabled backside illumination and series-connected high-voltage output. Although this architecture improves optical absorption and electrical isolation, it suffers from a limited active-area ratio, non-uniform backside illumination, and fill-factor degradation caused by substrate thinning and etching imperfections. Yan et al. [22] proposed a MEMS-based MPG employing patterned p^+/n^+ silicon junctions with interdigitated electrodes and an oxide-passivated surface to suppress recombination. The device was realized using CMOS–MEMS steps, including ion implantation, low-pressure chemical vapor deposition (LPCVD) polysilicon, dry etching, and Si_3N_4 anti-reflection coatings. Backside texturing and back-surface field implantation further enhanced absorption. Nonetheless, performance was constrained by minority-carrier recombination, wafer-lifetime limitations, and reduced efficiency under backside illumination. Sichao et al. [23] developed a back-contact silicon MPG integrated beneath a thermoelectric module. The structure utilized doped p/n regions formed through MEMS processes such as implantation, oxidation, plasma-enhanced chemical vapor deposition (PECVD) passivation, and gold metallization. Although illumination from both sides was possible, backside excitation produced lower efficiency due to recombination and optical losses. Moreover, the fabrication flow was complex and highly sensitive to substrate quality. Sun et al. [24] designed a MPG based on patterned silicon p–n junctions with interdigitated electrodes for efficient photocarrier collection. The device was fabricated using CMOS- and MEMS-compatible techniques including oxidation, implantation, LPCVD polysilicon, Si_3N_4 passivation, and gold contacts. Despite supporting dual-sided illumination, backside efficiency was degraded by recombination, substrate dependence, and optical interference from thermocouple structures. Arima et al. [25] proposed a MPG that employs two CMOS-compatible photodiodes, which have an N-well/P^+ diode and a P-substrate/N^+ diode, connected in series to increase the output voltage. Both junctions are formed naturally during the standard 0.35 μm CMOS source–drain implantation, eliminating the need for any additional processing steps. Illumination is applied from the chip surface, although the metal interconnect layers introduce shading over the CMOS circuitry. While the structure provides straightforward integration and produces 0.6–0.83 V, its conversion efficiency is low and the required device area is large, which significantly limits practical applicability. Law et al. [26] realized a MPG using stacked n-well/p^+ photodiodes to achieve high-voltage output within a standard CMOS process. Each photodiode was formed by ion implantation and CMOS-compatible layering, enabling two- and three-stage voltage multiplication. However, parasitic n-well/p-sub diodes diverted photocurrent, necessitating large parallel compensation arrays that reduced effective active area and significantly lowered overall conversion efficiency. Sultana et al. [27] demonstrated a heterojunction MPG combining p-CuO films with n-ZnO nanowires grown via chemical bath deposition, followed by indium tin oxide and Ag electrode formation. Although ZnO nanowires improved absorption due to their large surface area, device performance was limited by recombination at CuO/ZnO interfaces and variability in nanowire morphology, resulting in inconsistent efficiency. Masui et al. [28] presented a CMOS-integrated MPG using a deep n-well/p-well photodiode structure, enabling series-connected cells for higher output

 https://doi.org/10.3390/mi17010048

voltage. Fabricated in a 0.18 μm CMOS process, the device incorporated p^+/n^+ implants and metal interconnects, producing 0.97 V and 8 μA from four 400 μm cells. However, parasitic photodiodes in the substrate absorbed part of the incident light, while the shallow junctions restricted photocurrent, leading to low overall power generation.

In this work, we present a MPG that incorporates several innovations enabled by its CMOS-based architecture. The mesh-patterned p–n junction configuration substantially increases the effective junction area compared with conventional interdigitated designs, resulting in enhanced photocurrent generation. The device is realized using a standard 0.18 μm CMOS process and further improved through a wet-etching step that removes the SiO_2 passivation layer, allowing direct illumination of the active junctions. This post-processing technique strengthens optical coupling, reduces reflection-induced losses, and increases the output current by roughly 20.1%. In addition, the use of lightly doped n-well regions suppresses leakage and enhances junction quality. Together, these design and fabrication innovations enable a demonstrated energy-conversion efficiency of 12.9%, confirming the suitability of the proposed microgenerator as a compact, CMOS-compatible, and high-performance solution for MEMS and IoT applications.

2. Design of the Micro Photovoltaic Generator

A MPG is an electronic structure that directly transforms light energy into usable electrical power by means of the photovoltaic phenomenon. Its operation relies on how incident radiation interacts with semiconductor materials to produce charge carriers that can be collected as current. In the present research, silicon serves as the active medium owing to its well-established electronic properties and compatibility with microfabrication processes. When sunlight—comprising streams of photons—impinges upon the silicon surface, each photon imparts a quantized amount of energy to the atomic lattice. If the photon energy exceeds the bandgap of silicon, electrons gain sufficient energy to escape from their covalent bonds, leaving behind positively charged vacancies known as holes. To guide this carrier movement, the silicon is intentionally doped to form two adjacent regions: an n-type zone enriched with electrons and p-type zone rich in holes. The interface between these regions, called the p–n junction, exhibits a built-in electric field caused by the diffusion of charge carriers across the boundary. When photons are absorbed within this region, the generated electrons are driven toward the n-type side and the holes toward the p-type side by the internal field. This spatial separation of charges prevents immediate recombination and results in an electrostatic potential difference across the junction. By connecting an external load, the electrons that accumulate on the n-type contact travel through the circuit and recombine with holes on the p-type side, establishing a continuous flow of current. The resulting electrical energy can operate microelectronic components, charge capacitors, or be stored in energy-management systems for later use. In modern micro-scale applications, this process not only enables renewable power generation but also facilitates integration with complementary circuits for autonomous sensing and signal processing.

Figure 1 presents the structure of the proposed MPG, revealing how various doped regions are arranged within the silicon substrate to form the device's active architecture. The structure is created by intentionally introducing multiple doping profiles—p-type regions, moderately doped n-well zones, and heavily doped n-type areas—thus establishing a network of semiconductor junctions that collectively determine the device's electrical behavior. Unlike traditional photovoltaic layouts that commonly rely on interdigitated finger structures, this study adopts a mesh-like geometric pattern. As shown in Figure 1, a large-area, lightly doped n-well is distributed across the silicon wafer. This lightly doped region is not merely a structural choice; it plays a crucial electronic role. A lower doping concentration helps suppress parasitic leakage paths, reduces junction capacitance, and

 https://doi.org/10.3390/mi17010048

expands the depletion width, all of which contribute to enhanced carrier collection efficiency. Embedded within this n-well is a patterned array of p-type doped segments arranged in a grid configuration. This mesh configuration increases the available junction perimeter and creates multiple pathways through which photogenerated carriers can separate and travel. The coexistence of two types of p–n junctions—those formed between the patterned p-type mesh and the lightly doped n-well, as well as the junctions at the interface between the p-type substrate and the n-well—results in a more distributed electric field profile. This enhances photon absorption and strengthens carrier extraction across the device. Together, these engineered design choices help boost the microgenerator's ability to convert incident light into electrical power. The mesh-pattern architecture supports improved uniformity of carrier generation, reduces resistive losses, and maximizes the effective active area. As a result, the MPG achieves superior performance.

Figure 1. Structure of the micro photovoltaic generator (MPG).

In this work, the behavior and performance of the MPG were investigated using Sentaurus TCAD, a comprehensive semiconductor device simulation platform developed by Synopsys. The numerical analysis follows a structured workflow, beginning with the construction of the device geometry, followed by mesh generation, and concluding with multi-physics computation. The first stage involves building a detailed three-dimensional representation of the microgenerator using the Sentaurus Structure Editor (SDE). Within this environment, the physical layout is defined with precise spatial coordinates, enabling accurate placement of features such as the p-type regions, n-well, heavily doped contacts, and metal electrodes. Doping concentrations are assigned to each region, and Gaussian implantation profiles are incorporated to emulate lateral and vertical dopant diffusion. This ensures that the simulated p–n junctions closely resemble those formed during manufacturing. Following structural definition, the second stage focuses on generating the computational mesh. The quality of the mesh strongly influences both accuracy and simulation time. Therefore, mesh refinement is applied selectively in areas where strong gradients in electric field, carrier density, or optical absorption are expected—such as junction interfaces and surface regions exposed to illumination. Coarser meshing is employed in

https://doi.org/10.3390/mi17010048

electrically inactive zones to optimize efficiency without sacrificing precision. The third stage involves configuring physical models and simulation conditions through command and parameter files. The command file defines external factors such as operating temperature, light intensity, and illumination spectrum. These conditions are crucial because photovoltaic performance is highly sensitive to environmental variations. By enabling these models, the simulation captures the complex interactions among carriers, photons, and the semiconductor lattice. Once these configurations are completed, the device is subjected to numerical analysis. The solver computes carrier transport, electrostatic potential distribution, recombination activity, and resulting current–voltage characteristics under prescribed illumination.

Figure 2 presents the simulated current–voltage (I–V) response of the MPG when subjected to different illumination intensities. In this simulation, the temperature was fixed at 25 °C. The illumination strength was progressively varied from 200 W/m^2 to 1000 W/m^2 in increments of 200 W/m^2, enabling a detailed examination of how photogenerated carrier density changes with incident optical power. At the lowest illumination level of 200 W/m^2, the device exhibits an open-circuit voltage (Voc) of approximately 0.47 V, while the short-circuit current reaches about 48 μA. These values reflect the limited number of photons available to generate electron–hole pairs under weak lighting conditions. As the irradiance is increased, both voltage and current rise, though at different rates. When the irradiance reaches 1000 W/m^2—five times the initial value—the predicted open-circuit voltage climbs to around 0.52 V, and the short-circuit current increases almost an order of magnitude to roughly 254 μA. This substantial growth in current is expected, as photocurrent is linearly proportional to the photon flux absorbed within the depletion region. The gradual enhancement in open-circuit voltage is comparatively modest because the open-circuit voltage exhibits a logarithmic dependence on photocurrent, whereas the short-circuit current scales nearly linearly with irradiance. This distinction is a fundamental characteristic of photovoltaic devices and is well captured in the simulation results. The observed trends confirm that the microgenerator is highly responsive to changes in optical power and that its internal junction structure efficiently converts additional light into proportionally increased current output.

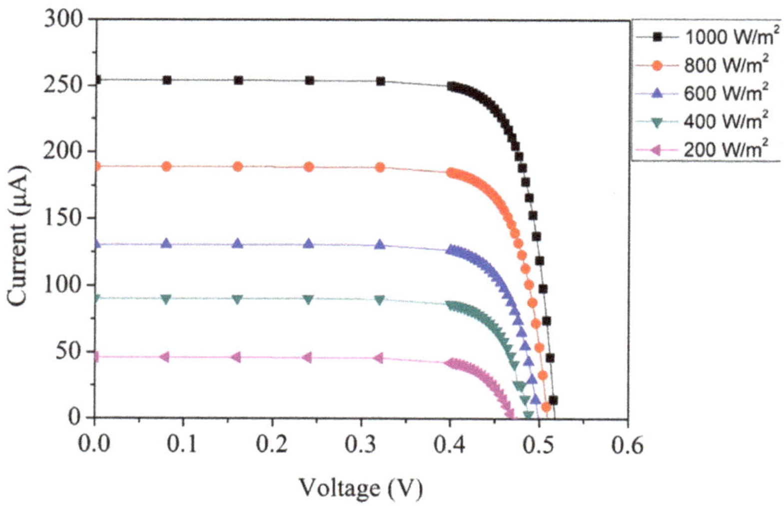

Figure 2. Simulated current–voltage response of the MPG.

https://doi.org/10.3390/mi17010048

TCAD simulations were employed in this study to support device design and to predict the global electrical behavior of the proposed photovoltaic microgenerator. The simulations focused on extracting current–voltage characteristics under various illumination conditions, which serve as the primary indicators of device performance and provide guidance for structural optimization prior to fabrication. While spatial distributions such as electric-field maps, carrier generation profiles, and recombination rate distributions are not explicitly presented, the key advantages of the mesh-patterned architecture can be inferred from its geometric configuration and carrier transport characteristics incorporated in the TCAD framework. Specifically, the mesh-patterned layout increases the junction perimeter density and distributes p–n junctions throughout the active area, resulting in a network of depletion regions that enhances carrier collection efficiency. Compared with conventional interdigitated or planar designs, this distributed junction geometry shortens the effective lateral carrier collection length and increases the probability that photogenerated carriers encounter nearby depletion regions, thereby promoting drift-assisted transport over long-range diffusion. These effects are reflected in the simulated current–voltage characteristics.

Output power is one of the most essential indicators for evaluating the effectiveness of a MPG, as it directly reflects how efficiently the device converts absorbed light into usable electrical energy. In general, the instantaneous electrical power delivered by such a device is determined by the product of its output voltage and current. This relationship is expressed mathematically as ref. [29]:

$$P_{out} = I_{out} V_{out} \tag{1}$$

where I_{out} denotes the current generated under illumination and V_{out} is the output voltage of the MPG. Once these two quantities are obtained—either experimentally or through numerical simulation—the output power can be readily computed using Equation (1).

Figure 2 provides the simulated current–voltage characteristics of the microgenerator at different illumination intensities. By applying the simulated voltage and current values to Equation (1), the output power profile as a function of irradiance can be determined. The resulting power curve is illustrated in Figure 3. At low irradiance, such as 200 W/m^2, the microgenerator is capable of delivering approximately 14 μW of power, corresponding to a simulated operating point of about 0.42 V. This modest power level is expected, as fewer photons are available to generate charge carriers. When the illumination is raised to 1000 W/m^2, the output voltage increases slightly to around 0.45 V, while the current grows substantially due to the higher photon flux. Consequently, the maximum output power reaches approximately 106 μW.

The energy conversion efficiency of a MPG quantifies how effectively the device transforms incoming optical energy into electrical power. In essence, it provides a direct measure of how much of the incident illumination is successfully harvested and converted into useful output. This parameter is particularly important for micro-scale energy harvesters, where the available surface area is limited and maximizing the power-per-unit area becomes crucial for enabling self-powered microsystems. The efficiency, commonly denoted as η, can be defined as the ratio between the maximum electrical power extracted from the device and the total optical power incident on its active area. Mathematically, this relationship is expressed as ref. [30]:

$$\eta = \frac{P_{pk}}{P_i} \times 100\% = \frac{P_{pk}}{A_e E_i} \times 100\% \tag{2}$$

where P_{pk} is the peak output power obtained from the microgenerator under illumination, P_i is the incoming light power, A_e is the effective active area of the MPG and E_i is the

 https://doi.org/10.3390/mi17010048

irradiance level of the incident light. Figure 3 illustrates the simulated power output characteristics. Under the highest simulated irradiance of 1000 W/m^2, the MPG achieves a maximum power of approximately 106 μW. With an active area of 0.7 mm^2, substituting these quantities into Equation (2) yields an estimated conversion efficiency of about 15%.

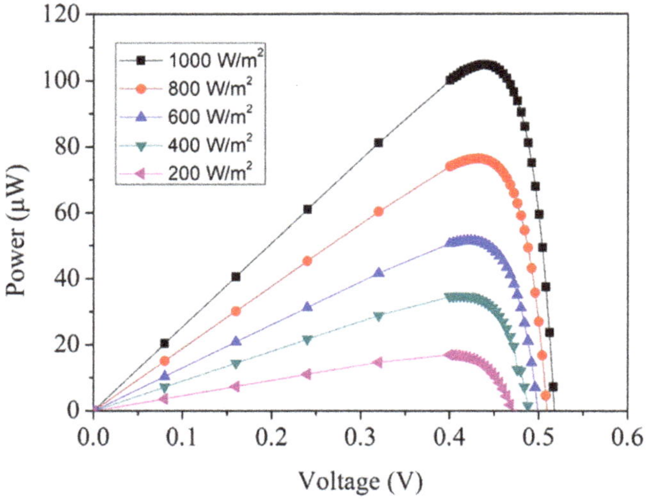

Figure 3. Simulated output power of the MPG.

3. Fabrication of the Micro Photovoltaic Generator

The MPG was fabricated using the 0.18 μm CMOS technology provided by Taiwan Semiconductor Manufacturing Company (TSMC). The foundry implemented the device layout—shown in Figure 1—using its standard CMOS process. A simplified overview of the fabrication sequence is presented in Figure 4. As with all CMOS-based devices, an oxide layer is formed during the process. This layer primarily functions as both an electrical insulator and a passivation coating, safeguarding the underlying semiconductor structures from moisture, ionic contaminants, and mechanical damage.

Figure 4a shows the cross-sectional profile of the device immediately after completion of the CMOS process. At this stage, a relatively thick silicon dioxide film covers the p–n junction regions. While beneficial from a protection standpoint, this dielectric layer can inadvertently reduce the amount of light reaching the active junction. Silicon dioxide possesses a refractive index mismatch relative to silicon, which can induce Fresnel reflections, unwanted scattering, and partial absorption—phenomena that diminish the number of photons penetrating into the depletion region. Consequently, the presence of this oxide layer can limit the efficiency of photovoltaic energy conversion. To address these optical losses, a post-CMOS modification was introduced [31]. As depicted in Figure 4b, the silicon dioxide layer above the active regions was selectively removed to expose the p–n junctions directly to incident illumination. The removal was performed using a silox vapox III etchant [32], chosen for its high selectivity and controlled etch rate. This post-processing step provides a clear optical pathway into the semiconductor, significantly enhancing the coupling of incident light into the depletion region. The resulting structure maximizes the absorption of photons and thus improves photocurrent generation, contributing to the higher measured efficiency of the device.

https://doi.org/10.3390/mi17010048

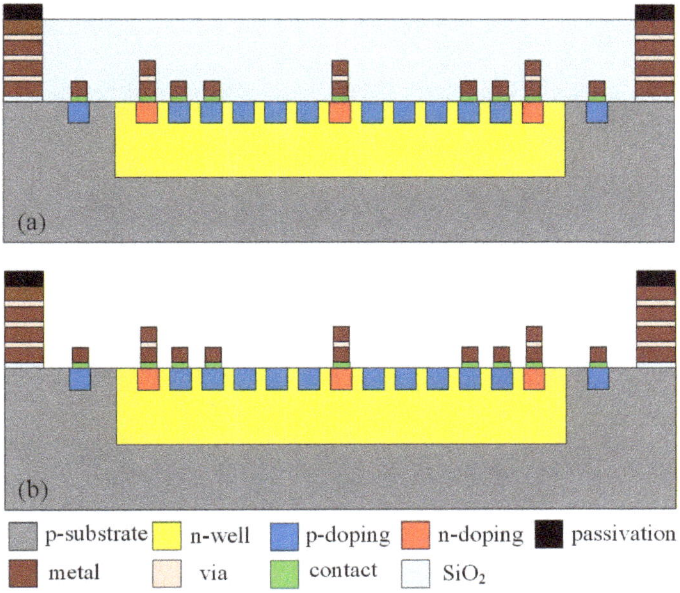

Figure 4. Fabrication flow of the MPG; (**a**) after the CMOS process, (**b**) after the post-processing.

Figure 5 presents an optical micrograph of the fabricated device before oxide removal, showing the overall pattern definition achieved through the CMOS process. In contrast, Figure 6 illustrates a scanning electron microscope (SEM) image captured after the post-etching process, where the recessed areas corresponding to the removed oxide layer are clearly visible. These SEM observations confirm that the etching process successfully exposed the underlying silicon while preserving the structural integrity of the patterned features. After fabrication and inspection, the MPG die was mounted onto a printed circuit board (PCB) for electrical characterization. Aluminum wire bonding was used to connect the device pads to the PCB traces. This packaging process provides reliable electrical connections for measurement equipment and enables stable, repeatable testing under controlled illumination conditions.

Figure 5. Optical image of the MPG after CMOS process.

https://doi.org/10.3390/mi17010048

Figure 6. Scanning electron microscope image of the MPG after the post-processing.

4. Results

To evaluate the operational characteristics of the MPG, a dedicated measurement setup was assembled consisting of a controlled illumination chamber, a calibrated optical power meter, and a precision digital multimeter. The illumination within the chamber was provided by a tungsten-filament lamp, which was selected due to its broad emission spectrum spanning roughly 380–780 nm—closely matching the visible wavelength range relevant to silicon-based photovoltaic devices. The lamp model used in this study (TFC 250 W, 120 V; Taiwan Fluorescent Lamp Co., Taipei, Taiwan) features adjustable brightness, enabling fine control of irradiance during testing. Prior to electrical measurements, the irradiance level was calibrated using the optical power meter to ensure consistent and repeatable illumination conditions. This calibration step is crucial because even small variations in light intensity can significantly influence photocurrent generation in micro-scale photovoltaic structures. Once calibration was completed, the MPG was mounted inside the chamber at a fixed distance from the lamp to maintain uniform exposure. The lamp's irradiance was adjusted using its integrated control mechanism, while the optical power meter continuously monitored the actual power density delivered to the device surface. The electrical output of the MPG—specifically its voltage and current—was measured using a high-accuracy digital multimeter. Current measurements required particular attention due to the relatively small magnitude of the short-circuit current produced by micro-scale generators. By synchronizing irradiance monitoring with electrical measurement, the experimental setup enabled reliable characterization of the device's response under various lighting conditions. In this study, a tungsten-filament lamp was employed as the illumination source for experimental characterization due to its stable output and broad continuous spectrum in the visible wavelength range. Although the spectral distribution of a tungsten lamp differs from the standard AM1.5G solar spectrum, the spectral overlap with silicon absorption is substantial within the visible range, which dominates photocarrier generation in silicon-based photovoltaic devices. The illumination intensity was carefully calibrated using an optical power meter, and all devices were measured under identical conditions to ensure consistent comparison of electrical performance.

To investigate how post-processing influences device behavior, both versions of the MPG—one retaining the original silicon dioxide passivation layer and the other subjected to oxide removal—were evaluated under identical illumination and temperature conditions. The resulting current–voltage characteristics are presented in Figure 7, where both devices were tested at room temperature under an irradiance of 1000 W/m^2.

https://doi.org/10.3390/mi17010048

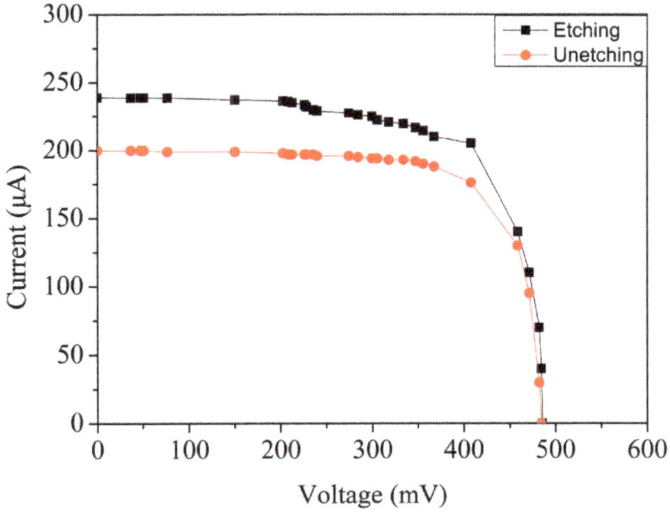

Figure 7. Measured current–voltage response of the MPGs with etched and unetched SiO_2 layers.

The contrast between the two curves is significant. The device with the unetched oxide layer generates a short-circuit current of approximately 199 μA, reflecting the partial attenuation of incoming photons caused by the thick silicon dioxide film. In comparison, the microgenerator from which the oxide layer was removed exhibits a higher short-circuit current, showing an improvement of roughly 20.1%. This enhancement is directly linked to the improved optical accessibility of the active p–n junctions. With the oxide layer still in place, part of the incident illumination is inevitably lost due to reflection at the air–oxide interface, refraction-induced angular scattering, and minor absorption within the dielectric. Removing the oxide eliminates these parasitic optical effects, allowing photons to penetrate directly into the depletion region without encountering an intermediate medium. The increased photon flux absorbed at the semiconductor surface results in greater electron–hole pair generation, which in turn elevates the short-circuit current.

The comparative results shown in Figure 7 clearly demonstrate that removing the silicon dioxide layer significantly enhances the MPG's electrical output. Because the etched device consistently exhibits superior photocurrent and improved I–V characteristics, all subsequent measurements in this study were performed using the oxide-removed version to ensure that the evaluation reflects the device's optimal operating condition. Figure 8 presents the measured current–voltage behavior of the microgenerator under a series of controlled irradiance levels. The illumination intensity was incrementally varied from 200 W/m^2 to 1000 W/m^2 in steps of 200 W/m^2, allowing a systematic assessment of the device performance across a broad range of lighting environments. At the lowest irradiance level of 200 W/m^2, the device produces an open-circuit voltage of approximately 0.46 V and a short-circuit current of 45 μA. These values correspond to the limited photon flux available under weak lighting and reflect the expected sublinear response in Voc observed in silicon-based photovoltaic devices. As the irradiance increases, a clear upward trend emerges. When the illumination reaches 1000 W/m^2, the open-circuit voltage rises to 0.49 V, while the short-circuit current increases substantially to 239 μA. The modest increase in voltage relative to the large increase in current is consistent with semiconductor photovoltaic theory. Because Voc has a logarithmic dependence on photocurrent, it changes only slightly even under large variations in irradiance. In contrast, the short-circuit current

https://doi.org/10.3390/mi17010048

is directly proportional to the number of absorbed photons, leading to a nearly linear growth as illumination intensifies.

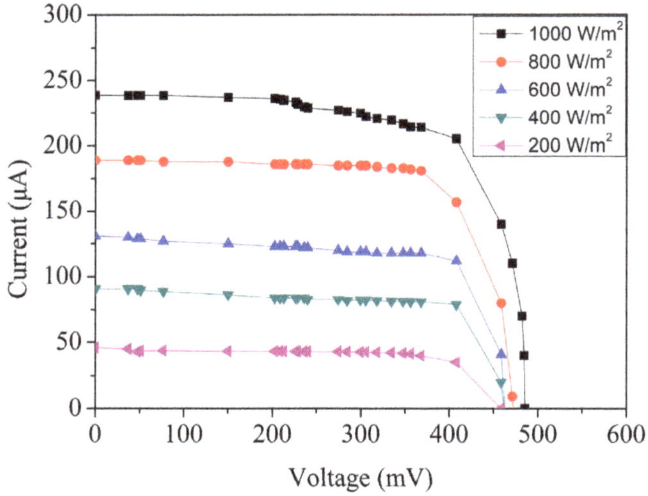

Figure 8. Measured current–voltage response of the MPG.

To further evaluate the electrical performance of the MPG, the measured current and voltage values obtained from Figure 8 were substituted into Equation (1) to evaluate the corresponding output power. The resulting power characteristics under different irradiance levels are presented in Figure 9. At an incident power density of 200 W/m^2, the device produces a maximum output power of approximately 13 μW, corresponding to an operating voltage of about 0.41 V. This relatively modest power generation is expected because fewer photons are available to create electron–hole pairs under low illumination conditions. Nevertheless, even at this level, the microgenerator generates a measurable and stable output, indicating good sensitivity to low-light environments. As the irradiance is increased, the output power rises significantly. When the illumination reaches 1000 W/m^2, the device's peak operating voltage increases slightly to 0.42 V, while the corresponding power output climbs to roughly 90 μW. At an irradiance of 1000 W/m^2, the simulated peak output power is slightly higher than the measured value. This discrepancy can be attributed to several practical non-idealities that are not fully captured in the TCAD simulation. First, optical shading caused by metal interconnects, contact pads, and routing reduces the effective illuminated junction area in the fabricated device, leading to a lower photocurrent than predicted. Second, contact resistance and series resistance, which include metal–silicon contact resistance, sheet resistance in lightly doped regions, and interconnect resistance, degrade the fill factor and thus reduce the extracted maximum power. Third, surface and edge recombination in the fabricated device may be higher than assumed in the simulation due to process-induced defects, interface states, and non-uniformities introduced during post-processing, which further reduce carrier collection efficiency. Finally, the TCAD model employs idealized assumptions such as uniform doping profiles, simplified optical boundary conditions, and perfect interfaces, and it does not account for packaging-related effects such as additional reflection or non-uniform illumination.

https://doi.org/10.3390/mi17010048

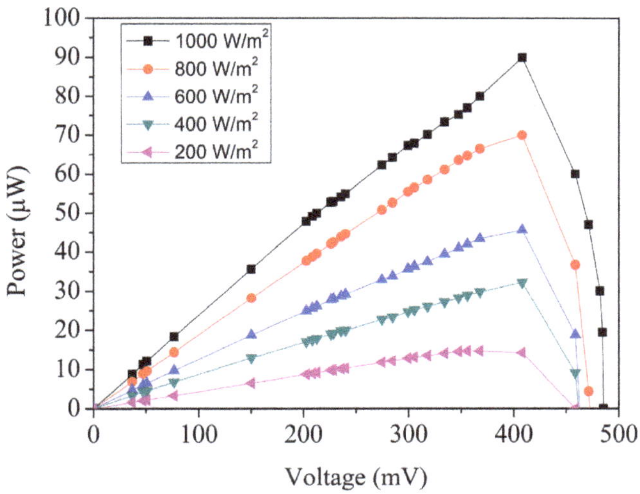

Figure 9. Measured output power of the MPG.

Based on the experimental data presented in Figures 8 and 9, the MPG demonstrates strong performance under high illumination. When exposed to an irradiance of 1000 W/m^2, the device produces an open-circuit voltage of approximately 0.49 V and a short-circuit current of 239 μA. At this same illumination level, the microgenerator delivers a peak electrical output of about 90 μW, confirming the effective extraction of photogenerated carriers under intense lighting conditions. Given that the active area of the device is 0.7 mm^2, its energy conversion efficiency can be calculated by substituting the measured performance metrics into Equation (2). The resulting efficiency is roughly 12.9%, which is a notable value for a CMOS-compatible MPG of such a small footprint. Achieving an efficiency above 10% in a sub-millimeter-scale device is particularly meaningful because microfabricated photovoltaic structures often suffer from limited absorption area, optical shading from metallization layers, and junction losses associated with shallow doping profiles. The measured efficiency therefore reflects not only the inherent quality of the p–n junctions but also the benefits of the post-etching process, which improves photon penetration and reduces reflective losses at the device surface. The energy conversion efficiency reported in this study is evaluated under an irradiance of 1000 W/m^2 using a tungsten-filament light source and should be interpreted as an experimental efficiency under broadband tungsten illumination rather than as a standard photovoltaic efficiency measured under AM1.5G conditions. Although spectral mismatch exists between a tungsten lamp and the standard solar spectrum, the tungsten source provides a stable and continuous spectrum with substantial overlap in the visible range, which dominates photocarrier generation in silicon-based devices. All measurements were performed under identical, calibrated illumination conditions, ensuring that the reported efficiency values are suitable for relative performance comparison and for assessing the effectiveness of the proposed design and post-processing strategy.

The performance improvement of the proposed photovoltaic microgenerator can be quantitatively attributed to the increased junction density enabled by the mesh-patterned layout. For a representative conventional interdigitated photovoltaic design reported in the literature [33], with an n-type finger width of 75 μm and a p-type finger width of 125 μm, the resulting finger pitch is approximately 200 μm. Using the standard geometric approximation for junction perimeter per unit area, the corresponding junction perimeter density is about 0.01 μm^{-1}. In contrast, the proposed mesh structure employs a minimum

https://doi.org/10.3390/mi17010048

unit cell size of 20 μm, yielding a junction perimeter density of approximately 0.10 μm^{-1}, which represents a substantially higher geometric junction density. In addition, the silox vapox III etchant removes the surface SiO_2 layer above the junction regions, enabling direct illumination of most mesh-distributed junction boundaries, whereas conventional interdigitated layouts are partially shielded by dielectric layers or metal fingers. Furthermore, the reduced mesh spacing limits the average lateral carrier collection length to approximately 10 μm, compared with about 100 μm in the interdigitated structure, thereby lowering recombination probability. The enhanced carrier collection of the proposed photovoltaic microgenerator originates from the mesh-patterned p–n junction architecture. Compared with conventional interdigitated layouts, the mesh structure increases the effective junction perimeter density, thereby enlarging the depletion boundary where photogenerated carriers are separated by the built-in electric field. The distributed mesh geometry also provides a more uniform spatial distribution of electric-field regions, allowing carriers generated throughout the active area to encounter nearby junctions. Furthermore, the reduced mesh spacing shortens the effective lateral carrier collection length, lowering recombination probability and promoting drift-assisted transport. These combined effects explain the experimentally observed enhancement in short-circuit current.

Table 1 presents a comparative overview of the reported performance of various micro power generators (MPGs). The short-circuit current achieved in this work is higher than those reported by Hung et al. [21], Arima et al. [25], Law et al. [26], and Masui et al. [28]. The MPGs reported in Refs. [22–28] were fabricated without any post-processing steps, whereas the proposed MPG incorporates a post-processing treatment to enhance device performance. Most previously reported MPGs, including those by Hung et al. [21], Yan et al. [22], Sun et al. [24], Arima et al. [25], Law et al. [26], and Sultana [27], exhibit energy conversion efficiencies below 5%. The device reported by Sichao et al. [23] shows a moderate improvement, achieving an efficiency of 5.45%. By contrast, the photovoltaic microgenerator developed in this work demonstrates a higher energy conversion efficiency of 12.9%. As a result, the proposed device outperforms the MPGs reported in Refs. [21–27] in terms of energy conversion efficiency.

Table 1. Summary of performance for various MPGs.

Authors	Open-Circuit Voltage (V)	Short-Circuit Current (μA)	Energy Conversion Efficiency (%)	Post-Processing
Hung [21]	0.5	11	0.5	Yes
Yan [12]	0.53	1.06×10^4	4.45	No
Sichao [23]	0.53	1.53×10^4	5.45	No
Sun [24]	0.53	1.06×10^4	4.5	No
Arima [25]	0.83	0.4	2.6	No
Law [26]	0.84	1.25×10^{-3}	0.3	No
Sultana [27]	1.1	1.1×10^4	4.19	No
Masui [28]	0.97	8	–	No
This work	0.49	239	12.9	Yes

While removal of the surface SiO_2 layer effectively enhances photocurrent by improving optical coupling, it may also introduce concerns related to surface recombination, environmental exposure, and long-term reliability. Exposed silicon surfaces and etch-induced interface states can increase surface recombination; however, in the proposed mesh structure, the short lateral carrier collection length reduces carrier residence time near the surface, partially mitigating this effect. In addition, oxide removal is confined to the active junction regions, while surrounding dielectrics and metal layers remain in-

https://doi.org/10.3390/mi17010048

tact and provide partial protection. For practical applications, additional passivation and packaging strategies, such as thin transparent passivation layers, hydrogen-based surface treatments, or transparent encapsulation, can be employed to balance optical transparency and surface stability.

The experimental results presented in this study are obtained from representative photovoltaic microgenerator devices fabricated using a standard 0.18-μm CMOS process followed by a uniform wet-etch post-processing step. Although a full statistical analysis across multiple wafers is beyond the scope of this work, the proposed mesh-patterned architecture is expected to exhibit good reproducibility and process tolerance. Unlike designs that rely on a single or narrowly defined junction, the mesh structure consists of a distributed network of p–n junctions, which reduces sensitivity to local variations in dopant concentration or junction geometry inherent to CMOS fabrication. The potential variations arising from CMOS doping fluctuations primarily affect local electric-field strength and series resistance but tend to average out over the large number of junction segments within the mesh. Similarly, post-etch non-uniformity is mitigated by the use of a global wet-etching process with controlled etch time, resulting in relatively uniform oxide removal across the active regions. As a result, the measured device performance is considered representative of the fabricated structures.

5. Conclusions

The MPG presented in this work was fabricated entirely using a commercial 0.18 μm CMOS process, followed by a post-processing step to improve its optical absorption characteristics. Prior to fabrication, TCAD simulations were conducted to predict device behavior. Under simulated conditions—specifically, a temperature of 25 °C and irradiance of 1000 W/m^2—the microgenerator was predicted to deliver an open-circuit voltage of approximately 0.52 V, a short-circuit current near 254 μA, and a maximum output power of roughly 106 μW. These simulated values established performance benchmarks for later experimental comparison. After completion of the CMOS manufacturing flow, a wet-etching post-process was applied to selectively remove the silicon dioxide layer overlying the active junction regions. Eliminating this dielectric layer allows incident photons to reach the depletion region with minimal reflection or scattering losses. As a result, the generation of electron–hole pairs is more efficient, leading directly to an improvement in photocurrent. Experimental measurements validate the benefit of this post-processing enhancement. When tested under the same irradiance of 1000 W/m^2, the device with the oxide layer removed exhibited a short-circuit current approximately 20.1% higher than its unetched counterpart. This substantial increase confirms that surface passivation layers—while useful for environmental protection—can significantly hinder optical coupling in photovoltaic microdevices. Further experimental characterization revealed that the device achieved an open-circuit voltage of 0.49 V, a short-circuit current of 239 μA, and a peak output power of 90 μW. These results translate to an overall energy conversion efficiency of 12.9%, which is competitive for CMOS-based MPGs. These findings demonstrate not only the accuracy of the TCAD modeling but also the effectiveness of combining CMOS compatibility with minimal post-processing to significantly enhance the performance of MPG.

Author Contributions: Conceptualization, C.-L.D.; Methodology, H.-W.C. and C.-L.D.; Software, H.-W.C.; Validation, H.-W.C., C.-Y.L. and C.-L.D.; Formal analysis, H.-W.C.; Investigation, H.-W.C.; Resources, C.-Y.L. and C.-L.D.; Data curation, C.-Y.L. and C.-L.D.; Writing—original draft, H.-W.C. and C.-L.D.; Writing—review & editing, C.-L.D.; Visualization, C.-Y.L.; Supervision, C.-L.D.; Project administration, C.-L.D. All authors have read and agreed to the published version of the manuscript.

 https://doi.org/10.3390/mi17010048

Funding: This research was funded by National Science and Technology Council of the Republic of China grant number NSTC 114-2221-E-005-067-MY2.

Data Availability Statement: The data presented in this study are available on request from the corresponding author.

Acknowledgments: The authors would like to thank the National Science and Technology Council of the Republic of China for financial support under the Grant: NSTC 114-2221-E-005-067-MY2. The authors also gratefully acknowledge the technical support provided by the Taiwan Semiconductor Research Institute (TSRI) and National Center for High-performance Computing (NCHC).

Conflicts of Interest: The authors declare no conflict of interest. Hung-Wei Chen is employees of Powerchip Semiconductor Manufacturing Corporation. The paper reflects the views of the scientists, and not the company.

References

1. Xia, L.; Ma, S.; Tao, P.; Pei, W.; Liu, Y.; Tao, L.; Wu, Y. A wind-solar hybrid energy harvesting approach based on wind-induced vibration structure applied in smart agriculture. *Micromachines* **2023**, *14*, 58. [CrossRef] [PubMed]
2. Lian, Q.; Han, P.; Mei, N. A review of converter circuits for ambient micro energy harvesting. *Micromachines* **2022**, *13*, 2222. [CrossRef]
3. Santos, M.S.; Marques Lameirinhas, R.A.; N. Torres, J.P.; P. Fernandes, J.F.; Correia, V.; Bernardo, C.P. Nanostructures for solar energy harvesting. *Micromachines* **2023**, *14*, 364. [CrossRef]
4. Rokonuzzaman, M.; Mishu, M.K.; Amin, N.; Nadarajah, M.; Roy, R.B.; Rahman, K.S.; Buhari, A.M.; Binzaid, S.; Shakeri, M.; Pasupuleti, J. Self-sustained autonomous wireless sensor network with integrated solar photovoltaic system for internet of smart home-building (IoSHB) applications. *Micromachines* **2021**, *12*, 653. [CrossRef]
5. Astigarraga, A.; Lopez-Gasso, A.; Golpe, D.; Beriain, A.; Solar, H.; del Rio, D.; Berenguer, R. A 21 m operation range RFID tag for "pick to light" applications with a photovoltaic harvester. *Micromachines* **2020**, *11*, 1013. [CrossRef]
6. Dai, Z.X.; Chen, C.Y.; Chiu, B.C.; Lee, C.Y. Manufacturing and measurement of CMOS-MEMS-based micro thermoelectric generators with long-length thermocouples. *J. Micromech. Microeng.* **2024**, *34*, 075001. [CrossRef]
7. Chen, Z.; Xie, Z. Investigation of the thermalconductance of MEMS contact switches. *Micromachines* **2025**, *16*, 872.
8. Dai, C.L.; Xiao, F.Y.; Juang, Y.Z.; Chiu, C.F. An approach to fabricating microstructures that incorporate circuits using a post-CMOS process. *J. Micromech. Microeng.* **2005**, *15*, 98. [CrossRef]
9. Shen, Y.X.; Tsai, Y.C.; Lee, C.Y.; Wu, C.C.; Dai, C.L. Thermoelectric energy micro harvesters with temperature sensors manufactured utilizing the CMOS-MEMS technique. *Micromachines* **2022**, *13*, 1258. [CrossRef]
10. Zhang, J.; Zhou, C.; Wu, Y. Research on pneumatic proportional pressure valve based on silicon microfluidic chip with V-shaped electrothermal microactuator. *Micromachines* **2025**, *16*, 566. [CrossRef]
11. Dai, Z.X.; Tsai, R.W.; Shih, Q.H. High-sensitivity magnetic field sensors manufactured utilizing CMOS–MEMS technology. *IEEE Sens. J.* **2025**, *25*, 9540–9549. [CrossRef]
12. Tsai, Y.C.; Huang, P.C.; Dai, C.L. Editorial for the special issue on MEMS/NEMS devices and applications, 2nd Edition. *Micromachines* **2025**, *16*, 189. [CrossRef] [PubMed]
13. Lee, C.Y.; Huang, Z.Y.; Hsu, J.Y.; Kuo, T.F.; Dai, Z.X. Measurement circuit design of 3-in-1 microsensor embedded in proton exchange membrane water electrolyzer. *Sens. Mater.* **2025**, *37*, 1537–1550. [CrossRef]
14. Wu, C.H.; Hsu, C.C.; Tsai, Y.C.; Lee, C.Y.; Dai, C.L. Design and measurement of microelectromechanical three-axis magnetic field sensors based on the CMOS technique. *Micromachines* **2023**, *14*, 1038. [CrossRef]
15. Dai, Z.X.; You, M.H. Modeling and measurement of CMOS-based three-axis magnetic field sensors with low cross-sensitivity. *IEEE Sens. J.* **2024**, *24*, 34217–34227. [CrossRef]
16. Lin, C.Y.; Hsu, C.C.; Dai, C.L. Fabrication of a micromachined capacitive switch using the CMOS-MEMS technology. *Micromachines* **2015**, *6*, 1645–1654. [CrossRef]
17. Li, H.; Hou, J.; Gong, Z.; Yu, H.; Liu, Y.; Shen, W. Hybrid actuation MEMS micromirror with decoupled piezoelectric fast axis and electromagnetic slow axis for crosstalk suppression. *Micromachines* **2025**, *16*, 1072. [CrossRef] [PubMed]
18. Kao, P.-H.; Dai, C.-L.; Hsu, C.-C.; Lee, C.-Y. Fabrication and characterization of a tunable in-plane resonator with low driving voltage. *Sensors* **2009**, *9*, 2062–2075. [CrossRef] [PubMed]
19. Shuaibu, A.H.; Rabih, A.A.S.; Blaquière, Y.; Nabki, F. Push–push electrothermal MEMS actuators with Si-to-Si contact for DC power switching applications. *Micromachines* **2025**, *16*, 977. [CrossRef]
20. Dai, C.L.; Peng, H.J.; Liu, M.C.; Wu, C.C.; Hsu, H.M.; Yang, L.J. A micromachined microwave switch fabricated by the complementary metal oxide semiconductor post-process of etching silicon dioxide. *Jpn. J. Appl. Phys.* **2005**, *44*, 6804. [CrossRef]

 https://doi.org/10.3390/mi17010048

21. Hung, Y.J.; Cheng, Y.C.; Cai, M.S.; Lu, C.H.; Su, H.W. High-voltage 12.5-V backside-Illuminated CMOS photovoltaic mini-modules. *IEEE J. Electron Devices Soc.* **2018**, *6*, 135–138. [CrossRef]
22. Yan, J.; Liao, X.; Ji, S.; Zhang, S. MEMS-based thermoelectric–photoelectric integrated power generator. *J. Microelectromech. Syst.* **2019**, *28*, 1–3. [CrossRef]
23. Sichao, J.; Liao, X. Researches on MEMS thermoelectric-photoelectric integrated energy harvester with metal heat sink. *Microelectron. J.* **2020**, *96*, 104702. [CrossRef]
24. Sun, M.; Liao, X. Theoretical and experimental investigation of a photoelectric-thermoelectric integrated power generator filled by Al layer. *IEEE Trans. Electron Devices* **2022**, *69*, 2462–2468. [CrossRef]
25. Arima, Y.; Ehara, M. On-chip solar battery structure for CMOS LSI. *IEICE Electron. Exp.* **2006**, *3*, 287–291. [CrossRef]
26. Law, M.K.; Bermak, A. High-voltage generation with stacked photodiodes in standard CMOS process. *IEEE Electron Device Lett.* **2010**, *31*, 1425–1427. [CrossRef]
27. Sultana, J.; Paul, S.; Saha, R.; Sikdar, S.; Karmakar, A.; Chattopadhyay, S. Optical and electronic properties of chemical bath deposited p-CuO and n-ZnO nanowires on silicon substrates: P-CuO/n-ZnO nanowires solar cells with high open-circuit voltage and short-circuit current. *Thin Solid Films* **2020**, *699*, 137861. [CrossRef]
28. Masui, Y. A CMOS temperature sensor with on-chip photovoltaic cells. *IEEJ Trans. Electr. Electron. Eng.* **2023**, *18*, 401–407. [CrossRef]
29. Dai, Z.X.; Chen, C.Y. Fabrication and evaluation of a CMOS-based energy harvesting chip integrating photovoltaic and thermoelectric energy harvesters. *Energy Convers. Manag. X* **2024**, *24*, 100748. [CrossRef]
30. Chen, C.Y.; Dai, Z.X. Fabrication and characterization of photovoltaic microgenerators using the complementary metal oxide semiconductor process. *Micromachines* **2023**, *14*, 2038. [CrossRef]
31. Dai, Z.X.; Li, W.H.; Liu, Y.N. Effect of post-processing techniques on the performance of CMOS-based photovoltaic energy harvesters with grid-like architectures. *J. Micromech. Microeng.* **2025**, *35*, 065005. [CrossRef]
32. Dai, C.L. A maskless wet etching silicon dioxide post-CMOS process and its application. *Microelectron. Eng.* **2006**, *83*, 2543–2550. [CrossRef]
33. Lammert, M.D.; Schwartz, R.J. The interdigitated back contact solar cell: A silicon solar cell for use in concentrated sunlight. *IEEE Trans. Electron Dev.* **1997**, *24*, 337–342. [CrossRef]

https://doi.org/10.3390/mi17010048

micromachines

Article

Experimental Study on the Performance of Light-Controlled Ion Drag Pump Based on PLZT Ceramic

Yujuan Tang [1,*], Yujie Shi [2], Zhen Lv [2], Zihao Guo [1] and Xinjie Wang [2]

[1] School of Intelligent Science and Control Engineering, Jinling Institute of Technology, Nanjing 211169, China; zhguo020528@163.com
[2] School of Mechanical Engineering, Nanjing University of Science and Technology, Nanjing 210094, China; yjshi@njust.edu.cn (Y.S.); zlv@njust.edu.cn (Z.L.); xjwang@njust.edu.cn (X.W.)
* Correspondence: yjtang@jit.edu.cn

Abstract

Light-controlled ion drag pumps have attracted considerable interest in soft robotics, biomedical engineering, and microelectromechanical systems (MEMS) due to their non-contact energy supply and high spatiotemporal controllability of light. However, experimental studies on their pumping performance and influencing factors remain limited. This study integrates the photoelectric effect with field emission phenomena to design and fabricate a light-controlled ion drag pump using lanthanum-modified lead zirconate titanate (PLZT) ceramic. The light-controlled pump enables non-contact energy transfer and fluid transport via high-energy laser irradiation. A series of experiments systematically investigate its pumping performance and key influencing factors. Results indicate that optimizing electrode structure and fluid channel design, along with increased light intensity, significantly enhances pumping performance. This work provides fundamental design guidelines for the application of light-controlled ion drag pumps in microfluidics, flexible robotics, and microdevice thermal management.

Keywords: electrohydrodynamics; PLZT ceramics; charge injection; ion drag pump; pumping performance

1. Introduction

The advancement of micro-electro-mechanical systems (MEMS) and micro-optical-electro-mechanical systems (MOEMS) has created a strong demand for compact, efficient, and wireless micro-actuation technologies [1]. Conventional micro-pumps, driven by electrostatic, piezoelectric, electrothermal, or electromagnetic principles, often involve moving parts, complex wiring, and susceptibility to electromagnetic interference [2]. In contrast, electrohydrodynamic (EHD) pumps, which operate on Coulomb forces without moving parts, offer advantages such as silent operation, simplicity, and high reliability [3]. Among EHD methods, ion drag pumps in which charges injected into a dielectric fluid drag the fluid from an emitter to a collector electrode show broad application potential in fluid actuation, particularly in microfluidics, soft robotics, and microdevice cooling [4,5].

However, a significant limitation persists: Ion drag pumps typically require an external high-voltage power supply (often kilovolts) [6,7], which complicates system integration, increases size, and generates electromagnetic noise, thereby hindering miniaturization and use in sensitive environments. Recent research has focused on optimizing electrode configurations [8], adopting interdigitated electrodes [9,10] to improve the performance of the ion drag pumps, but the fundamental reliance on wired, high-voltage sources remains unaddressed. The integration of high-voltage power supply units complicates the overall pumping system and is prone to generating electromagnetic interference, which hinders the development of injection pump systems in terms of miniaturization, high integration, and electro-magnetic interference resistance.

Parallelly, lanthanum-modified lead zirconate titanate (PLZT) ceramics represent a distinctive photoelectric smart material. Under ultraviolet (UV) excitation, they exhibit anomalous photovoltaic, thermoelectric, photothermal, thermal expansion, and piezoelectric effects [11]. From a photophysical perspective, the kilovolt-level photovoltage generated when UV light irradiates the PLZT polarization direction is highly significant for practical applications, e.g., hybrid photo-vibrational energy harvesting and micro-drive. Compared with conventional solar cells, which typically generate low-voltage direct current requiring power conditioning circuits for most applications, PLZT ceramics can produce a high photovoltage (on the order of kilovolts) directly, eliminating the need for external voltage amplifiers. Their operation generates minimal electromagnetic interference, making them suitable for noise-sensitive environments such as biomedical or precision measurement systems. This property has been exploited in lunar dust removal devices [12], novel actuators [13], and micro gripper [14] and damping systems [15], suggesting great potential for clean, light-driven micro-systems. Thus, the rapid photovoltage generated by PLZT ceramics under illumination can provide energy for injection pumps, thereby enabling light-controlled pumping of fluids.

In contrast to previous ion drag pumps that rely on external high-voltage supplies, this work introduces a fundamentally distinct driving strategy by integrating the photoelectric effect of PLZT ceramics with charge injection pumping. We present, for the first time, a fully light-controlled ion drag pump where the high photovoltage generated by PLZT ceramics under UV illumination directly powers the pump electrodes, enabling wireless and non-contact fluid actuation. Beyond the conceptual integration, this study provides a systematic experimental investigation into the key performance-influencing factors (light intensity, electrode geometry, and channel dimensions), establishing practical design guidelines. This approach not only addresses the miniaturization and electro-magnetic interference challenges associated with traditional high-voltage sources but also opens new avenues for optically addressable microfluidics and soft robotic systems.

2. Mathematical Modelling of the Light-Controlled Ion Drag Pump Based on PLZT Ceramic

Figure 1 illustrates the schematic of the light-controlled PLZT ion drag pump. The system comprises a PLZT ceramic providing high voltage and an ion drag pump chip, interconnected via wires. The pump chip features a four-layer sandwich structure (substrate, electrode, channel, and cover layers). The staggered electrode structure within the electrode layer generates the required electric field upon receiving voltage from the PLZT ceramic, driving dielectric fluid flow in the channel layer. A single-sided serrated electrode structure is adopted for the pump chip. To prevent dielectric breakdown, dielectric materials are used for the substrate, channel, and cover layers. In this design, UV light (wavelength ~365 nm) irradiates the PLZT ceramic, inducing a rapidly increasing photovoltage via the photoelectric effect. This photovoltage is applied to the electrode layer of the pump chip. As shown in Figure 1, the pump chip's positive and negative electrodes function as the collector and emitter, respectively. When the photoelectric field strength surpasses the energy barrier, field emission occurs, injecting electrons from the emitter into the dielectric fluid. Neutral molecules capture these electrons, forming ions. Under the photoelectric field, ions migrate toward the collector, releasing electrons and thereby driving fluid flow.

https://doi.org/10.3390/mi17010045

Upon UV cessation, the PLZT ceases photovoltage generation, the inter-electrode field gradually dissipates, and flow stops. Thus, powered solely by PLZT-generated photovoltage, this pump operates without external power sources, offering significant miniaturization advantages. Key dimensional parameters of the pump chip are labeled in Figure 1. For the ion drag pump chip, the channel height h_c, electrode spacing g_e, and gap between the electrode pairs g_p, are the primary design variables investigated in this study. Their values are varied within the range of 0.3 mm to 2 mm in subsequent experiments to systematically evaluate their influence on pumping performance.

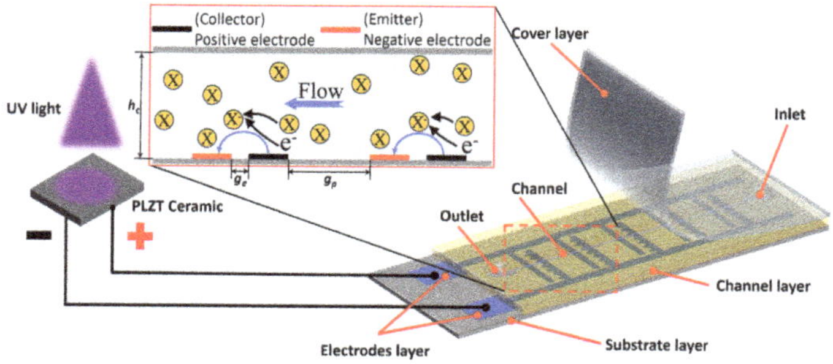

Figure 1. Structure and working mechanism of the light-controlled ion drag pump based on PLZT ceramic.

According to related literature, PLZT ceramics can be represented as an equivalent circuit consisting of a current source I_p in parallel with a resistor R_p and a capacitor C_p. When PLZT ceramics are connected to the ion drag pump, the equivalent capacitance and resistance of the pump are effectively placed in parallel with resistor R_p and a capacitor C_p in parallel. The photovoltage V_p of the light-controlled ion drag pump is expressed as follows [15]:

$$V_p = I_p R_{eq}\left(1 - e^{-\frac{t}{\tau}}\right) \tag{1}$$

where R_{eq} is the equivalent resistance of the light-controlled ion drag pump $R_{eq} = \frac{R_p R_{pump}}{R_p + R_{pump}}$, and τ is the time constant $\tau = \frac{(C_p + C_{pump}) R_p R_{pump}}{R_p + R_{pump}}$; and C_{pump} and R_{pump} represent the equivalent capacitance and resistance of the pump.

Additionally, according to the Korteweg–Helmholtz formula, the electric volume force exerted on a unit volume of fluid in an electric field F_d is:

$$F_d = \rho_d E - \frac{1}{2} E^2 \nabla \varepsilon + \frac{1}{2} \nabla \left[E^2 \rho \left(\frac{\partial \varepsilon}{\partial \rho} \right)_T \right] \tag{2}$$

where ρ_d, E, ε, and ρ denote the space charge density, electric field strength, dielectric constant of the fluid, and density of the fluid, respectively.

Equation (2) comprises three components: the Coulomb force (electrophoretic term) acting on free charges, the dielectrophoretic force arising from permittivity gradients, and the electrocapillary force relevant in compressible dielectrics. For an incompressible fluid, the third term vanishes.

Based on the law of conservation of mass, the continuity equation for fluids can be derived:

$$\frac{\partial \rho}{\partial t} + \nabla \cdot \rho v = 0 \tag{3}$$

where t is time and v is the flow velocity.

https://doi.org/10.3390/mi17010045

Equation (3) rests on the continuum assumption, which is valid for the flow regimes considered in this EHD analysis. According to the principle of conservation of momentum, fluid motion in a flow field satisfies the Navier–Stokes momentum equation:

$$\rho\left[\frac{\partial u}{\partial t}+(v\cdot\nabla)v\right]=-\nabla p+\eta\nabla^{2}v+F_{d} \quad (4)$$

where η and p are the fluid viscosity and the pressure, respectively.

According to the electrostatic Poisson's equation, the relationship between the electric field and photovoltage is:

$$E=-\nabla V=-\nabla\left[I_{p}\left(\frac{R_{p}R_{pump}}{R_{p}+R_{pump}}\right)\left(1-e^{-\frac{t}{(C_{p}+C_{pump})R_{p}R_{pump}}(R_{p}+R_{pump})}\right)\right] \quad (5)$$

The space charge density ρ_d follows charge conservation:

$$\frac{\partial\rho_{d}}{\partial t}+\nabla\cdot J=0 \quad (6)$$

where J is current density, representing total charge passing per unit area per unit time. Current density J is expressed as:

$$J=\mu_{d}\rho_{d}E+D\nabla\rho_{d}+\rho_{d}u+\sigma_{d}E \quad (7)$$

where μ_d, D, and σ_d represent the electron mobility coefficient, molecular diffusion coefficient, and electrical conductivity coefficient, respectively.

Additionally, the output performance of this light-controlled micro-injection pump is defined by its output pressure and output flow rate. The output pressure can be calculated using Equation (5), while the flow rate Q is obtained by integrating the velocity profile over the channel cross-section A:

$$Q=\int_{A}v\,dA \quad (8)$$

3. Experimental Analysis of Influencing Factors on the Output Performance of the Light-Controlled Ion Drag Pump

3.1. Experimental Platform for the Output Performance of Light-Controlled PLZT Ion Drag Pump

Figure 2 shows the experimental platform for evaluating the output performance of the light-controlled ion drag pump. During testing, the PLZT ceramic is continuously irradiated with adjustable-intensity continuous-wave UV light. A pressure gauge (PS4, Shanghai Pengzan Biotechnology Co., Ltd., Shanghai, China) and a flowmeter (MicroFluidics, Dolymate Ltd., Royston, UK) measure output pressure and flow rate, respectively. Illumination duration is 300 s, with data collection continuing for 400 s post-illumination. The cantilever configuration of the PLZT ceramic enables the UV light to be easily aligned vertically with the top surface of the PLZT ceramic and reduces parasitic capacitive coupling and unintended charge leakage paths of the photo-induced voltage. A standard ion drag pump chip (channel height: 1 mm, electrode spacing: 0.3 mm, gap between electrode pairs: 0.6 mm, 25 electrode pairs) and a PLZT ceramic patch (15 mm × 5 mm × 1 mm) are used in this paper.

https://doi.org/10.3390/mi17010045

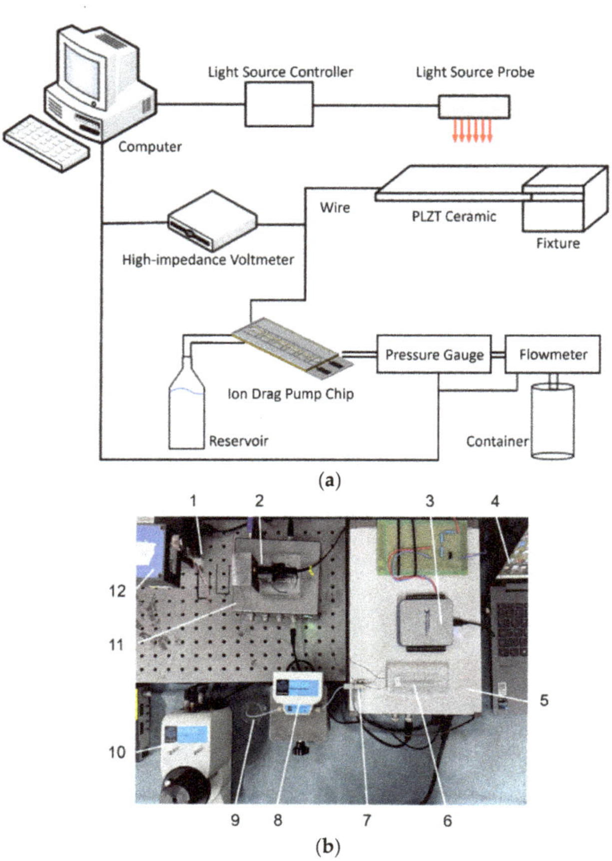

Figure 2. Experimental platform of the performance of light-controlled ion drag pump: (**a**) Schematic diagram; (**b**) Experimental platform: 1—PLZT ceramic, 2—high-impedance voltmeter probe, 3—data acquisition card, 4—computer, 5—high-impedance voltmeter, 6—micro-injection pump chip, 7—pressure gauge, 8—flow meter, 9—beaker, 10—flow meter controller, 11—pressure gauge controller, and 12—continuous-wave UV light probe.

PLZT ceramics generate different photovoltages under varying UV intensities. To characterize pump performance under different light intensities, pressure and flow rate are tested at 50 mW/cm^2 and 100 mW/cm^2, as shown in Figure 3. Comparison of Figure 3a,b shows that increasing light intensity from 50 mW/cm^2 to 100 mW/cm^2 shortens the pump's response time from 150 s to 53 s. The difference in response time is fundamentally governed by the photo-charging dynamics of the PLZT ceramic. As described by Equation (1), the photovoltage Vp builds up over time upon illumination. A higher light intensity (100 mW/cm^2) leads to a higher photovoltage Vp and smaller time constant, which in turn charges the equivalent circuit (Rp//Cp) more rapidly. Consequently, the threshold electric field required for charge injection and pump activation is reached in a shorter time (~70 s) compared to the lower intensity case (50 mW/cm^2, ~150 s). Peak output pressure and flow rate increase from 0.17 kPa and 280 μL/min to 0.25 kPa and 430 μL/min, representing increases of 47% and 53%, respectively. These results validate the performance and feasibility of the light-controlled ion drag pump. Furthermore, the performance of the pump can be adjusted by changing the light intensity.

https://doi.org/10.3390/mi17010045

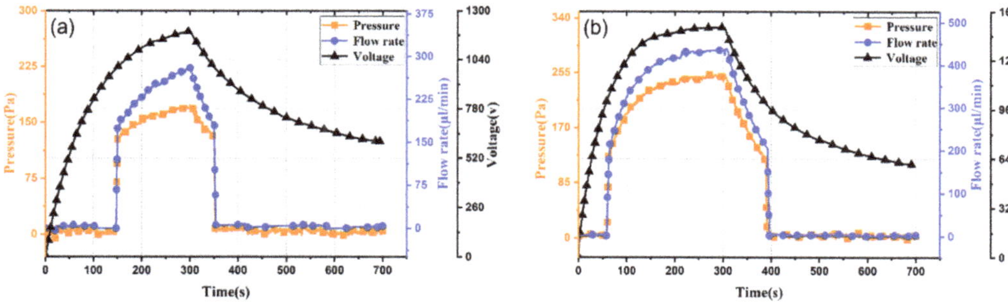

Figure 3. Dynamic response curves of the light-controlled ion drag pumps under different light intensities: (**a**) 50 mW/cm^2; (**b**) 100 mW/cm^2.

Subsequent sections investigate key factors affecting output performance: microchannel height, electrode spacing, number of electrode pairs, and gap between electrode pairs. Experimental data are compared with numerical simulations based on the mathematical model.

3.2. Effect of Microchannel Height

Figure 4 presents experimental results for light-controlled ion drag pumps with channel heights of 1 mm, 1.5 mm, and 2 mm. As shown in Figure 4a–c, channel height has negligible impact on output response time, which remains about 55 s for all heights, determined by the PLZT photovoltaic voltage characteristics, which are unchanged across channel heights. Thus, channel height does not affect response speed.

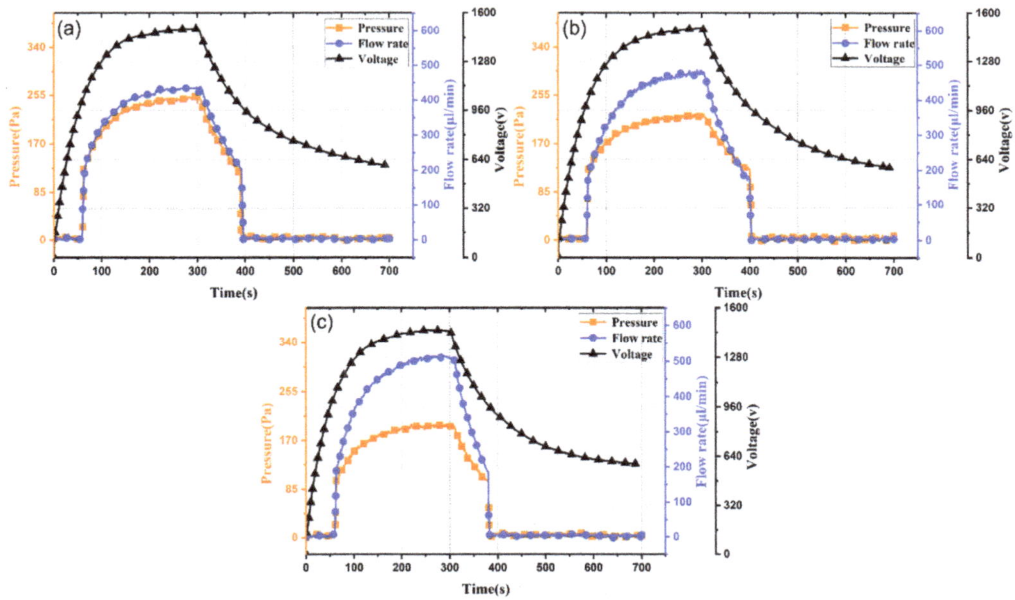

Figure 4. Time-dependent performance curves of light-controlled ion drag pumps at different channel heights: (**a**) 1 mm; (**b**) 1.5 mm; and (**c**) 2 mm.

Comparing output pressure and flow rate reveals that as channel height increases, peak output pressure decreases, while output flow rate increases. At 100 mW/cm^2, increasing channel height from 1 mm to 2 mm reduces maximum output pressure from 250 Pa to 200 Pa, while maximum flow rate increases from 430 μL/min to 510 μL/min (18.6%

https://doi.org/10.3390/mi17010045

increase). Therefore, increasing channel height effectively enhances output flow rate but reduces output pressure.

The inverse relationship between the channel height and the output pressure, as well as the positive impact on the flow rate, is consistent with the basic principles of fluid dynamics. A larger cross-sectional area reduces the flow resistance, thereby achieving a higher volumetric flow rate. However, for a given charge injection density and electric field strength, the momentum transferred by the unit volume of fluid decreases as the channel expands, resulting in a lower achievable pressure. This trade-off between pressure and flow rate is crucial for adjusting the pump according to specific application requirements, such as seeking a higher flow rate for cooling or a higher pressure to overcome system resistance.

To validate the mathematical model from Section 2, Table 1 compares simulated and experimental maximum output pressures for different channel heights at 100 mW/cm^2. The close agreement confirms model reliability.

Table 1. Comparison of numerical simulation and experimental results for the maximum output pressure at different channel heights.

	1 mm	1.5 mm	2 mm
Simulation (Pa)	265 Pa	236 Pa	213 Pa
Experiment (Pa)	250 Pa	222 Pa	200 Pa

3.3. *Effect of Electrode Spacing*

Electric field strength distribution, crucial for fluid driving capability, is significantly influenced by positive–negative electrode spacing. Experiments investigated pumps with electrode spacings of 0.3 mm, 0.4 mm, and 0.5 mm. Figure 5 shows time-dependent performance curves of the light-controlled ion drag pumps.

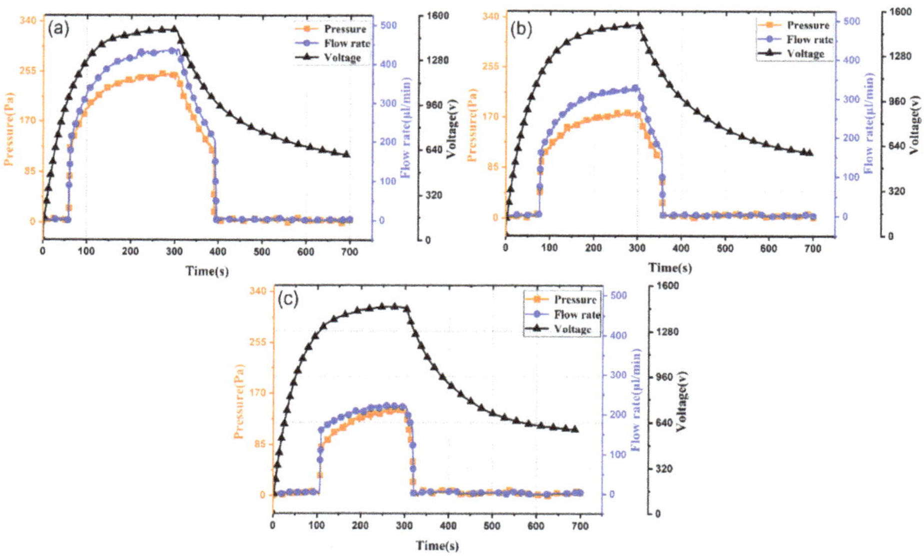

Figure 5. Time-dependent output performance curves of light-controlled ion drag pumps with different electrode spacings: (**a**) 0.3 mm; (**b**) 0.4 mm; and (**c**) 0.5 mm.

Comparing Figure 5a–c, increasing electrode spacing from 0.3 mm to 0.5 mm prolongs output response time from 57 s to 109 s. This is attributed to differing times to reach the

 https://doi.org/10.3390/mi17010045

threshold activation electric field: smaller spacing yields higher field strength under identical voltage, enabling faster response. Moreover, increasing electrode spacing from 0.3 mm to 0.5 mm reduces peak output pressure and flow rate from 0.25 kPa and 430 μL/min to 0.14 kPa and 220 μL/min (decreases of 78% and 95%, respectively). Thus, reducing electrode spacing effectively enhances output flow rate, pressure, and response speed.

The significant sensitivity of pump performance to electrode spacing highlights the crucial role of electric field intensity in controlling the ion drag mechanism. As the electrode spacing increases, the response time significantly increases, which underscores the key limitations and design insights of the light-controlled ion drag pump: the activation of the pump depends on the photovoltage reaching the threshold electric field intensity. A larger electrode spacing requires a higher absolute voltage to reach this threshold, thereby prolonging the charging time of the equivalent RC circuit of the PLZT pump. This result strongly indicates that reducing the electrode size is the main strategy for achieving faster activation of light-driven EHD pumps.

Table 2 compares simulated and experimental maximum output pressures for different electrode spacings. Errors are within 7.1%.

Table 2. Comparison of numerical simulation and experimental results for the maximum output pressure at different electrode spacings.

	0.3 mm	0.4 mm	0.5 mm
Simulation (Pa)	265	182	149
Experiment (Pa)	250	170	140

3.4. Effect of Number of Electrode Pairs

The effect of the number of planar interdigitated electrode pairs on output performance is investigated using pumps with 20, 25, and 30 pairs, as shown in Figure 6. Figure 6 shows that as the number of electrode pairs increases, output response time remains nearly constant (~57 s), with response voltage around 100 V. However, increasing electrode pairs from 20 to 30 raises output pressure amplitude from 0.21 kPa to 0.29 kPa and flow rate from 380 μL/min to 490 μL/min (increases of 38% and 29%, respectively). Thus, increasing electrode pairs effectively enhances maximum output pressure and flow rate.

The linear enhancement of output performance with an increasing number of electrode pairs demonstrates the effective summation of electrohydrodynamic forces along the flow path. Each additional electrode pair gradually increases the contribution to the total pressure head, as it expands the area where the electric volume force is applied. Under different numbers of electrode pairs, the response time of the light-controlled pump remains almost unchanged, which is an important finding. It indicates that the excitation dynamics are controlled by the local field conditions at the first few electrode pairs rather than the total system capacitance. Once the threshold field is reached locally, pumping will almost simultaneously start throughout the channel.

Table 3 compares simulated and experimental maximum output pressures for different numbers of electrode pairs. Errors are within 7.6%.

Table 3. Comparison of numerical simulation and experimental results for the maximum output pressure with different numbers of electrode pairs.

	20 pairs	25 pairs	30 pairs
Simulation (Pa)	225	265	308
Experiment (Pa)	210	250	290

 https://doi.org/10.3390/mi17010045

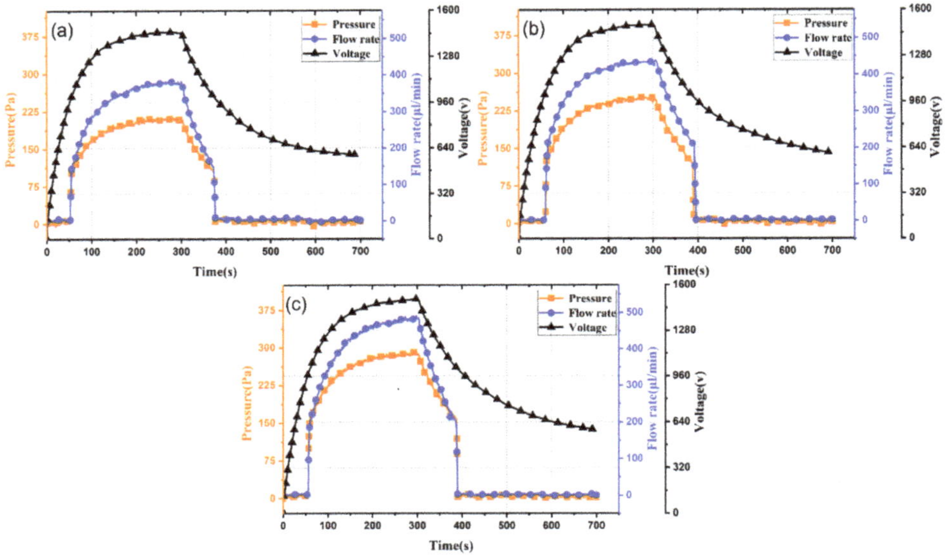

Figure 6. Time-dependent output performance curves of light-controlled ion drag pumps with different numbers of electrode pairs: (**a**) 20 pairs; (**b**) 25 pairs; and (**c**) 30 pairs.

3.5. *Effect of Gap Between the Electrode Pairs*

The operating mechanism of the light-controlled ion drag pump involves an opposing electric field between adjacent electrode pairs, counteracting the driving field and impeding fluid flow. To investigate the effect of the electrode pair gap, experiments were conducted with gaps of 0.6 mm, 0.8 mm, 1 mm, and 1.2 mm. Results are shown in Figure 7.

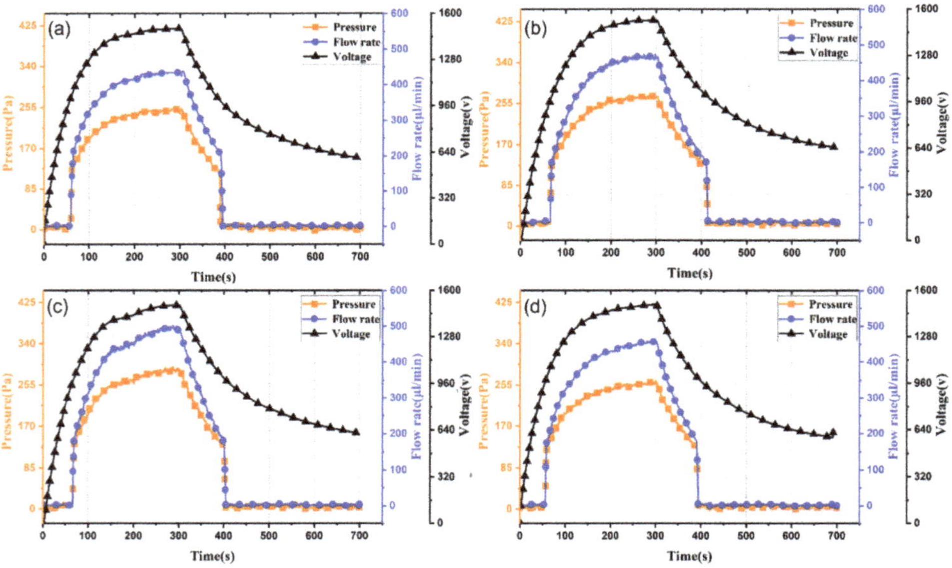

Figure 7. Time-dependent output performance curves of light-controlled ion drag pumps with different gaps between the electrode pairs: (**a**) 0.6 mm; (**b**) 0.8 mm; (**c**) 1 mm; and (**d**) 1.2 mm.

https://doi.org/10.3390/mi17010045

Varying the gap between electrode pairs has negligible effect on output response time, which remains about 55 s, determined by the PLZT photovoltaic voltage characteristics. Comparing output flow rate and pressure across gaps (Figure 7) shows that as the gap increases from 0.6 mm to 1.2 mm in 0.2 mm increments, both output pressure and flow rate initially rise then fall. Increasing the gap reduces the opposing electric field strength, enhancing output performance to some extent. However, further gap increase diminishes fluid kinetic energy transfer along the channel and, with fixed electrode pairs, lengthens the channel, increasing friction losses. Consequently, overall output performance decreases. Experimentally, optimal output performance occurs at a 1 mm gap, with maximum output pressure of 300 Pa and flow rate of 510 μL/min. Thus, rational gap selection effectively improves output performance.

In terms of output performance, the non-monotonic trend observed with the change in the gap between adjacent electrodes reveals the complex interplay between competitive electrical effects and flow effects. There exists an optimal gap (in this study, 1 mm), which can balance the two factors: (1) A larger gap will weaken the undesired reverse electric field between adjacent electrodes of the same polarity, thereby allowing for more effective net positive pumping. (2) When the gap is excessively large, the flow force generated by one electrode pair will decay before reaching the next electrode pair, thereby reducing the effective coupling of forces. Moreover, for a fixed number of electrode pairs, a longer total channel for a fixed number of pairs increases frictional losses. The existence of this optimal gap highlights the importance of the overall design, which requires considering the spatial distribution of the electric field and the resulting flow field morphology, rather than focusing on a single geometric parameter.

Table 4 compares simulated and experimental maximum output pressures for different gaps between electrode pairs. Errors are within 6.2%.

Table 4. Comparison of numerical simulation and experimental results for the maximum output pressure with different gaps between the electrode pairs.

	0.6 mm	0.8 mm	1 mm	1.2 mm
Simulation (Pa)	265	288	318	270
Experiment (Pa)	250	272	300	255

3.6. *Stall Pressure Test of the Light-Controlled Ion Drag Pump*

Considering that fluid pumps must overcome hydraulic loads present in fluid flow during practical applications, such as flow resistance in channels, micro pumps require a certain load-driving capability. The metric for evaluating a fluid micro pump's load-driving capability is stall pressure, defined as the maximum output pressure achievable when the pump's output flow rate is zero. Figure 8 presents a schematic diagram of the load capacity testing platform for the light-controlled PLZT ion drag pump. Select the optimal set of parameters from the aforementioned study—namely an ion drag pump chip with a channel height of 1 mm, electrode spacing of 0.3 mm, electrode pair spacing of 1 mm, and 30 electrode pairs—for experimentation. Figure 9 illustrates the temporal variation in liquid level height at the pump outlet channel under different light intensities, revealing the static response characteristics of the ion drag pump. During the initial phase, when illumination begins, the photovoltaic voltage has not yet reached the threshold activation electric field of the ion drag pump, causing the outlet liquid level height to remain unchanged. As the photovoltaic voltage reached and exceeded the threshold activation electric field while continuing to increase, the liquid level height also rose continuously. After illumination ceased, the photovoltaic voltage began to decrease. When the internal photovoltaic electric field of the ion drag pump fell below the threshold activation electric field, its static output

https://doi.org/10.3390/mi17010045

pressure dropped to zero. As shown in the experimental results of Figure 9, the liquid level inside the test tube continued to rise with prolonged irradiation time.

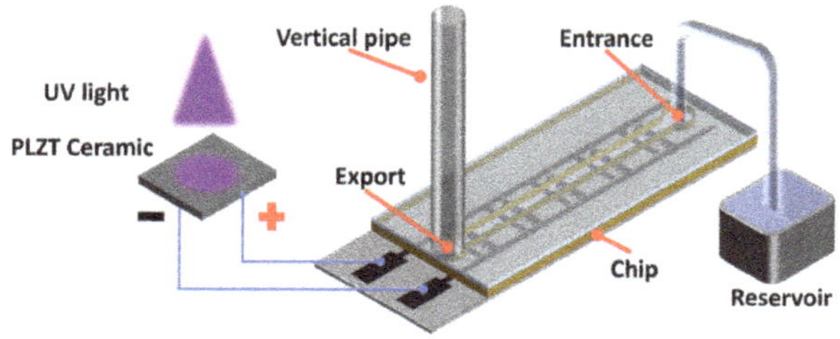

Figure 8. Schematic diagram of stall pressure test of the light-controlled ion drag pump.

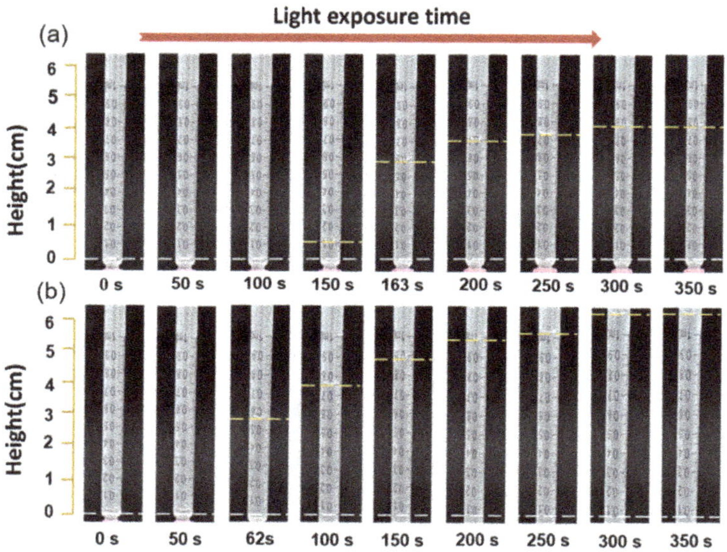

Figure 9. Stall pressure of the light-controlled ion drag pump under different light intensities: (**a**) 50 mW/cm^2; (**b**) 100 mW/cm^2.

In practical applications, fluid pumps must overcome hydraulic loads such as flow resistance. Stall pressure, the maximum output pressure at zero flow rate, evaluates a micro pump's load-driving capability. Figure 8 shows the stall pressure testing platform schematic. The optimal parameter set (channel height: 1 mm, electrode spacing: 0.3 mm, electrode pair gap: 1 mm, 30 electrode pairs) is selected. Figure 9 shows the temporal variation of liquid level height at the pump outlet under different light intensities, revealing static response characteristics. Initially, upon illumination, before the photovoltaic voltage reaches the threshold activation electric field, the outlet liquid level remains unchanged. As the voltage exceeds and continues to increase, the liquid level rises. When the photovoltage reaches saturation, the liquid level remains stable. Figure 9 shows the liquid level rising with prolonged irradiation. We used the pressure formula:

$$p = \rho_d g h \quad (9)$$

where h is the liquid length in the vertical pipe.

At 50 mW/cm^2 and 100 mW/cm^2, the stall pressures of the light-controlled pump are about 0.7 kPa and 1 kPa, respectively.

 https://doi.org/10.3390/mi17010045

4. Conclusions

This study presents a light-controlled ion drag pump based on PLZT ceramic by integrating the photoelectric effect with field emission. An EHD model for the pump was developed. Experimental studies on the photoelectric response characteristics of PLZT ceramics integrated into ion drag pump chips established a foundation for their application. An experimental platform was constructed to investigate key factors influencing output performance, including electrode shape, channel height, electrode spacing, number of electrode pairs, and gaps between electrode pairs. Under constant light intensity, output pressure and flow rate were measured to compare the effects of fluid channel dimensions and electrode structure sizes. Experimental results show that fluid channel height, number of electrode pairs, and electrode spacing monotonically affect output performance: Output performance is proportional to channel height and number of electrode pairs, and inversely proportional to electrode spacing. Notably, output pressure and flow rate first increase then decrease as the gap between electrode pairs widens.

This work establishes a basic framework for the light-controlled ion drag pump and verified its viability through systematic experimentation. The unique driving mode—utilizing light to both power and control the pump—opens up unique application potential that is different from the limitations faced by traditional wired, high-voltage charge, gradient-driven pumps. Specifically, this pump is particularly suitable for scenarios that require wireless, low electromagnetic interference, and spatially selective liquid driving. Promising application areas include:

(1) Optically Addressable Microfluidic Networks: In integrated lab-on-a-chip systems, different pump units on a single substrate can be independently activated by focused light beams, enabling complex and reconfigurable control of fluid paths without the need for a matrix structure composed of physical wires or connectors. This facilitates highly multiplexed and programmable biological detection.

(2) Closed-Environment or Implantable Devices: For systems where electrical conduction is not allowed or is not suitable (such as sealed containers used for sensitive chemical reactions, or possible future implantable drug delivery systems), light becomes a clean and safe medium for the transmission of energy and control signals.

(3) Microfluidics in Electromagnetically Sensitive Environments: For operations where the complete absence of generated electromagnetic interference is critical. This includes the microfluidic preparation, processing, or online detection of flammable chemicals and explosive materials, or operations carried out in systems integrated with high-precision electromagnetic sensors. By eliminating the arc discharge and broadband electronic noise inherent in traditional high-voltage drives, it offers a fundamentally safer and more compatible driving solution for these high-reliability and safety-critical applications.

(4) Distributed Drive in Soft Robotics: As a wireless pressure source, it can be embedded in soft robotic structures. Selective illumination of different pump units allows for localized, model-free control of fluidic channels, thereby achieving bending or shape deformation. This simplifies the design and control architecture of the robot.

Future efforts aimed at significantly reducing the response time and further miniaturizing the system are crucial for translating these conceptual advantages into practical applications. The current performance parameters (e.g., response time on the order of tens of seconds, pressure output of ~1 kPa) have already made it possible for application in microfluidics and soft robotics, where speed is not the main limiting factor and wireless control is of critical importance.

https://doi.org/10.3390/mi17010045

Author Contributions: Conceptualization, Y.T. and X.W.; methodology, Y.T. and Y.S.; software, Z.G. and Z.L.; validation, Y.T. and Z.L.; formal analysis, Y.T. and Y.S.; investigation, Y.T.; resources, Z.G.; data curation, Z.L.; writing—original draft preparation, Y.T. and Y.S.; writing—review and editing, Y.T. and X.W.; visualization, Z.G.; supervision, Y.T. and X.W.; project administration, Y.T.; funding acquisition, Y.T. All authors have read and agreed to the published version of the manuscript.

Funding: This research was funded by the Natural Science Foundation of Jiangsu Higher Education Institutions (No. 22KJA460005 and No. 24KJA460004).

Data Availability Statement: All data are presented within the paper.

Conflicts of Interest: The authors declare no conflicts of interest.

References

1. Sudhawiyangkul, T.; Yoshida, K.; Eom, S.I.; Kim, J.-W. A study on a hybrid structure flexible electro-rheological microvalve for soft microactuators. *Microsyst. Technol.* **2020**, *26*, 309–321. [CrossRef]
2. Wang, Q.; Jia, Y.; Li, J. Liquid transport strategies in wearable and implantable microfluidic systems. *Lab A Chip* **2025**, *25*, 4252–4272. [CrossRef] [PubMed]
3. Chen, J.; Meng, F.; Feng, Z.; Gao, W.; Liu, C.; Zeng, Y. Matching the optimal operating mode of polydimethylsiloxane check valves by tuning the resonant frequency of the resonator in a piezoelectric pump for improved output performance. *Micromachines* **2022**, *14*, 15. [CrossRef] [PubMed]
4. Xin, Y.; Zhou, X.; Tan, M.R.J.; Chen, S.; Huang, P.; Jiang, Y.; Wu, W.; Gao, D.; Lv, J.; Magdassi, S.; et al. 3D-printed electrohydrodynamic pump and development of anti-swelling organohydrogel for soft robotics. *Adv. Mater.* **2025**, *37*, 2415210. [CrossRef] [PubMed]
5. Peng, Y.; Li, D.; Yang, X.; Ma, Z.; Mao, Z. A review on electrohydrodynamic (EHD) pump. *Micromachines* **2023**, *14*, 321. [CrossRef] [PubMed]
6. Smith, M.; Cacucciolo, V.; Shea, H. Fiber pumps for wearable fluidic systems. *Science* **2023**, *379*, 1327–1332. [CrossRef] [PubMed]
7. Cacucciolo, V.; Shintake, J.; Kuwajima, Y.; Maeda, S.; Floreano, D.; Shea, H. Stretchable pumps for soft machines. *Nature* **2019**, *572*, 516–519. [CrossRef] [PubMed]
8. Gao, X.L.; Bao, X.D.; Pang, S.J.; Wu, J.; Luo, K.; Yi, H.L. The electrohydrodynamic enhancement of heat transfer on interdigitated electrodes by a charge injection pump. *Phys. Fluids* **2024**, *36*, 033612. [CrossRef]
9. Gao, X.L.; Lu, C.L.; Chen, D.L.; Wu, J.; Yi, H.L.; Luo, K. Mechanism of charge injection-based electrohydrodynamic pump with interdigitated electrodes. *Phys. Fluids* **2023**, *35*, 033610. [CrossRef]
10. Lin, S.C.; Huang, B.L.; Lai, F.C. Evaluation of swirl-flow generation by two-stage EHD pumps with three configurations using anchored electrodes. *J. Electrost.* **2025**, *134*, 104043. [CrossRef]
11. Huang, J.H.; Wang, X.J.; Wang, J. A mathematical model for predicting photo-induced voltage and photostriction of PLZT with coupled multi-physics fields and its application. *Smart Mater. Struct.* **2016**, *25*, 025002. [CrossRef]
12. Lu, Y.; Jiang, J.; Yan, X.; Wang, L. A new photovoltaic lunar dust removal technique based on the coplanar bipolar electrodes. *Smart Mater. Struct.* **2019**, *28*, 085010. [CrossRef]
13. Liu, Y.; Wang, X.; Wang, J.; Shao, Y.; Huang, J. Investigation on influence factors of opto-electrostatic hybrid driving torsion actuator based on PLZT ceramic. *Mater. Sci. Eng. B* **2021**, *263*, 114798. [CrossRef]
14. Lv, Z.; Liu, Z.; Tang, Y.; Wang, X. Photo-induced electrostatic actuators towards micro gap regulator and micro gripper based on lead lanthanum zirconate titanate ceramics. *Mater. Des.* **2023**, *232*, 112078. [CrossRef]
15. Liu, Z.; Lv, Z.; Tang, Y.; Shao, Y.; Wang, X.; Wang, J. Multiphysics modeling and simulation of a light-controlled variable damping system. *Materials* **2023**, *16*, 3194. [CrossRef] [PubMed]

 https://doi.org/10.3390/mi17010045

micromachines

MDPI

Article

Behavior of Electrothermal Actuator Analyzed by Polynomial Point Interpolation Collocation Method

Yujuan Tang [1], Aidong Qi [2], Yuanhu Gu [3,4,5], Yinfa Zhu [6,7], Haojie Li [8], Dao Gu [9] and Hao Chen [6,*]

1 School of Intelligent Science and Control Engineering, Jinling Institute of Technology, Nanjing 211169, China; yjtang@jit.edu.cn
2 Zhejiang Jiali (Lishui) Industrial Co., Ltd., Lishui 323000, China; lishuijiali@vip.163.com
3 Artificial Intelligence School, Lishui University, Lishui 323000, China; 18921884431@189.com
4 State Key Laboratory of Fluid Power and Mechatronic Systems, Zhejiang University, Hangzhou 310027, China
5 Lishui Key Laboratory of High Power Density Intelligent Drive System, Lishui 323000, China
6 Engineering School, Lishui University, Lishui 323000, China; lszhuyinfa@163.com
7 State Key Laboratory of Precision Manufacturing for Extreme Service Performance, Central South University, Changsha 410000, China
8 Lishui Zhe Li Digital Resources Management Co., Ltd., Lishui 323000, China; miaojia527@163.com
9 Faculty of Engineering and Information Technology, The University of Technology Sydney, Ultimo, NSW 2007, Australia; gudao901@gmail.com
* Correspondence: 17766106120@163.com

Abstract: This paper presents a novel implementation of the Polynomial Point Interpolation Collocation Method (PPCM) for analyzing the coupled electrothermal and thermomechanical behavior of V-shaped microactuators. Within the PPCM framework, the governing equations for heat transfer and structural mechanics are discretized over the computational domain. The resulting discrete electrothermal system is solved in a fully coupled manner via an incremental load method to determine the temperature field. Subsequently, the displacement field is computed by solving the discrete mechanical equation, which incorporates terms from the natural boundary conditions. The MQ radial basis function behaves well in convergence when its parameters p_a and p_q are 1 and 1.8. Under a 6 V voltage, the difference between the PPCM and FEM temperature values is less than 1 °C. Meanwhile, the discrepancy between the PPCM and experimental temperature values is approximately 20 °C, corresponding to an approximate error of 10%. Furthermore, the displacement error between the PPCM and FEM is as low as approximately 2 μm under an applied voltage of 12 V. These results validate the PPCM for predicting the driving characteristics of V-shaped microactuators.

Keywords: polynomial point interpolation collocation method; thermo-mechanical response; microactuator meshfree

1. Introduction

Microactuators serve as fundamental components in micro- and nano-electromechanical systems (MEMS and NEMS), enabling diverse applications such as micro-grippers [1], high-precision positioning systems [2,3], and bistable or multistable micro-switches [4,5]. These devices operate by converting electrical signals into mechanical motion through multi-physical interactions. Depending on their operational principles, microactuators are commonly classified into piezoelectric [6], electrostatic [7], electrothermal [8], and electromagnetic [9] types. In particular, electrothermal actuators have garnered significant interest owing to their ability to generate large displacements and high driving forces, along with their compatibility with standard microfabrication processes.

The past few decades have witnessed the fast development of advanced computational techniques for small-scale structural analysis, which continues to attract significant and growing interest within the research community. Ouakad et al. [10]. Predicted the steady-state behavior of an electrostatically actuated and initially curved actuator arrangement with the meshless Galerkin decomposition method. Jankowski et al. [11] considered the vibration of a piezoelectrically actuated smart nanoactuator via nonlocal strain gradient theory on the basis of the meshless differential quadrature method. The work of Chen et al. [12] featured the implementation of the local radial point interpolation method (LRPIM) for electrothermal structure dynamics, where temperature data served to validate the results. Xia et al. [13] completed an electroelastic analysis of piezoelectric structures based on the meshfree generalized finite difference method (GFDM). Nourmohammadi et al. [14] numerically solved the shear deformation problem of a functionally graded piezoelectric plate using the radial point interpolation technique. In addition, the emerging field of data-driven numerical computation integrates data science with traditional computational physics. Its feature is the fusion of physical principles with data to create methods that are less reliant on explicit, closed-form equations. This is often achieved through machine learning, which can reveal hidden patterns and complex nonlinear relationships, leading to a transformative shift in how we simulate and understand complex systems. Machine learning techniques have demonstrated remarkable potential in diverse interdisciplinary applications, including biomedical engineering and soft robotics. For instance, machine learning-enhanced soft robotic systems inspired by biological functions have been developed for medical diagnostics and rehabilitation [15]. While these applications differ in scale and domain from microactuator analysis, they exemplify the transformative potential of data-driven approaches in solving complex, nonlinear engineering problems. This cross-fertilization of ideas motivates the exploration of novel computational frameworks, such as PPCM, that can bridge different physical scales and application domains. Lao et al. [16] employed a Physics-Informed Neural Networks (PINNs) enhanced Deep Reinforcement Learning (DRL) framework for high-precision vibration control of piezoelectric cantilever beams. In order to predict the minimum gap dynamics of an electrostatically actuated microbeam, Barbulescu et al. [17] proposed a data-driven method based on recurrent neural networks.

The classification of meshless methods, according to their solution formulation, falls into two primary types: the strong form and the weak form. The implementation of meshless methods founded on weak formulations [18,19], such as the Galerkin-based approaches or LRPIM, is usually challenging. This is primarily due to the requirement for cumbersome numerical integration over a background mesh and the intricate procedures needed to impose essential boundary conditions, a consequence of their non-interpolating shape functions. Compared with the weak form, the strong formulation, despite its advantages of lower computational expense and a more direct implementation, comes with the drawback of sensitivity to node irregularities and the need for calculation stabilization. This issue is particularly pronounced in the Generalized Finite Difference Method (GFDM) [20,21], where the coefficient matrix from the least-squares fit is highly prone to ill-conditioning. Meanwhile, the PINN, another strong form method, relies on large-scale iterative training and is better adapted for solving inverse problems that incorporate known observational data [22,23]. In addition, the Polynomial Point Interpolation Collocation Method (PPCM) is also a strong-form meshless technique that constructs polynomials to exactly pass through all nodes within local support domains. Given the relatively regular geometry of the electrothermal microactuator, the adoption of PPCM not only ensures numerical stability during computation but also leads to high computational efficiency.

Considering the cumbersome implementation and high computational costs associated with the weak form, as well as addressing limitations such as low accuracy in the Generalized Finite Difference Method (GFDM) and the time-consuming training requirements of Physics-Informed Neural Networks (PINNs), this research focuses on the electrothermal–thermomechanical analysis of V-shaped actuators by employing PPCM to construct an efficient and robust convergent numerical scheme. The paper is structured as follows: Section 2 outlines the fundamental principles of the PPCM and presents a flowchart for the actuator analysis. Section 3 derives the discretized equations by integrating the PPCM with the incremental load method, based on a pre-established electrothermal coupling model. In Section 4, actuator deflection is estimated using the PPCM within a thermo-elastic framework. Section 5 compares the thermal and mechanical results obtained from the PPCM, finite element method (FEM), and experimental measurements and assesses the accuracy and convergence of the PPCM. Finally, Section 6 provides concluding remarks and suggests future research directions.

2. Polynomial Point Interpolation Collocation Method

To facilitate comprehension, the shape function constructed by the radial basis polynomial is first derived here. A 2D problem domain Ω is discretized into N points. Provided that N^l discrete points are identified in the local field of the point $\mathbf{x}_I$, (x_I, y_I), the field variable vector $\mathbf{u}$ can be expressed as

$$\mathbf{u} = \mathbf{Ra} + \mathbf{Qb} \tag{1}$$

where position vector $\mathbf{x}_I$ contains the x and y, $\mathbf{a}$ and $\mathbf{b}$ are the vectors of undetermined coefficients,

$$\mathbf{u}_s = [u_1, u_2, \ldots, u_{N^l}]^T \tag{2}$$

and

$$\mathbf{R} = \begin{bmatrix} R_1(\mathbf{x}_1) & R_2(\mathbf{x}_1) & \ldots & R_{N^l}(\mathbf{x}_1) \\ R_1(\mathbf{x}_2) & R_2(\mathbf{x}_2) & \ldots & R_{N^l}(\mathbf{x}_2) \\ \ldots & \ldots & \ldots & \ldots \\ R_1(\mathbf{x}_{N^l}) & R_2(\mathbf{x}_{N^l}) & \ldots & R_{N^l}(\mathbf{x}_{N^l}) \end{bmatrix} \tag{3}$$

where the MQ radial basis function (RBF) is

$$R_i(\mathbf{x}_j) = \begin{cases} (1 + (p_a * r_{ij}/d)^{p_q}, & r_{ij} \le d \\ 0, & r_{ij} > d \end{cases} \tag{4}$$

where the distance $r_{ij} = |\mathbf{x}_i - \mathbf{x}_j|_2$ and d is the radius of the support domain. To enhance the smoothness of the shape function, the constant and linear terms are added so that

$$\mathbf{Q} = \begin{bmatrix} 1 & 1 & \ldots & 1 \\ x_1 & x_2 & \ldots & x_{N^l} \\ y_1 & y_2 & \ldots & y_{N^l} \\ x_1y_1 & x_2y_2 & \ldots & x_{N^l}y_{N^l} \end{bmatrix}^T \tag{5}$$

Further, the equation can be rewritten as

$$\widetilde{\mathbf{u}_s} = \begin{bmatrix} \mathbf{u}_s \\ 0 \end{bmatrix} = \begin{bmatrix} \mathbf{R} & \mathbf{Q} \\ \mathbf{Q}^T & 0 \end{bmatrix} \begin{bmatrix} \mathbf{a} \\ \mathbf{b} \end{bmatrix} = \mathbf{G}\mathbf{a}_0 \tag{6}$$

For any point $\mathbf{x}_I$ (x_I, y_I), at the current local domain, its field variable u_I can be expressed as

$$u_I = \mathbf{\Phi}_I \widetilde{\mathbf{u}_s} \tag{7}$$

where vector $\mathbf{\Phi}_I$ is with $1 \times (N^l + 4)$.

$$\mathbf{\Phi}_I = [\ R_1(\mathbf{x}_I) \quad R_2(\mathbf{x}_I) \quad \cdots \quad R_{N^l}(\mathbf{x}_I) \quad 1 \quad \cdots \quad x_I y_I]\ \mathbf{G}^{-1} \tag{8}$$

For the convenience of calculation, the last four elements are removed from the vector $\mathbf{\Phi}_I$ to construct the shape function vector $\mathbf{N}_I$, in which the dimension is $1 \times N^l$. Hence, a more unified compact form of Equation (7) can be expressed as

$$u_I = \mathbf{N}_I \mathbf{u}_s \tag{9}$$

For a 2D boundary value problem, the derivative terms—both within the governing PDE and the accompanying boundary conditions—can be supplanted by differentiating the corresponding linear Equation (9).

3. Electrothermal Analysis

The geometric model of the V-shaped actuator is shown in Figure 1a. Given the characteristics of the temperature distribution on the actuator, the computational model of the electrothermal analysis can be simplified as a 1D model. Thus, the design of the discrete domain is shown in Figure 1b, where its support domain will be a segment, just related to the x-coordinate. According to the temperature distribution from the 1D model, the discrete node distribution in the thermomechanical analysis is presented in Figure 1c.

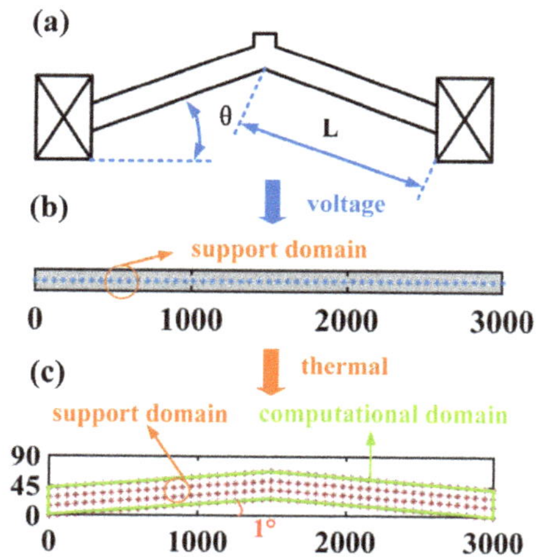

Figure 1. (**a**) Geometric schematic of the V-shaped actuator. (**b**) Discrete node distribution in the electrothermal analysis. (**c**) Discrete node distribution in the thermomechanical analysis.

According to our previous studies, the coupling equation between the electrical and thermal fields can be derived as

$$k(T)\frac{\partial^2 T}{\partial x^2} - (k_v \frac{w+2h}{wh} + \frac{S}{h})T + \frac{U_0^2 \rho(T)}{(\int_0^{2l} \rho(T)\,\mathrm{d}x)^2} = 0 \tag{10}$$

where thermal conductivity of polysilicon k = 210658(T − 30)$^{-1.2747}$ W·m^{-1}·°C^{-1}, width w and thickness h of the actuator are 42 and 100 μm, electrical resistivity ρ = 0.0002(1 + 0.003(T − 30)) Ω·m, the parameters of the heat dissipate rate k_v and S are 70 and 0.1(T − 30) W·m^{-2}, and U_0 is the voltage. At both ends of the actuator, the temperatures are forced to be the room temperature. According to the PPCM above, the matrix **Q** in Equation (5) is rewritten as

$$\mathbf{Q} = \begin{bmatrix} 1 & 1 & \dots & 1 \\ x_1 & x_2 & \dots & x_{N^I} \end{bmatrix}^{\mathrm{T}} \tag{11}$$

Then the vector $\mathbf{\Phi}_I$ in Equation (7) is with 1 × (N^I + 2). Hence, the last two columns should be removed from the vector $\mathbf{\Phi}_I$ to construct the shape function vector $\mathbf{N}_I$. For Equation (8), its discretization form at the interior point x_I should satisfy

$$\mathbf{K}^I\mathbf{T}^I = p^I \tag{12}$$

where

$$\mathbf{K}^I = k(T_I)\mathbf{N}_{I,xx} - (k_{\mathrm{v}}\frac{w+2h}{wh} + \frac{S}{h})\mathbf{N}_I \tag{13}$$

and

$$p^I = -U_0^2\rho(T_I)/(\frac{2l}{N}\sum_{j=1}^{N-1}\rho(T(x_j))) \tag{14}$$

where T_I denotes the temperature at node x_I. $\mathbf{N}_{I,x}$ and $\mathbf{N}_{I,xx}$ denote the first- and second-order derivatives at the x_I versus x, respectively. They can be separately obtained from $\mathbf{\Phi}_{I,x}$ and $\mathbf{\Phi}_{I,xx}$. Here, $\mathbf{\Phi}_{I,x}$ and $\mathbf{\Phi}_{I,xx}$ can be expressed as

$$\mathbf{\Phi}_{I,x} = \begin{bmatrix} \frac{\mathrm{d}R_1(x_I)}{\mathrm{d}x} & \frac{\mathrm{d}R_2(x_I)}{\mathrm{d}x} & \dots & \frac{\mathrm{d}R_{N^I}(x_I)}{\mathrm{d}x} & 0 & 1 \end{bmatrix}\mathbf{G}^{-1} \tag{15}$$

$$\mathbf{\Phi}_{I,xx} = \begin{bmatrix} \frac{\mathrm{d}^2R_1(x_I)}{\mathrm{d}x^2} & \frac{\mathrm{d}^2R_2(x_I)}{\mathrm{d}x^2} & \dots & \frac{\mathrm{d}^2R_{N^I}(x_I)}{\mathrm{d}x^2} & 0 & 0 \end{bmatrix}\mathbf{G}^{-1} \tag{16}$$

Thus, this linear vector Equation (12) will be assembled into the total matrix Equation (17)

$$\mathbf{KT} = \mathbf{P} \tag{17}$$

where the matrix **K** has dimension of $N^I \times N^I$, vectors, **T** and **P**, are with N^I × 1. For any element $\mathbf{K}^I_{1,j}$ in the $\mathbf{K}^I$, the global index assigned to the point x_I is denoted by I, while the local index assigned to the point x_j is denoted by j. The global index J corresponding to the local index j is obtained through a mapping. Hence, the element $\mathbf{K}^I_{1,j}$ is to be added to the global stiffness matrix K at Ith row and Jth column. Similarly, the element $\mathbf{T}^I_j$ in the $\mathbf{T}^I$ is to be added in the Jth row of the vector **T**. The p_I is assembled into the Ith row of vector **P**. Once assembly is complete, the essential boundary condition is imposed at the two boundary points. Because of the highly nonlinear parameters ρ and k varying with the temperature, the incremental load iteration needs to be applied to solve Equation (16). In the iteration computation, an equal increment of the input power, U = $\sqrt{U_0^2/N^U}$, is constantly implemented for each iteration step. In it, N^U is the number of iterations. The procedure of the computational iteration is shown in Figure 2, and the detailed iteration procedure is as follows:

(i) Set the initial values for the nonlinear material parameters and then calculate the matrix K, inverse matrix $\mathbf{K}^{-1}$, and vector $\mathbf{P}^1$. From these, the temperature $\mathbf{T}^1$ at the first iteration is computed by Equation (17).

(ii) Update the applied voltage U^n and material parameters based on the temperature of the previous iterative step, where U^n = $\sqrt{nU_0^2/N^U}$. After that, the matrixes **K**, $\mathbf{K}^{-1}$, and vector $\mathbf{P}^n$ are obtained based on the temperature from the previous iteration T^{n-1}. To avoid the accumulation of computational error, the transition temperature $\mathbf{T}^t$ is determined from Equation (18).

$$\mathbf{T}^t = \mathbf{K}^{-1}(\mathbf{P}^n - \mathbf{K} * \mathbf{T}^{n-1}) + \mathbf{T}^{n-1} \tag{18}$$

(iii) Further refine the transition temperature by

$$\mathbf{T}^\theta = (1-\theta)\mathbf{T}^n + \theta\mathbf{T}^t, \theta = 0.9 \tag{19}$$

(iv) Renewed the matrix $\mathbf{K}^{-1}$ and vector $\mathbf{P}^n$ using $\mathbf{T}^\theta$. Finally, the temperature $\mathbf{T}^n$ at the nth iteration is computed via Equation (20), from which the next iteration proceeds.

$$\mathbf{T}^n = \mathbf{K}^{-1}(\mathbf{P}^n - \mathbf{P}^{n-1}) + \mathbf{T}^{n-1} \tag{20}$$

(v) Termination condition: If the current iteration count reaches or exceeds the maximum allowed, terminate the iteration. Otherwise, proceed to step (ii).

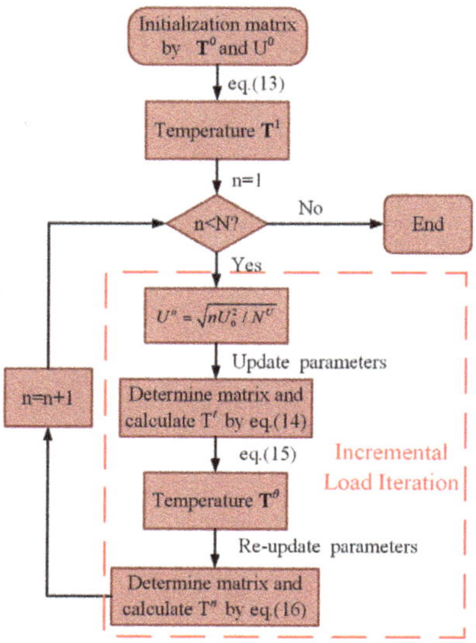

Figure 2. Iteration computation on the electrothermal analysis by the PPCM.

4. Thermomechanical Analysis

Combining the isotropic thermal expansion coefficient with the temperature distribution obtained from the above analysis, the thermal strain profile can be further calculated based on the discrete domain in Figure 1c. Hence, the steady-state thermoelastic equation is further presented as

$$\begin{cases} \frac{E}{1-\mu^2}\left(\frac{\partial^2 u}{\partial x^2} + \frac{1-\mu}{2}\frac{\partial^2 u}{\partial y^2} + \frac{1+\mu}{2}\frac{\partial^2 v}{\partial x \partial y} - \frac{\partial \varepsilon_x^t}{\partial x} - \mu\frac{\partial \varepsilon_y^t}{\partial x}\right) = 0 \\ \frac{E}{1-\mu^2}\left(\frac{\partial^2 v}{\partial y^2} + \frac{1-\mu}{2}\frac{\partial^2 v}{\partial x^2} + \frac{1+\mu}{2}\frac{\partial^2 u}{\partial x \partial y} - \mu\frac{\partial \varepsilon_x^t}{\partial y} - \frac{\partial \varepsilon_y^t}{\partial y}\right) = 0 \end{cases} \tag{21}$$

where E is the Young's modulus, μ is the Poisson's ratio, and u and v are the corresponding displacement components in the x- and y-directions. The $\varepsilon_x{}^t$, $\alpha(T) \times T$, denotes the thermal strain along the x-direction. The thermal expansion coefficient α, $-4 \times 10^{-12}T^2 + 8 \times 10^{-9}T + 4 \times 10^{-7}$, varies with the temperature. The natural boundary condition is

$$\begin{cases} \frac{E}{1-\mu^2}\left(n_x\frac{\partial u}{\partial x} + \mu n_x\frac{\partial v}{\partial y} + \frac{1-\mu}{2}n_y\left(\frac{\partial u}{\partial y} + \frac{\partial v}{\partial x}\right) + n_x(\varepsilon_x^t + \mu\varepsilon_y^t)\right) = 0 \\ \frac{E}{1-\mu^2}\left(n_y\frac{\partial v}{\partial y} + \mu n_y\frac{\partial u}{\partial x} + \frac{1-\mu}{2}n_x\left(\frac{\partial u}{\partial y} + \frac{\partial v}{\partial x}\right) + n_y(\mu\varepsilon_x^t + \varepsilon_y^t)\right) = 0 \end{cases} \tag{22}$$

where the components of the normal vector are defined as n_x and n_y in the x- and y-directions. Additionally, essential boundary conditions are applied at the two ends of the actuator. For Equation (10), the discrete point $\mathbf{x}_I$ (x_I, y_I) determines the elements in the matrix $\mathbf{K}^I$and vector p^I.

Case I: If the point $\mathbf{x}_I$ is in the internal computational domain, $\mathbf{K}^I$ and p^I should be derived based on Equation (21). The matrix $\mathbf{K}^I$ has the dimension of $2 \times 2N^l$.

$$\begin{cases} \mathbf{K}^I_{1,2*j-1} = \frac{E}{1-\mu^2}(\mathbf{N}_{I,xx} + \frac{1-\mu}{2}\mathbf{N}_{I,yy}) \\ \mathbf{K}^I_{1,2*j} = \frac{E}{2(1-\mu)}\mathbf{N}_{I,xy} \end{cases} \tag{23}$$

$$\begin{cases} \mathbf{K}^I_{2,2*j-1} = \frac{E}{2(1-\mu)}\mathbf{N}_{I,xy} \\ \mathbf{K}^I_{2,2*j} = \frac{E}{1-\mu^2}(\mathbf{N}_{I,yy} + \frac{1-\mu}{2}\mathbf{N}_{I,xx}) \end{cases} \tag{24}$$

where $\mathbf{N}_{I,xx}$, $\mathbf{N}_{I,xy}$, and $\mathbf{N}_{I,\,yy}$ can also be obtained from the $\mathbf{\Phi}_{I,xx}$, $\mathbf{\Phi}_{I,xy}$, and $\mathbf{\Phi}_{I,xy}$ based on Equation (8). For example, the $\mathbf{\Phi}_{I,xy}$ can be expressed as

$$\mathbf{\Phi}_{I,xy} = \begin{bmatrix} \frac{\mathrm{d}^2 R_1(\mathbf{x}_I)}{\mathrm{d}x\mathrm{d}y} & \frac{\mathrm{d}^2 R_2(\mathbf{x}_I)}{\mathrm{d}x\mathrm{d}y} & \cdots & \frac{\mathrm{d}^2 R_{N^l}(x_I)}{\mathrm{d}x\mathrm{d}y} & 0 & 0 & 0 & 1 \end{bmatrix}\mathbf{G}^{-1} \tag{25}$$

and

$$\begin{cases} p^I_1 = \frac{E}{1-\mu^2}(\varepsilon^t_{Ix,x} + \mu\varepsilon^t_{Iy,x}) \\ p^I_2 = \frac{E}{1-\mu^2}\left(\mu\varepsilon^t_{Ix,y} + \varepsilon^t_{Iy,y}\right) \end{cases} \tag{26}$$

where $\varepsilon^t_{Ix,x}$ and $\varepsilon^t_{Iy,x}$ are the first derivative, with respect to x, of the thermal strain component (in the x- and y-directions) at the specific point $\mathbf{x}_I$. Based on the temperature distribution, their value can be calculated by the central finite difference method. $\varepsilon^t_{Ix,y}$ and $\varepsilon^t_{Iy,y}$ both can be regarded as 0 in the light of the 1D model established in the electrothermal analysis.

Case II: If the point $\mathbf{x}_I$ belongs to the natural boundary, $\mathbf{K}^I$ and p^I should be derived based on Equation (22).

$$\begin{cases} \mathbf{K}^I_{1,2*j-1} = \frac{E}{1-\mu^2}(n_{I_x}\mathbf{N}_{I,x} + \frac{1-\mu}{2}n_{I_y}\mathbf{N}_{I,y}) \\ \mathbf{K}^I_{1,2*j} = \frac{E}{1-\mu^2}(\mu n_{I_x}\mathbf{N}_{I,y} + \frac{1-\mu}{2}n_{I_y}\mathbf{N}_{I,x}) \end{cases} \tag{27}$$

$$\begin{cases} \mathbf{K}^I_{2,2*j-1} = \frac{E}{1-\mu^2}(\mu n_{I_y}\mathbf{N}_{I,x} + \frac{1-\mu}{2}n_{I_x}\mathbf{N}_{I,y}) \\ \mathbf{K}^I_{2,2*j} = \frac{E}{1-\mu^2}(n_{I_y}\mathbf{N}_{I,y} + \frac{1-\mu}{2}n_{I_x}\mathbf{N}_{I,x}) \end{cases} \tag{28}$$

where the formula of $\mathbf{\Phi}_{I,x}$, adopted to reconstruct $\mathbf{N}_{I,x}$, is

$$\mathbf{\Phi}_{I,x} = \begin{bmatrix} \frac{\mathrm{d}R_1(\mathbf{x}_I)}{\mathrm{d}x} & \frac{\mathrm{d}R_2(\mathbf{x}_I)}{\mathrm{d}x} & \cdots & \frac{\mathrm{d}R_{N^l}(x_I)}{\mathrm{d}x} & 0 & 1 & 0 & y_I \end{bmatrix}\mathbf{G}^{-1} \tag{29}$$

Other first-order derivatives of the shape function vector can be obtained by the same method, according to Equation (8).

$$\begin{cases} p^I_1 = \frac{E}{1-\mu^2}n_{Ix}(\varepsilon^t_{Ix} + \mu\varepsilon^t_{Iy}) \\ p^I_2 = \frac{E}{1-\mu^2}n_{Iy}(\mu\varepsilon^t_{Ix} + \varepsilon^t_{Iy}) \end{cases} \tag{30}$$

where n_{Ix} and n_{Iy} are components of the normal vector in the x- and y-directions at point $\mathbf{x}_I$.

Case III: Otherwise, point $\mathbf{x}_I$ belongs to the Dirichlet boundary. To meet the coercive boundary condition, matrix $\mathbf{K}^I$ and p^I are directly rewritten as

$$\mathbf{K}^I_{1,1} = \mathbf{K}^I_{2,2} = 1 \tag{31}$$

and

$$p_1^I = p_2^I = 0 \tag{32}$$

where the remaining entries in the local matrix $\mathbf{K}^I$ are zero. Finally, Equations (23), (24), (26)–(28), and (30)–(32) must be assembled into a global system of equations, analogous to Equation (12). In this global system, the stiffness matrix **K** and the force vector P have dimensions of $2N^l \times 2N^l$ and $2N^l \times 1$, respectively. For the element $\mathbf{K}^I_{1,2*j-1}$, I represents the global index and j represents the local index. The corresponding global index J is defined for the point $\mathbf{x}_j$ in the whole problem domain. Accordingly, the element $\mathbf{K}^I_{1,2*j-1}$ is placed in the $(2I-1)^{\text{th}}$ row and $(2J-1)^{\text{th}}$ column of the global stiffness matrix **K**. Similarly, the element $\mathbf{K}^I_{2,2*j-1}$ belongs in the $(2I)^{\text{th}}$ row and $(2J-1)^{\text{th}}$ column. The p_1^I and p_2^I are assembled into the $(2I-1)^{\text{th}}$ and $(2I)^{\text{th}}$ row of the **P**. For the purposes of this analysis, the resulting global matrix equation can be solved efficiently using the direct iteration method.

5. Numerical Results and Discussions

A. Thermal measurement

For precise temperature acquisition, an infrared thermal microscope system with high spatial resolution was utilized. The system records thermal infrared images automatically at a frame rate of 200 Hz, each with a resolution of 392 × 258 pixels. A spatial calibration, performed with reference to the known dimensions of the ETA in the field of view, indicates that one pixel represents a physical distance of approximately 3.75 µm. As in the previous study, the high temperature existing in the middle area has a decisive impact on the performance of the actuator. Hence, the thermal image around the middle area should be preferentially captured in the experiment. Temperature distributions around the middle area under different voltages are measured and shown in Figure 3. As shown in the figure, the temperature decreases from the middle to either side of the actuator. When a voltage of 6 V is applied, the maximum temperature of about 210 °C lies in the central area. The thermal radiation may happen because the temperature near the actuator is higher than the room temperature.

B. Thermal prediction

To evaluate the performance of the incremental load (IL) method within the PPCM framework, its results were compared with those from the direct iteration (DI) method for solving Equation (10). Both sets of results were benchmarked against reference solutions obtained from the finite element method (FEM). Figure 4 presents the comparative results. As shown, the temperature difference calculated by the DI method increases significantly with applied voltage, reaching approximately 40 °C at 6 V. In contrast, the IL method demonstrates superior convergence within the PPCM. The temperature difference between the PPCM (IL) and FEM solutions remains within 1 °C across all voltage levels. Therefore, the integration of the IL method into the PPCM yields more accurate results in this study. This superiority can be attributed to the IL method's enhanced capability in handling the material nonlinearity, where the thermal conductivity decreases as the temperature rises, while the resistivity has an opposite relationship with the temperature. These factors generate high nonlinearity in the equation. In addition, another interesting phenomenon is that the temperature result from FEM with DI is valid. That is to say, the FEM can be more stable than PPCM in numerical calculations.

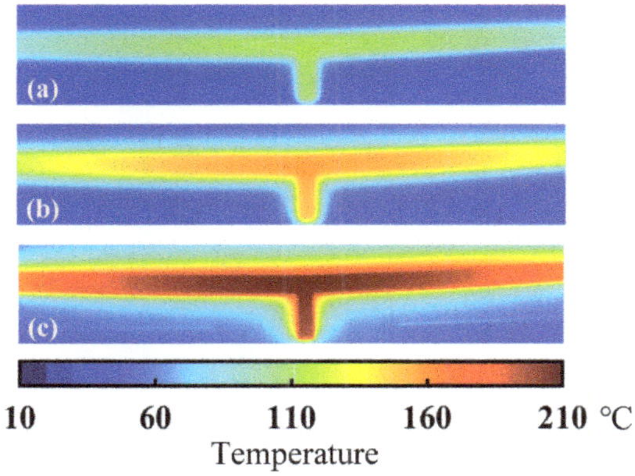

Figure 3. Experimental temperature distribution around the middle of the actuator under different voltages (**a**) −4 V, (**b**) −5 V, (**c**) −6 V.

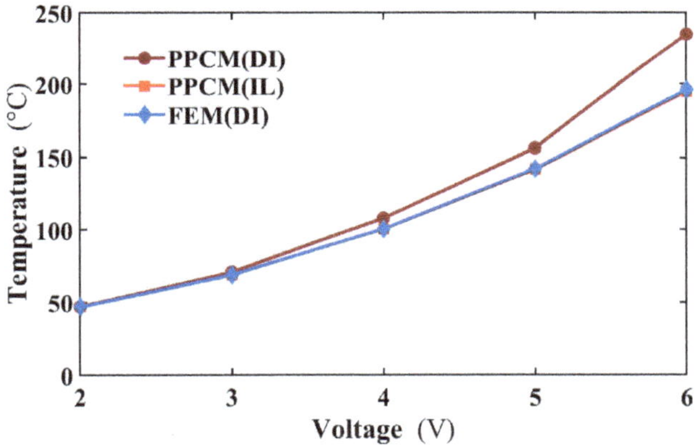

Figure 4. The middle temperatures of the actuator obtained by these three methods and FEM at different voltages.

To further explore the calculation stability of the PPCM (IL), the convergence test was performed. Three other types of RBF were employed in the PPCM to solve the equation, and the maximum temperature at 6 V is presented in Table 1. From the table, these three kinds of RBF have such poor performance that they may be unable to converge or solve the stagnation at the initial condition in the PPCM. The MQ RBF is best for this study. Based on the expression of the MQ RBF, its convergence analysis on two parameters, p_a and p_q was further conducted. The results are shown in Figures 5 and 6, respectively. When pa varies over a wider scale range (10^{-1}–10^3), a better computational stability can be seen from Figure 5. However, the calculation diverges rapidly with p_a close to zero. For p_q in Figure 6, the parameter has a narrow range of computational convergence. Its calculation diverges rapidly when p_a is close to 2. Moreover, a numerical singularity exists at the point $p_q = 1$, in which MQ degenerates into a simple polynomial. In this case, its performance is similar to that of the cubic RBF. Thus, the value of p_q should be far from 1. Finally, p_a and p_q are 1 and 1.8 for the electrothermal analysis.

Table 1. Temperature obtained from different kinds of RBF in the PPCM.

Name	Expression	Temperature/°C
MQ	$(1+p_a r^2)^{p_q}$	195
Cubic	$1-6r^2+8r^3-3r^4$	30.15
Spline	$\exp(-cr^2)$	30.18
CSRBF (Wl-C4)	$(1-r)^6(3+18r+35r^2)$	30.09

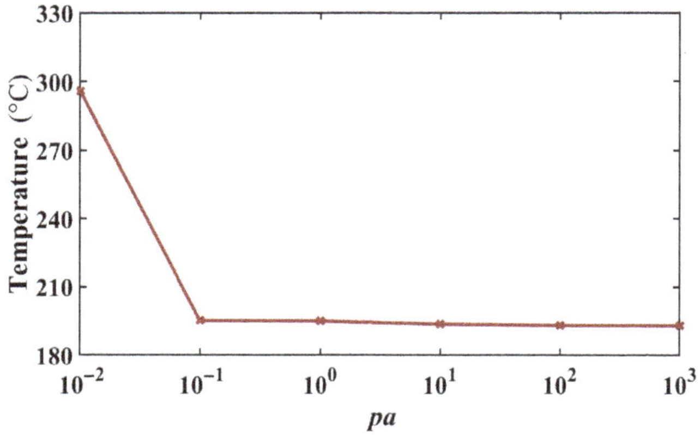

Figure 5. Maximum temperatures under 6 V varying with p_a in MQ RBF.

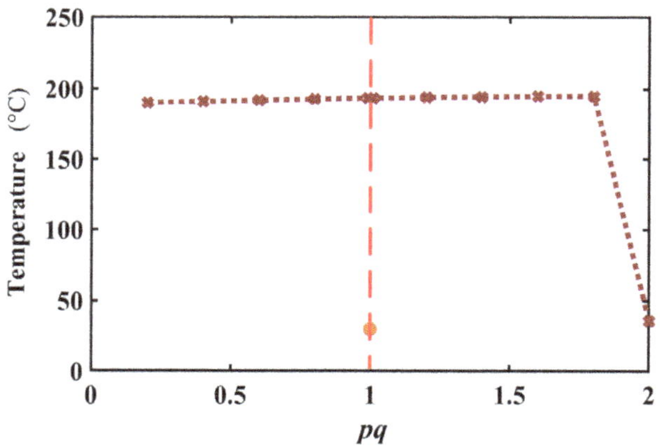

Figure 6. Maximum temperatures under 6 V vary with p_q in MQ RBF.

When the actuator is loaded with different voltages, its temperatures separately obtained from PPCM and FEM were compared with those from the experiment in Figure 7. The temperatures are extracted from the middle area of the actuator at the steady state. In this figure, the predicted temperature curve is in strong agreement with the FEM one. Furthermore, the PPCM results show reasonable agreement with experimental data, with a maximum temperature error of approximately 13 °C. For a more detailed evaluation, temperature profiles along the actuator length were extracted from thermal images. As shown in Figure 8, the temperature profiles predicted by PPCM agree well with FEM results at both 4 V and 6 V. The maximum discrepancy between PPCM and FEM is about 7 °C, located near the three-quarter point of the actuator at 6 V. However, compared to the experimental profile, the PPCM prediction shows a slight deviation. Specifically, a temperature difference emerges midway along the actuator and gradually increases towards both ends. As estimated from the curve at 6 V, the discrepancy between the PPCM and experimental temperature values is approximately 20 °C, corresponding to an approximate error of 10%. This can mainly be ascribed to the uncertain material parameters and excessive simplification of the boundary condition, where the Dirichlet boundary condition is employed.

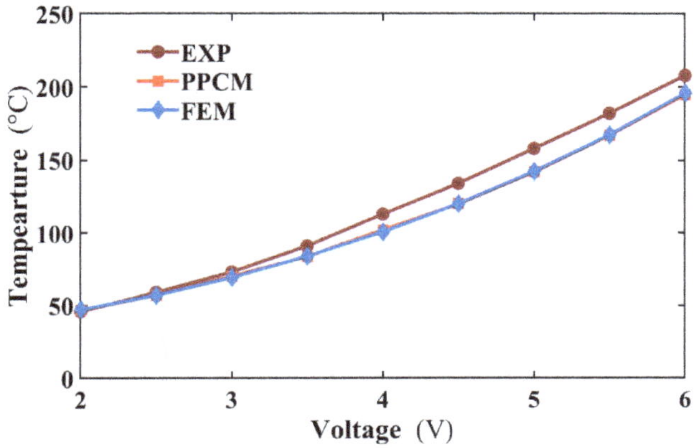

Figure 7. Middle temperatures comparison between the experiment, PPCM, and FEM from 2 to 6 V.

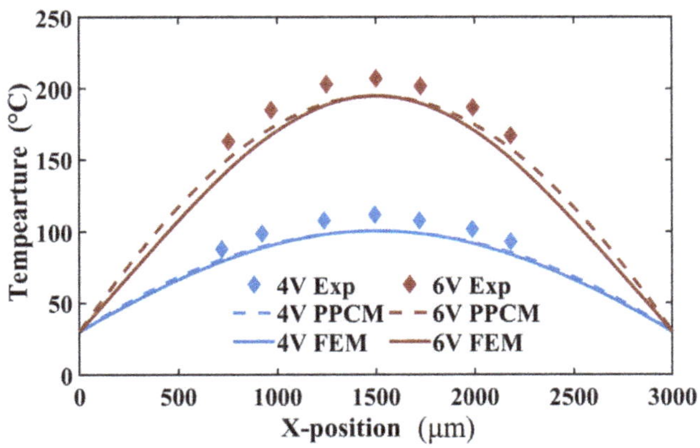

Figure 8. Comparison of the temperature distribution between the experiment, GFDM, and FEM at 4 and 6 V.

C. Mechanical prediction

Figure 1c shows the node distribution of the 2D computational domain of the actuator. Here, the calculation stability of the PPCM for the thermomechanical analysis is explored. It was found that p_a and p_q are 1 and 1.03. Moreover, the radius of the support domain d = 46 μm is employed in the calculation. Under different voltages, the displacements in the middle area of the actuator, where the maximum displacement occurs, are calculated by the PPCM and FEM, are presented in Figure 9. In the figure, though a striking difference can be observed at lower voltages, the displacement curve of the PPCM is overall in good agreement with that of the FEM. The error of the displacement is merely about 2 μm when 12 V is applied. In our previous study [24], strong agreement was demonstrated between the same FEM model and experimental displacement data. When the V-shaped actuator is loaded with 8 V and 12 V, the displacement distribution along the length is presented in

Figure 10. Based on the PPCM, the deflection at the sample points is in good agreement with that in the FEM curves. Given its good performance in the mechanical prediction, the PPCM can be employed to solve the thermomechanical model of the actuator.

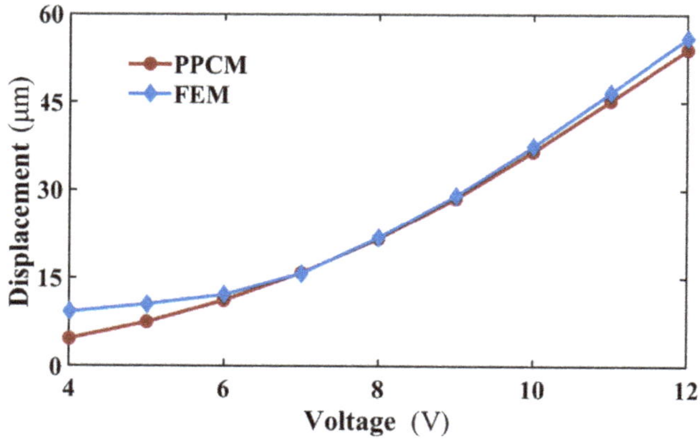

Figure 9. Maximum displacement of the actuator by the PPCM and FEM loaded with different voltages.

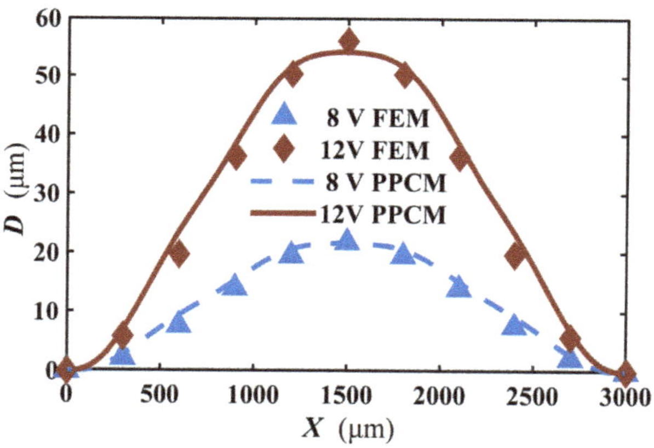

Figure 10. Comparison of displacement distribution along the length at 8 V and 10 V.

6. Conclusions

This study presents a successful application of the Polynomial Point Interpolation Collocation Method (PPCM) to the coupled electrical, thermal, and mechanical analysis of a V-shaped microactuator. Using the PPCM, the thermal and mechanical behaviors of the actuator were evaluated independently. The temperature distribution obtained via PPCM shows good agreement with both finite element method (FEM) results and experimental data. Through computational stability analysis, the incremental load method is identified as particularly suitable for handling the strong nonlinearities inherent in the electrothermal model. Further comparison reveals, however, that the boundary conditions used in this study require refinement, as indicated by increasing temperature deviations from the middle toward both ends of the actuator. In terms of mechanical response, the displacement field computed with PPCM also aligns well with FEM predictions. These findings collectively demonstrate that the PPCM-based scheme provides a convenient, intuitive, and efficient implementation while avoiding high computational costs associated with the weak form of the meshless method. It is also validated that this scheme is more stable and accurate than other strong forms, such as GFDM and PINN, when dealing with multi-physics problems. The further acquisition and comparison of experimental displacement field data for this microactuator will be performed in the next step of our research.

In the context of topology optimization, PPCM completely avoids the computational instabilities and accuracy degradation associated with mesh distortion or remeshing in conventional finite element methods and strong-form meshfree approaches. PPCM also enables efficient and precise multi-physics coupling optimization by solving multiple governing equations in a unified manner within a single node-based framework, using distinct local approximating polynomials. As a node-based discretization method, PPCM generates structured data such as nodal field variables that are highly compatible with machine learning models, thereby facilitating the development of a hybrid intelligent optimization paradigm. In summary, PPCM provides a novel approach to topology optimization that overcomes mesh-related limitations and exhibits significant potential for applications in complex multi-physics systems and data-driven integration.

Author Contributions: Conceptualization, Y.T.; methodology, Y.T.; software, H.L.; validation, Y.T.; formal analysis, A.Q.; investigation, A.Q.; resources, Y.T. and Y.G.; data curation, Y.Z.; writing—original draft preparation, Y.T. and A.Q.; writing—review and editing, D.G. and H.C.; visualization, Y.Z.; supervision, D.G. and H.C.; project administration, H.L. and H.C.; funding acquisition, Y.G. All authors have read and agreed to the published version of the manuscript.

Funding: This research was supported by the Natural Science Foundation of Jiangsu Higher Education Institutions (No. 22KJA460005), the Joint Funds of the Zhejiang Provincial Natural Science Foundation of China (No. LLSSY24E070001), the Open Research Fund of State Key Laboratory of Precision Manufacturing for Extreme Service Performance (No. Kfkt2024-13), the National Key R&D Program of China (No. 2024YFF0504900), and the Open Foundation of the State Key Laboratory of Fluid Power and Mechatronic Systems (No. GZKF-202433).

Data Availability Statement: All data are presented within the paper.

Conflicts of Interest: Author Aidong Qi was employed by the company Zhejiang Jiali (Lishui) Industrial Co., Ltd. Author Haojie Li was employed by the company Lishui Zhe Li Digital Resources Management Co., Ltd. The remaining authors declare that the research was conducted in the absence of any commercial or financial relationships that could be construed as a potential conflict of interest.

References

1. Zhou, H.; Zhang, S.; Liu, Z.; Chi, B.; Li, J.; Wang, Y. Untethered microgrippers for precision medicine. *Small* **2024**, *20*, 2305805. [CrossRef] [PubMed]
2. Peng, L.; Sun, Y.; Qi, Y.; Ru, X. Motion Characteristics of Self-Sensing Piezoelectric Actuator for Yarn Micro-Gripper. *Appl. Sci.* **2024**, *14*, 2620. [CrossRef]
3. Al-Rawashdeh, Y.M.; Al Saaideh, M.; Heertjes, M.F.; Oomen, T.; Al Janaideh, M. Model-free control for an industrial long-stroke motion system with a nonlinear micropositioning actuator. *Mechatronics* **2024**, *104*, 103257. [CrossRef]
4. Li, J.; Li, Z.; Chen, H. Design and Contact Performance Analysis of 3D-Printed Alloy Metal Inertial Micro Switch. *Micromachines* **2025**, *16*, 560. [CrossRef]
5. Wang, L.; Zhao, J.; You, Z. Optimizing MEMS inertial switches for efficient event-based environmental interaction: Motivation, approaches, and purposes. *Microsyst. Nanoeng.* **2025**, *11*, 150. [CrossRef]
6. Xia, G.; Fang, J.; Shou, D.; Wang, X. Phase structure deciphering for pure polymers with a giant piezoelectric response. *Prog. Mater. Sci.* **2024**, *146*, 101340. [CrossRef]
7. Xu, Y.; Wen, J.; Burdet, E.; Taghavi, M. Monolithic electrostatic actuators with independent stiffness modulation. *Nat. Commun.* **2025**, *16*, 1174. [CrossRef]
8. Ho, Y.N.; Michael, A.; Kwok, C.Y.; Pulikkaseril, C. Parallel in-plane electrothermal actuators. *J. Microelectromech. Syst.* **2024**, *33*, 305–307. [CrossRef]
9. Pérez-San Lázaro, R.; Mendoza-Bautista, K.J.; Fuentes-Aguilar, R.Q.; Chairez, I. State-restricted adaptive control of a multilevel rotating electromagnetic mechanical flexible device using electromagnetic actuators. *ISA Trans.* **2024**, *155*, 346–360. [CrossRef]

10. Ouakad, H.M.; Krzysztof Kamil, U.R. On the snap-through buckling analysis of electrostatic shallow arch micro-actuator via meshless galerkin decomposition technique. *Eng. Anal. Bound. Elem.* **2022**, *134*, 388–397. [CrossRef]
11. Jankowski, P.; Żur, K.K.; Farajpour, A. Analytical and meshless DQM approaches to free vibration analysis of symmetric FGM porous nanobeams with piezoelectric effect. *Eng. Anal. Bound. Elem.* **2022**, *136*, 266–289. [CrossRef]
12. Chen, H.; Sun, X.; Chen, M.; Kong, X. Dynamic response of a MEMS electrothremal actuator by the local radial point interpolation method. *Nonlinear Dyn.* **2025**, *113*, 11215–11225. [CrossRef]
13. Xia, H.; Gu, Y. The generalized finite difference method for electroelastic analysis of 2D piezoelectric structures. *Eng. Anal. Bound. Elem.* **2021**, *124*, 82–86. [CrossRef]
14. Nourmohammadi, H.; Behjat, B. Static analysis of functionally graded piezoelectric plates under electro-thermo-mechanical loading using a meshfree method based on RPIM. *J. Stress Anal.* **2020**, *4*, 93–106.
15. Mao, Z.; Suzuki, S.; Nabae, H.; Miyagawa, S.; Suzumor, K.; Maeda, S. Machine learning-enhanced soft robotic system inspired by rectal functions to investigate fecal incontinence. *Bio-Des. Manuf.* **2025**, *8*, 482–494. [CrossRef]
16. Lao, J.; Chen, Y.; Li, H. Vibration Control of Piezoelectric Cantilever Beam With Physics-Informed Neural Networks. *IEEE Access* **2025**, *13*, 145442–145450. [CrossRef]
17. Barbulescu, R.; Ciuprina, G.; Duca, A.; Machado, P.; Silveira, L.M. Physics-informed neural networks for a highly nonlinear dynamic system. *J. Math. Ind.* **2025**, *15*, 7. [CrossRef]
18. Shao, Y.; Duan, Q.; Chen, R. Adaptive meshfree method for fourth-order phase-field model of fracture using consistent integration schemes. *Comput. Mater. Sci.* **2024**, *233*, 112743. [CrossRef]
19. Demkowicz, L.F.; Gopalakrishnan, J. An overview of the discontinuous Petrov Galerkin method. In *Recent Developments in Discontinuous Galerkin Finite Element Methods for Partial Differential Equations: 2012 John H Barrett Memorial Lectures*; Springer: Cham, Switzerland, 2014; pp. 149–180.
20. Xu, B.; Zhang, R.; Yang, K.; Yu, G.; Chen, Y. Application of generalized finite difference method for elastoplastic torsion analysis of prismatic bars. *Eng. Anal. Bound. Elem.* **2023**, *146*, 939–950. [CrossRef]
21. Sun, W.; Qu, W.; Gu, Y.; Zhao, S. Meshless generalized finite difference method for two-and three-dimensional transient elastodynamic analysis. *Eng. Anal. Bound. Elem.* **2023**, *152*, 645–654. [CrossRef]
22. Cuomo, S.; Di Cola, V.S.; Giampaolo, F.; Rozza, G.; Raissi, M.; Piccialli, F. Scientific machine learning through physics–informed neural networks: Where we are and what's next. *J. Sci. Comput.* **2022**, *92*, 88. [CrossRef]
23. Guo, R.; Sui, F.; Yue, W.; Wang, Z.; Pala, S.; Li, K.; Lin, L. Deep learning for non-parameterized MEMS structural design. *Microsyst. Nanoeng.* **2022**, *8*, 91. [CrossRef]
24. Chen, H.; Wang, X.J.; Wang, J.; Xi, Z.W. Analysis of the dynamic behavior of a V-shaped electrothermal microactuator. *J. Micromech. Microeng.* **2020**, *30*, 085005. [CrossRef]

micromachines

MDPI

Article

Design and Evaluation of a Capacitive Micromachined Ultrasonic Transducer(CMUT) Linear Array System for Thickness Measurement of Marine Structures Under Varying Environmental Conditions

Changde He, Mengke Luo, Hanchi Chai, Hongliang Wang, Guojun Zhang, Renxin Wang, Jiangong Cui, Yuhua Yang, Wendong Zhang and Licheng Jia *

State Key Laboratory of Extreme Environment Optoelectionic Dynamic Measurement Technology and Instrument, North University of China, Taiyuan 030051, China; hechangde@nuc.edu.cn (C.H.); s202306010@st.nuc.edu.cn (M.L.); s202306026@st.nuc.edu.cn (H.C.); wanghongliang@nuc.edu.cn (H.W.); zhangguojun1977@nuc.edu.cn (G.Z.); wangrenxin@nuc.edu.cn (R.W.); jgcui@nuc.edu.cn (J.C.); yangyuhua@nuc.edu.cn (Y.Y.); wdzhang@nuc.edu.cn (W.Z.)

* Correspondence: jialicheng@nuc.edu.cn

Abstract: This paper presents the design, fabrication, and experimental evaluation of a capacitive micromachined ultrasonic transducer (CMUT) linear array for non-contact thickness measurement of marine engineering structures. A 16-element CMUT array was fabricated using a silicon–silicon wafer bonding process, and encapsulated in polyurethane to ensure acoustic impedance matching and environmental protection in underwater conditions. The acoustic performance of the encapsulated CMUT was characterized using standard piezoelectric transducers as reference. The array achieved a transmitting sensitivity of 146.82 dB and a receiving sensitivity of −229.55 dB at 1 MHz. A complete thickness detection system was developed by integrating the CMUT array with a custom transceiver circuit and implementing a time-of-flight (ToF) measurement algorithm. To evaluate environmental robustness, systematic experiments were conducted under varying water temperatures and salinity levels. The results demonstrate that the absolute thickness measurement error remains within ±0.1 mm under all tested conditions, satisfying the accuracy requirements for marine structural health monitoring. The results validate the feasibility of CMUT-based systems for precise and stable thickness measurement in underwater environments, and support their application in non-destructive evaluation of marine infrastructure.

Keywords: MEMS; CMUT array; sensitivity; time of flight (ToF); thickness detection

1. Introduction

In marine environments [1], ships [2], offshore engineering structures, and subsea pipelines are continuously subjected to factors such as seawater corrosion [3], pressure fluctuations [4], and biological fouling [5], where changes in material thickness can directly affect their safety, reliability, and service life. Accurate thickness measurement therefore represents a critical inspection technique for ensuring the structural integrity of marine assets. Compared with alternative methods such as X-ray [6], laser-based techniques, and mechanical contact testing, ultrasonic technology [7–9] has emerged as one of the most important non-destructive testing (NDT) solutions. This prominence is attributed to its distinct advantages: non-invasive operation, high measurement precision, adaptability to complex environments, compatibility with a wide range of materials, rapid testing capabilities, and the ability to penetrate opaque media. These features make ultrasonic methods indispensable for evaluating thickness variations in submerged or otherwise inaccessible structures [10–12].

Ultrasonic transducer arrays [13], serving as the core components for acoustic–electric signal conversion in ultrasonic thickness measurement systems, play a pivotal role in determining the overall performance of such systems. Compared with traditional piezoelectric transducers (e.g., Lead Zirconate Titanate (PZT)), microelectromechanical ultrasonic transducers (MUTs) exhibit significant advantages in underwater detection: First, MUTs reduce reliance on coupling agents through the thin-film vibration mechanism, with an acoustic impedance matching error of <5% within the 1–10 MHz frequency band [14]. Second, the influence of frequency on resolution is notable: MUTs can achieve an axial resolution of <300 μm at high frequencies (e.g., 10 MHz), which is superior to the >1 mm resolution of PZT at low frequencies (<5 MHz) [15]. Third, the integration capability with complementary metal-oxide semiconductors (CMOSs) enables the system power consumption to be reduced to 3.8 mW and the chip area to be controlled within 4 mm^2, far outperforming the discrete integration scheme of PZT. In contrast, capacitive micromachined ultrasound transducers (CMUTs), a new generation of ultrasonic transducers based on microelectromechanical system (MEMS) technology, emerge as a promising alternative. The integration capability of MEMS technology is a defining advantage, enabling the fabrication of 2D arrays with thousands or even tens of thousands of micro-scale elements and facilitating monolithic integration with signal processing circuits [16]. This not only expands the detection coverage but also significantly enhances data acquisition precision. In addition, CMUTs offer superior features, such as wide bandwidth, excellent underwater adaptability eliminating the need for coupling agents, and low power consumption. As CMUT technology continues to advance, its potential for enabling high-performance, scalable, and environmentally robust thickness measurement systems makes it an indispensable solution for non-destructive evaluation in offshore engineering structures, poised to drive transformative advancements in marine asset integrity management [14,17–20].

Compared with single-element CMUT transducers, CMUT linear array transducers enable simultaneous waveform emission from multiple elements, leveraging signal energy superposition to enhance ultrasonic transmission intensity. This augmentation significantly improves penetration through thick or highly attenuating materials, ensuring robust reflection signal strength during reception and thereby enhancing measurement feasibility in challenging scenarios. Meanwhile, the array configuration allows for data averaging or fusion across multiple elements [21], capitalizing on the randomness of individual measurement errors to mitigate overall errors and improve measurement precision [22]. Additionally, the spatial distribution of array elements expands the effective detection area, enabling efficient coverage of large-scale structures or multi-point thickness measurements. This feature drastically reduces inspection time and operational workload compared to sequential single-element scanning. Given these combined advantages in signal strength, measurement accuracy, and detection efficiency, CMUT linear array transducers are selected in this study for thickness measurement applications.

In ultrasonic measurement techniques, the ToF method [23], as one of the earliest developed approaches, has been widely adopted across diverse ultrasonic measurement applications [24]. In 2016, Zhang et al. [25] demonstrated ultrasonic measurement of oil film thickness using piezoelectric transducers via the ToF principle. More recently, in 2022, B. Bühling et al. introduced a non-contact air-coupled ultrasonic ToF measurement method based on laser Doppler vibrometry, achieving a distance measurement error of less than 1.5% [?]. However, despite these advancements, the application of CMUT linear array transducers to construct thickness detection systems for offshore engineering structures using the ToF method remains unexplored in the current literature, representing an unaddressed research gap in ultrasonic non-destructive evaluation for marine environments.

Addressing the application requirements of ultrasonic transducers in offshore engineering structure thickness detection systems, this study designed and fabricated a high-density integrated CMUT linear array transducer, followed by comprehensive performance characterization of the as-prepared device. The developed detection system utilizes the CMUT linear array to emit and receive ultrasonic signals, determining object thickness via the ToF method by recording the time delay between reflected signals from the front and back interfaces. Environmental impacts on measurement accuracy—specifically temperature and salinity variations—were systematically evaluated to assess the system's robustness in complex marine conditions. The thickness detection system constructed using CMUT transducers demonstrates substantial application potential for precise, reliable thickness measurement in offshore engineering structures, addressing the critical need for non-destructive evaluation in harsh marine environments.

2. CMUT Linear Array Design and Fabrication

2.1. Design

Figure 1a illustrates the CMUT cell structure, which consists of a movable top electrode diaphragm, a fixed lower electrode layer, and an insulating dielectric layer sandwiched between the two electrodes. The movable upper electrode diaphragm is constructed from a deformable single-crystal silicon membrane integrated with a patterned metal electrode. This diaphragm structure undergoes mechanical deformation and generates vibrations under voltage excitation for ultrasonic emission, or conversely, vibrates in response to incident acoustic pressure signals during reception, enabling bidirectional acoustic–electric conversion.

A vacuum cavity is formed between the movable single-crystal silicon membrane and the monocrystalline silicon substrate, with a silicon dioxide insulating layer retained at the bottom of the vacuum cavity. This design prevents the movable membrane from collapsing and making direct contact with the substrate, thereby avoiding high-voltage breakdown within the device due to electrode short-circuiting. The cross-sectional scanning electron microscope (SEM) image performed by the TESCAN MIRA3 of a sensing cell in the MEMS hydrophone sensor is shown in Figure 1b. Table 1 lists the structural parameters of the CMUT cell, where the thickness of the movable single-crystal silicon membrane is 2.83 μm and the vacuum gap height is 0.65 μm.

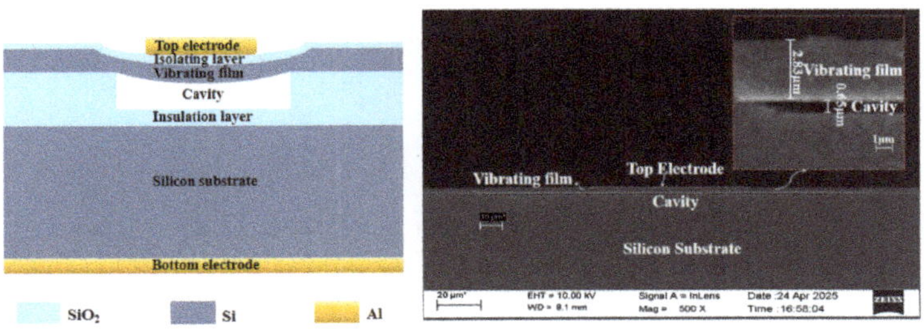

(a) (b)

Figure 1. CMUT linear array transducers. (**a**) Working principle diagram of a sensing cell. (**b**) SEM image of the cross-section of the CMUT array (20 μm corresponds to the scale bar in the SEM image).

Table 1. Key parameters for CMUT array.

Parameters	Value (μm)
Monocrystalline silicon thickness	2.83
Cavity diameter	180
Cavity height	0.65
Cavity insulation thickness	0.15
Oxide thickness	1

During operation, CMUTs involve bidirectional conversion between electric field forces and mechanical vibrations, with mechanical motions coupled to ultrasonic waves during transmission or inducing mechanical responses from incident ultrasonic waves during reception. The equivalent circuit models for the CMUT ultrasonic transducer sensing cell [27] are depicted in Figure 2, where Figure 2a represents the model under transmission mode and Figure 2b illustrates the model under reception mode.

In transmission mode, an applied excitation voltage V_{in} is imposed across the CMUT electrodes, generating electrostatic forces that induce vibration of the CMUT membrane and subsequently produce ultrasonic signals in the coupling medium. In the electrical domain, C_0 represents the static capacitance, and energy conversion from the electrical to mechanical domain is achieved via an ideal transformer with a turns ratio of $1:\eta$. In the mechanical domain, the model incorporates the effective mass m_m and effective stiffness k_m, which characterize the sensing cell dynamic behavior for a specific vibration mode. In the acoustical domain, P_{out} denotes the output acoustic pressure, and Z_a represents the acoustic impedance, which is coupled to the water through an area conversion ratio of $A_{eff}:1$. The overall input impedance in the frequency domain is expressed as:

$$Z_{in}(s) = \frac{1}{sC_0} + \frac{1}{\eta^2}\left(sm_m + \frac{1}{sk_m} + \frac{Z_a}{A_{eff}^2}\right) \quad (1)$$

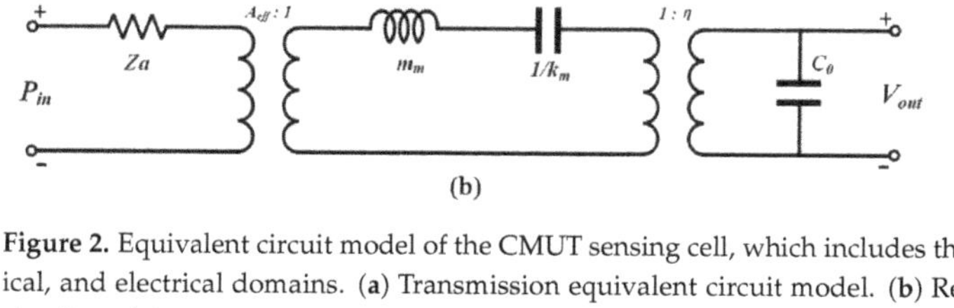

Figure 2. Equivalent circuit model of the CMUT sensing cell, which includes the acoustic, mechanical, and electrical domains. (**a**) Transmission equivalent circuit model. (**b**) Receiving equivalent circuit model.

In reception mode, the CMUT surface is subjected to an external ultrasonic acoustic pressure P_{in}, which generates an equivalent driving force causing the membrane to undergo mechanical displacement. This displacement induces charge variations between the electrodes, thereby generating an output voltage V_{out}. In the acoustic domain, the incident acoustic pressure P_{in} acting on the sensing cell effective area A_{eff} is converted into a mechanical force $F = A_{eff} \cdot P_{in}$, which drives mechanical vibrations in the mechanical domain. Z_a is the acoustic impedance. The mechanical displacement X(s) is converted into electrical charge through electromechanical transformation with a turns ratio η. In the electrical domain, C_0 denotes the static capacitance of the sensing cell, and V_{out} represents the output voltage generated by the charge variation on the sensing surface. In the frequency domain, the relationship between the output voltage and the input acoustic pressure can be characterized by a transfer function.

$$V_{out}(s) = \frac{\eta A_{eff}}{C_0\left(sm_m + \frac{1}{sk_m} + \frac{Z_a}{A_{eff}^2}\right)} P_{in}(s) \tag{2}$$

where s denotes the Laplace variable in the frequency domain.

2.2. Fabrication

When operating in a vacuum, the resonant frequency of a CMUT is denoted as f_0 [28]. In contrast, when the CMUT is immersed in a liquid environment with density ρ_L, its resonant frequency shifts to f_1 [29].

$$f_0 = \frac{0.467h}{a^2\sqrt{\frac{E}{\rho(1-\sigma^2)}}} \tag{3}$$

$$f_1 = \frac{0.467h}{a^2\sqrt{\frac{E}{\rho(1-\sigma^2)}}} / \sqrt{1 + 0.67\rho_L a/\rho h} \tag{4}$$

where h is the thickness of the membrane, a is the radius of the membrane, E is the Young's modulus of the membrane material, ρ is the material density, and σ is the Poisson's ratio of the thin film. From the equation, it can be observed that the resonant frequency of the membrane is exclusively determined by its geometric dimensions and material properties. For a given membrane material, a larger radius or smaller thickness leads to a higher resonant frequency, reflecting the inverse relationship between frequency and membrane size under fixed material conditions [30].

To ensure sufficient detection depth for underwater operation, the CMUT in this study was designed to have a resonant frequency of 1 MHz. Previous studies have indicated that excessively thin membranes are prone to vibration fatigue, leading to structural damage, whereas overly thick membranes necessitate higher driving voltages to achieve effective actuation. Balancing these trade-offs, a membrane thickness of 2.83 µm was determined as the optimal compromise, ensuring both mechanical durability and operational efficiency under the target frequency.

This study presents a MEMS microfabrication process for CMUT linear array transducers, with silicon–silicon wafer bonding as the core technique to fabricate the CMUT array device [31]. The detailed fabrication process of the CMUT array is illustrated in Figure 3. An oxidized silicon wafer with an 0.8 µm thick thermal oxide layer and a silicon-on-insulator (SOI) wafer were first prepared. After photolithographically patterning and etching 0.65 µm deep cavities on the front side of the oxidized silicon wafer, it was pre-bonded with the SOI wafer at room temperature under vacuum conditions. It is critical to ensure sufficient vacuum level and surface cleanliness to prevent the formation of unbonded regions or micro-voids. The bonded wafers were then annealed in a high-temperature oxidation furnace, with uniform temperature distribution during annealing to avoid stress discrepancies caused by inconsistent bonding strength.

The substrate of the SOI wafer was thinned via chemical mechanical polishing (CMP), taking care to avoid membrane rupture. The remaining silicon substrate was etched away using a silicon etchant, followed by removal of the buried oxide layer of the SOI wafer with buffered oxide etchant (BOE) to form a free-standing silicon membrane structure. Silicon etching and silicon dioxide (SiO_2) deposition were subsequently performed; during etching, over-etching (which would cause membrane thickness to exceed the designed value) and under-etching (which would leave residual substrate) must be avoided to prevent yield degradation.

A metal layer was sputtered onto the SiO_2 layer and patterned to form the top electrode of the capacitor. Precise alignment during patterning is essential to prevent short circuits between the top and bottom electrodes. On the substrate side, a 1 μm thick metal layer was deposited and annealed to achieve ohmic contact between the substrate and metal, serving as the bottom electrode of the capacitor.

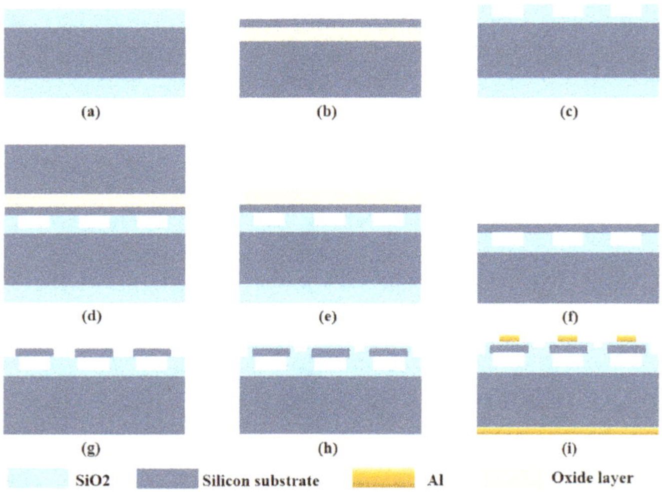

Figure 3. Processing flowchart of CMUT based on wafer bonding. (**a**) Thermal oxidation of the silicon wafer. (**b**) SOI wafer. (**c**) Etching of the cavity and alignment markers. (**d**) Wafer bonding of the SOI wafer. (**e**) Removal of silicon substrate. (**f**) Removal of the oxide layer. (**g**) Etching isolation grooves and marking grooves. (**h**) Chemical vapor deposition oxide layer. (**i**) Sputtering on the upper and lower electrodes.

Figure 4 presents a top-view image of the CMUT array, where the CMUT linear array transducer is composed of 16 elements. Each element is configured with 90 × 9 (totaling 810) micro-capacitive units arranged in parallel, forming a high-density integrated structure to enhance acoustic signal sensitivity and array uniformity.

Figure 4. Optical microscope image of the CMUT manufactured in this work.

3. Development and Operational Principle of Thickness Detection System

To prevent direct contact between CMUT chip surfaces and water—as such contact could induce electrical short circuits or corrosion of microstructures—CMUT arrays require packaging. In non-contact ultrasonic thickness measurement using water as the coupling medium, the encapsulation layer material of the transducer plays a decisive role in enabling effective acoustic energy coupling and propagation. The encapsulation material must not only exhibit high acoustic transparency but also possess acoustic impedance characteristics matched to water, the coupling medium, to minimize interfacial reflection and enhance measurement sensitivity and resolution. This dual requirement ensures minimal energy loss at the transducer–medium interface, thereby optimizing the signal-to-noise ratio and enabling precise thickness detection in non-destructive evaluation applications [32]. Acoustic impedance is determined by the material density ρ and the longitudinal wave velocity ν and can be expressed as follows:

$$Z = \rho\nu \tag{5}$$

The acoustic impedance of water is approximately $1.48 \times 10^6 (\text{kg}/\text{m}^2\text{s})$, while that of commonly used polyurethane materials ranges from 1.2 to $1.7 (\text{kg}/\text{m}^2\text{s})$, depending on their curing density, degree of molecular crosslinking, and additive composition. This range closely matches the acoustic impedance of water, thereby creating a low reflectivity at the water–polyurethane interface. This matching minimizes energy loss due to reflection, enabling efficient transmission of ultrasonic energy into the target structure and back to the transducer, which is critical for maintaining high-fidelity signal transduction in non-contact thickness measurement systems.

According to the reflection coefficient formula

$$R = \left(\frac{Z_2 - Z_1}{Z_2 + Z_1}\right)^2, \tag{6}$$

when the acoustic impedance of polyurethane, Z_2, closely matches that of water, Z_1, the reflection coefficient R approaches zero. This indicates minimal energy reflection at the interface, allowing the majority of ultrasonic waves to propagate through unimpeded. Such impedance matching is critical in non-contact ultrasonic thickness measurement—especially in dual-interface systems involving water, the encapsulation layer, and the target material—as it significantly enhances the quality of echo signals and the system's signal-to-noise ratio (SNR). Beyond impedance matching, polyurethane exhibits additional advantageous properties: it combines excellent flexibility with mechanical strength [33,34], providing robust physical protection for the ultrasonic transducer. Furthermore, its low water absorption and hydrolysis resistance ensure stable performance during long-term underwater operation [35,36], minimizing performance drift. Given these attributes, the 16-element CMUT linear array developed in this study was encapsulated with polyurethane, rendering it well-suited for non-contact ultrasonic thickness measurement in aquatic environments.

A thickness measurement system for marine engineering structures was developed using a 16-element CMUT linear array transducer, with its conceptual diagram illustrated in Figure 5a. The experimental test platform is shown in Figure 5b. The target specimen and the 16-element linear array transducer were vertically positioned and rigidly fixed within a water tank, maintaining a 10 cm acoustic path between them. The transducer was connected to a transmit–receive integrated circuit, where a 30V DC bias voltage and a 22V amplitude AC excitation signal were applied to its electrodes to generate ultrasonic waves. A transmit–receive integrated circuit is used to drive the CMUT for transmitting and receiving waveforms, as shown in Figure 6a. The MAX14808 high-voltage pulse excitation chip is employed as the core component of the transmit circuit to drive the CMUT transducer, as illustrated in Figure 6b. The ultrasonic receiving circuit is constructed using the AFE5818 chip. When detecting CMUT echo signals, the AFE chip is configured as follows: the echo signal passes through the TR switch of the MAX chip, flows out from the LVOUT pin, and enters the input pins INP and INM of the AFE chip. Upon transmission, the received echo signals were sent to a PC for real-time waveform display, enabling systematic characterization of the transducer's performance in water-based non-destructive thickness measurement.

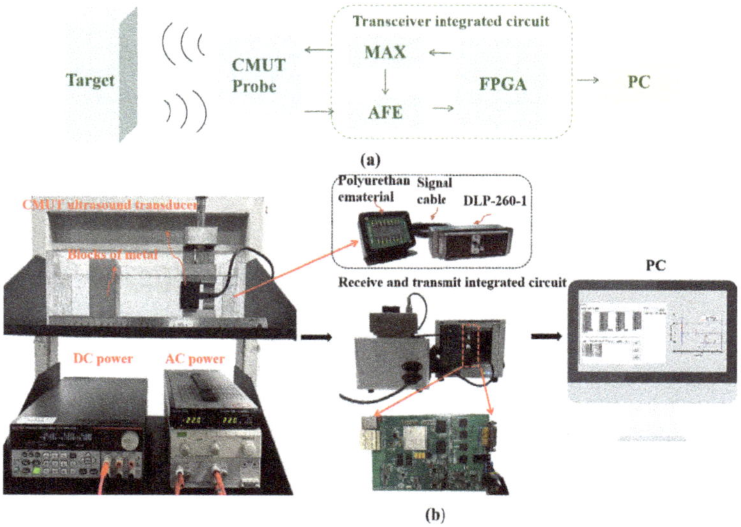

Figure 5. Schematic diagram of the CMUT-based thickness detection system. (**a**) Functional block. (**b**) Testing process.

The core of this system relies on the ToF method to estimate the thickness of the target object, combined with high-resolution signal processing techniques to enhance time delay extraction accuracy. This process involves recording the echo signals reflected from the target interfaces and extracting the time delay between the primary peaks of the front and back surface waveforms—denoted as ToF. Assuming the sound velocity within the specimen is v, the thickness d can be calculated using the following equation:

$$d = \frac{v * ToF}{2} \tag{7}$$

The division by 2 in the equation accounts for the round-trip travel of the ultrasonic wave through the specimen. In this study, echo signals acquired from the 16 array elements were first averaged to reduce noise, followed by envelope processing using an absolute-value low-pass filtering method to highlight the signal amplitude variations. A quadratic parabola interpolation technique was then employed to achieve sub-sample level correction for the envelope peak positions, enabling a time precision of 0.01 µs in ToF extraction. The specific interpolation procedure is as follows.

$$\delta = \frac{1}{2} \cdot \frac{y_1 - y_3}{y_1 - 2y_2 + y_3} \tag{8}$$

$$t_{\text{fine}} = (i + \delta) \cdot \Delta t \tag{9}$$

where, y_1, y_2, and y_3 represent three consecutive sampled values around the envelope peak, y_2, at the peak position, Δ_t denotes the sampling interval (0.1 μs), and i represents the integer sampling point index adjacent to the peak point of the envelope signal, which is an integer corresponding to the position of the original sampling point. δ represents the correction.

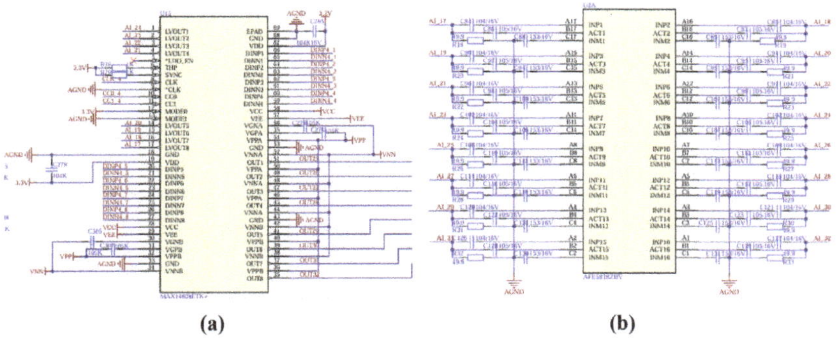

Figure 6. Schematic diagrams of the transmit–receive integrated circuit. (**a**) Schematic diagram of the transmit circuit. (**b**) Schematic diagram of the receive circuit.

The processed waveform is shown in Figure 7, where T_1 and T_2 indicate the primary peak positions of the front and back surface reflection signals from the specimen, respectively. These peaks are identified after envelope processing and sub-sample interpolation, providing precise time markers for calculating the ToF with sub-microsecond accuracy.

In the marine engineering field, structures such as ship hulls, seawater pipelines, offshore platforms, and wind turbine substructures are subjected to harsh environments characterized by high salinity, humidity, and wave impacts during long-term service. The selected metallic materials must exhibit excellent corrosion resistance, a balanced strength–toughness ratio, and good weldability. Carbon steel, valued for its low cost, reliable mechanical properties, and ease of welding, is widely used in the main structures of offshore platforms and ship hulls. Stainless steel, particularly 316 L, offers superior resistance to pitting, crevice corrosion, and stress corrosion cracking, making it ideal for pipeline systems and critical connection components in direct seawater contact [37]. Aluminum alloys, prized for their high strength-to-weight ratio, are essential in weight-sensitive marine structures such as high-speed craft, boarding platforms, and buoys.

To validate the CMUT-based thickness measurement system, three representative materials were selected as test specimens: 6061 aluminum alloy, 316 L stainless steel, and AH36 carbon steel, with thicknesses of 20 mm, 40 mm, 60 mm, and 80 mm. Prior to thickness measurement, sound velocity calibration for each material was performed using the pulse-echo method in accordance with the international standard GB/T 23900-2009 [38], ensuring measurement accuracy through systematic characterization of acoustic propagation properties. The calibration process involved transmitting ultrasonic pulses through the specimens and recording round-trip travel times, from which material-specific sound velocities were derived to enable precise thickness calculations. Table 2 shows the acoustic velocities of the calibration blocks: the acoustic velocity of 6061 aluminum alloy is 6299.2 m/s, that of 316L stainless steel is 5743 m/s, and that of AH36 carbon steel is 5873.1 m/s.

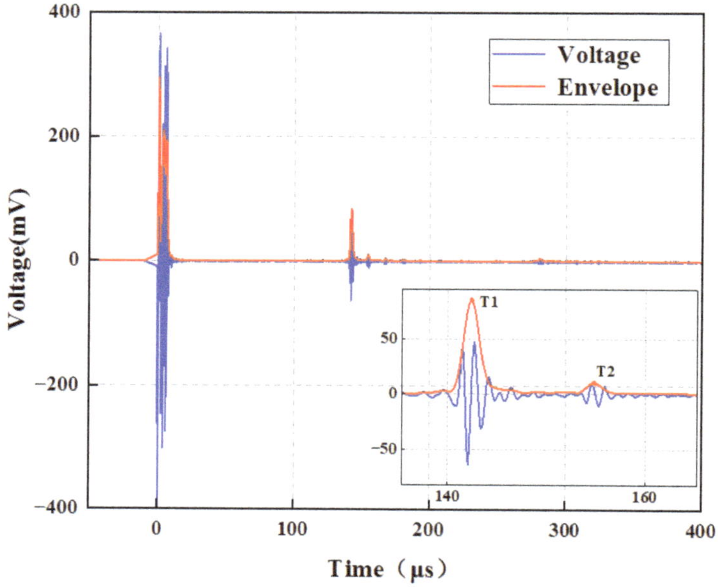

Figure 7. Transmitted and received signals collected by the system.

Table 2. Sound velocity of calibration block.

Serial No.	Material of Calibration Block(m/s)	Sound Velocity
1	6061 aluminum alloy	6299.2
2	316 L stainless steel	5743.0
3	AH36 carbon steel	5873.1

4. Experimental Results and Discussion

The transmitting sensitivity and receiving sensitivity are critical parameters for evaluating the transmission and reception performance of CMUT linear array transducers. A dedicated test platform was constructed, as shown in Figure 8, to characterize these metrics. The transmitting sensitivity of the CMUT array was measured in a water tank at a distance of approximately 30 cm across a frequency range of 0.8 MHz–1.2 MHz. As shown in Figure 9a, the reflection response spectrum and final emission response spectrum of the standard transducer show a transmit voltage response (TVR) of 146.82 dB at 1 MHz, indicating effective generation of sound waves under the specified excitation conditions. For receiving sensitivity characterization, a reciprocal calibration method was employed using a standard 1 MHz piezoelectric transducer as the reference. This approach involved measuring the electrical output of the CMUT element in response to known acoustic pressures across the same frequency range. As shown in Figure 9b, the receiving sensitivity of the standard transducer and the tested CMUT is determined to be −229.55 dB at 1 MHz, demonstrating the transducer's ability to convert incident acoustic energy into high fidelity electrical signals. These measurements validate the CMUT array's suitability for high-precision thickness detection in marine environments, where both robust signal transmission and sensitive reception are essential for reliable non-destructive evaluation.

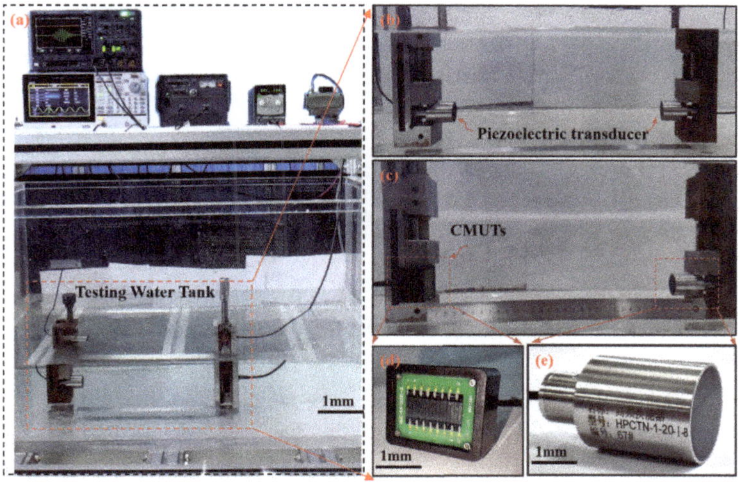

Figure 8. Experimental testing platform. (**a**) Testing process. (**b**) Calibration of piezoelectric transducer under receiving sensitivity test. (**c**) Transmit voltage response test. (**d**) Polyurethane material. (**e**) Standard 1 MHz piezoelectric transducer.

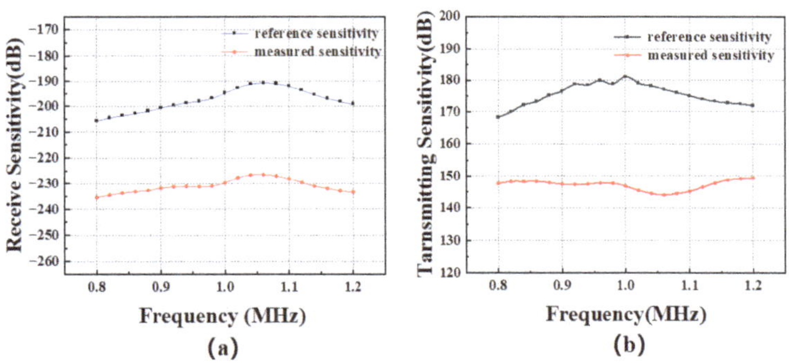

Figure 9. The performance parameters. (**a**) Transmitting sensitivity curve of the CMUT linear array transducer and standard transducer. (**b**) Receiving sensitivity curve of the CMUT linear array transducer and standard transducer.

The impedance of the CMUT array measured in air by an impedance analyzer (Keysight E499oA) is shown in Figure 10. As can be seen from the impedance frequency spectrum, the resonant frequency in the air for the CMUT array is 1311 kHz and the anti-resonant frequency is 1330 kHz. As can be seen from the figure, the phase curve shows an overall increasing trend, but its maximum value remains negative, indicating that the capacitive property of the CMUT dominates. The Z curve exhibits an overall decreasing trend, suggesting that the capacitive reactance decreases with increasing frequency.

To analyze the impact of environmental factors in marine environments on the accuracy of CMUT-based thickness measurement, thickness measurements were carried out over a temperature range of 10 °C–30 °C (in increments of 5 °C) and a salinity range of practical salinity units of 0.1–0.35% (in increments of 0.05%). This experimental design systematically evaluates how temperature and salinity variations affect measurement precision, critical for ensuring reliable non-destructive testing in harsh offshore conditions where water properties can fluctuate significantly.

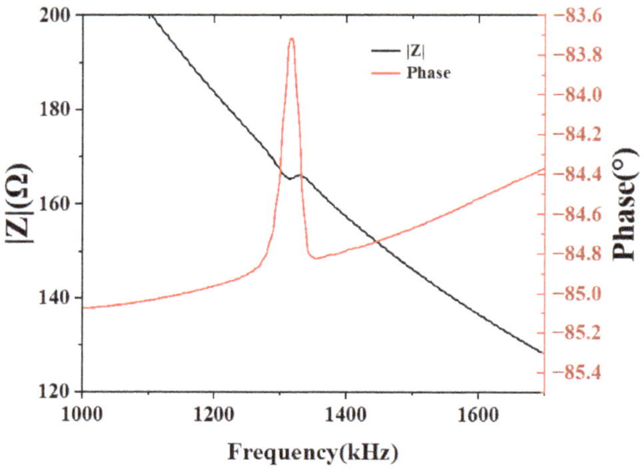

Figure 10. Impedance curve of the CMUT array in air.

To investigate the influence of temperature variations on the thickness measurement accuracy of the CMUT system, the following experimental protocol was implemented: A sealed container filled with pure water was first cooled to 5 °C using a DHC-05 low-temperature constant-temperature water bath. The cooled water was then transferred to the test water tank containing the CMUT linear array transducer. A MJ-R10 constant-temperature heater was used to incrementally raise the water temperature in 5 °C steps up to 30 °C. At each target temperature, the water was stabilized within ±0.2 °C of the setpoint for 3 min before measurements were initiated. Water temperature was continuously monitored in real-time using a high-precision thermocouple thermometer (Type k; measurement accuracy: ±0.1°C). Each temperature condition was tested five times, with the average value used for the analysis. Figure 11 illustrates the relationship between specimen thickness and temperature variations across different materials. The results demonstrate that, over the 10 °C–30 °C temperature range, thickness measurement precision remained within ±0.1 mm, and relative errors were confined to within ±0.5% of the target thickness. These findings confirm that temperature fluctuations do not significantly impact the measurement accuracy of the thickness detection system, satisfying the error tolerance requirements for marine engineering structural thickness testing.

In the experiment investigating the effect of salinity on thickness measurement accuracy, salt solutions with mass fractions of 0%, 0.1%, 0.15%, 0.20%, 0.25%, 0.30%, and 0.35% were prepared by dissolving pure NaCl in deionized water according to the following formula. This method ensures precise control over the salinity gradient, with each solution meticulously formulated to achieve the target concentration by systematically varying the mass of NaCl added while maintaining a constant total solution volume. The use of deionized water as the base medium minimizes impurities that could interfere with acoustic propagation, allowing for isolated evaluation of salinity effects on measurement performance.

$$m_l = s * \frac{m_h}{1000 - s} \tag{10}$$

In this equation, m_l represents the mass of solute required for preparation, s denotes the salinity concentration of the solution, and m_h is the mass of water in the salt solution. For each target concentration, the required solute was weighed using an electronic balance, dissolved in deionized water through thorough stirring to ensure homogeneity, and calibrated using a salinity meter (measurement accuracy: ±0.1%). Under each salinity condition, specimens of different materials and thicknesses were fully immersed in the salt solution, with measurements repeated five times per specimen to calculate average values for analysis. Figure 12. depicts the relationship between specimen thickness and salinity variations across different materials. The results show that in pure water, the thickness measurement precision of the system remained within ±0.1mm. As the salinity concentration was varied, the absolute error band stayed within ±0.1 mm relative to the standard thickness, with relative errors confined to ±0.5% of the measured thickness. These error margins fall within the controllable range, confirming that salinity fluctuations do not significantly affect the measurement accuracy of the thickness detection system. This performance meets the error tolerance requirements for thickness testing of marine engineering structures, validating the system's reliability in saline environments. Overall, the marine engineering structural thickness detection system developed using a 16-element CMUT linear array transducer with the ToF method exhibits measurement accuracy unaffected by environmental factors such as temperature and salinity, demonstrating deterministic and stable performance across the tested conditions.

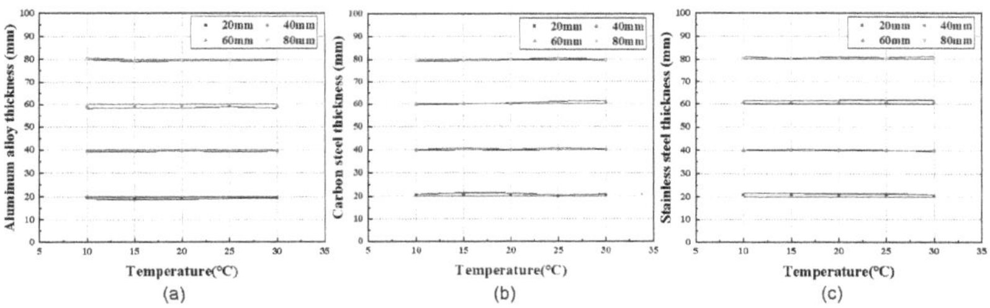

Figure 11. Measured data of specimens with different thicknesses at different temperatures. (**a**) Aluminum alloy. (**b**) Carbon steel. (**c**) Stainless steel.

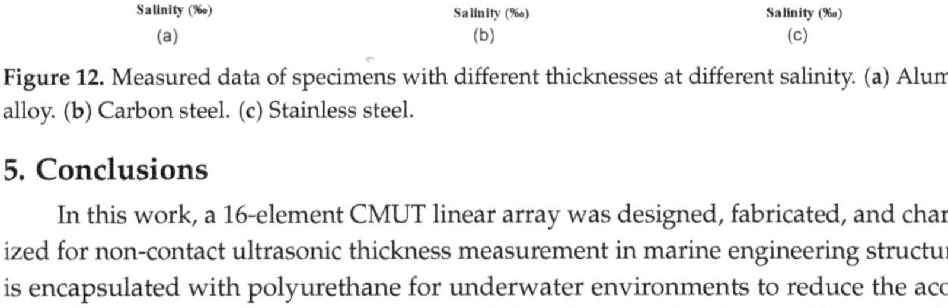

Figure 12. Measured data of specimens with different thicknesses at different salinity. (**a**) Aluminum alloy. (**b**) Carbon steel. (**c**) Stainless steel.

5. Conclusions

In this work, a 16-element CMUT linear array was designed, fabricated, and characterized for non-contact ultrasonic thickness measurement in marine engineering structures. It is encapsulated with polyurethane for underwater environments to reduce the acoustic impedance mismatch rate between the CMUT and water, thereby significantly reducing signal loss. The transmit voltage response level is 146.82 dB@1 MHz, and the receive sensitivity is −229.55 dB@1 MHz [39]. When integrated with a custom transmit–receive circuit, the system enhances the signal-to-noise ratio (SNR) and receive gain [40], and employs the time-of-flight (ToF) method to perform accurate thickness measurements without requiring physical contact. The influence of environmental parameters, specifically water temperature (10 °C–30 °C) and salinity (0–0.35%), was experimentally evaluated. The results showed that the system maintained a measurement accuracy within 0.1 mm in all conditions tested, meeting the practical requirements for offshore structural monitoring. These findings validate the robustness and reliability of the CMUT array in variable marine environments and highlight its potential for scalable, high-precision, and non-invasive thickness evaluation in demanding offshore applications.

Author Contributions: Conceptualization—C.H. and L.J.; methodology—C.H., M.L. and H.C.; validation—M.L. and H.C.; formal analysis—G.Z., R.W., J.C., Y.Y., H.W. and L.J.; data curation—C.H., M.L. and H.C.; original draft preparation—C.H. and M.L.; review and editing—G.Z., R.W., J.C., Y.Y., H.W. and L.J.; funding acquisition—C.H. and W.Z. All authors have read and agreed to the published version of the manuscript.

Funding: This work was supported in part by the National Key Research and Development Project of China under Grant 2022YFB3205400.

Data Availability Statement: All data supporting this study are included in the article.

Conflicts of Interest: The authors declare no conflicts of interest.

References

1. Liu, Y.; Jing, B.; Zhang, G.; Pei, J.; Jia, L.; Geng, Y.; Bai, Z.; Zhang, J.; Guo, Z.; Wang, J.; et al. Design and Algorithm Integration of High-Precision Adaptive Underwater Detection System Based on MEMS Vector Hydrophone. *Micromachines* **2024**, *15*, 514 [CrossRef]
2. Vicen-Bueno, R.; Carrasco-Álvarez, R.; Jarabo-Amores, M.P.; Nieto-Borge, J.C.; Alexandre-Cortizo, E. Detection of Ships in Marine Environments by Square Integration Mode and Multilayer Perceptrons. *IEEE Trans. Instrum. Meas.* **2021**, *60*, 712–724. [CrossRef]
3. Chen, B.-Q.; Zhang, X.; Guedes-Soares, C. The effect of general and localized corrosions on the collapse pressure of subsea pipelines. *Ocean Eng.* **2022**, *247*, 110719. [CrossRef]
4. Lu, Y.; Fu, W.; Xue, D. Deformation Characteristics of Soft Marine Soil Tested under Cyclic Loading with Low Frequency. *Adv. Civ. Eng.* **2020**, *2020*, 8875315. [CrossRef]
5. Liu, H.; Chen, W.; Tan, Y.; Meng, G.; Liu, H.; Cheng, Y.; Liu, H.W. Characterizations of the biomineralization film caused by marine Pseudomonas stutzeri and its mechanistic effects on X80 pipeline steel corrosion. *J. Mater. Sci. Technol.* **2022**, *125*, 15–28. [CrossRef]
6. Cao, Z.; Cao, P. Research Progress on Low-Surface-Energy Antifouling Coatings for Ship Hulls: A Review. *Biomimetics* **2023**, *8*, 502. [CrossRef]
7. Shi, L.; Jia, L.; Liu, C.; Sun, C.; Liu, S.; Wu, G. A Miniaturized Ultrasonic Sugar Concentration Detection System Based on Piezoelectric Micromachined Ultrasonic Transducers. *IEEE Trans. Instrum. Meas.* **2022**, *71*, 7503509. [CrossRef]
8. Shomaji, S.; Dehghanzadeh, P.; Roman, A.; Forte, D.; Bhunia, S.; Mandal, S. Early Detection of Cardiovascular Diseases Using Wearable Ultrasound Device. *IEEE Consum. Electron. Mag.* **2019**, *8*, 12–21. [CrossRef]
9. Luo, Y.; Hu, D.; Yang, F.; Zhou, H. UPVnet: A Neural Network for First-Arrival Picking in Ultrasonic Pulse Velocity Testing on Rock Samples. *IEEE Trans. Geosci. Remote Sens.* **2025**, *63*, 5906117. [CrossRef]
10. Chen, L.; U, Y.W.; Zhang, T.Z. PAUT+TOFD inspection qualification and application of girth welds of submarine pipelines. *Nondestruct. Test.* **2023**, *45*, 18–21.
11. Zima, B. Guided ultrasonic wave technique for corrosion monitoring and thickness variability analysis. *Measurement* **2025**, *245*, 116584. [CrossRef]
12. Zhang, J.; Cho, Y.; Kim, J.; Malikov, A.K.U.; Kim, Y.H.; Yi, J.-H.; Li, W. Non-Destructive Evaluation of Coating Thickness Using Water Immersion Ultrasonic Testing. *Coatings* **2015**, *11*, 1421 [CrossRef]
13. Song, H.; Popovics, J.S.; Park, J. Development of an automated contactless ultrasonic scanning measurement system for wavefield imaging of concrete elements. In Proceedings of the 2017 IEEE International Ultrasonics Symposium (IUS), Washington, DC, USA, 6–9 September 2017; pp. 1–3.
14. Li, Z.; Zhao, L.; Li, J.; Zhao, Y.; Xu, T.; Liu, Z.; Luo, G.; Zhang, S.; Hu, K.; Hoffman, T.; et al. Nonlinear behavior analysis of electrostatically actuated multilayer anisotropic microplates with residual stress. *Compos. Struct.* **2021**, *255*, 112964. [CrossRef]
15. Li, Z.; Zhao, Y.; Barbruni, G.L.; Li, J.; Li, Z.; Yuan, J.; Yang, P.; Zhao, L.; Jiang, Z.; Carrara, S. Development of an Integrated CMUTs-Based Resonant Biosensor for Label-Free Detection of DNA with Improved Selectivity by Ethylene-Glycol Alkanethiols. *Engineering* **2024**, *41*, 231–241. [CrossRef]
16. Zheng, Z.; Yao, Y.; Liu, J.A.; Sun, Y.; Yeow, J.T.W. Highly sensitive CMUT-based humidity sensors built with nitride-to-oxide wafer bonding technology. *Sensors Actuators B Chem.* **2019**, *294*, 123–131. [CrossRef]

17. Brenner, K.; Ergun, A.S.; Firouzi, K.; Rasmussen, M.F.; Stedman, Q.; Khuri–Yakub, B. Advances in Capacitive Micromachined Ultrasonic Transducers. *Micromachines* **2019**, *10*, 152 [CrossRef]
18. Huang, Y.; Zhuang, X.; Haeggstrom, E.O.; Ergun, A.S.; Cheng, C.H.; Khuri-Yakub, B.T. Capacitive micromachined ultrasonic transducers with piston-shaped membranes: Fabrication and experimental characterization. *IEEE Trans. Ultrason. Ferroelectr. Freq. Control* **2009**, *56*, 136–145. [CrossRef]
19. Jiao, D.; Ni, Z.; Wang, J.; Li, X. High Fill Factor Array of Piezoelectric Micromachined Ultrasonic Transducers with Large Quality Factor. *Sensors Mater.* **2020**, *32*, 1785. [CrossRef]
20. Park, S.; Yoon, L.; Lee, S.; Kim, H.; Seo, J.; Chung, Y.Y.; Unger, A.; Kupnik, M.; Lee, H. CMUT-based resonant gas sensor array for VOC detection with low operating voltage. *Sensors Actuators B Chem.* **2018**, *273*, 1556–1563. [CrossRef]
21. Zhang, R.; Zhang, W.D.; He, C.H.; Song, J.L.; Mu, L.F.; Cui, J.; Zhang, Y.M.; Xue, C.Y. Design of capacitive micromachined ultrasonic transducer (CMUT) linear array for underwater imaging. *Sens. Rev.* **2016**, *36*, 77–85. [CrossRef]
22. Peller, S.; Zankl, T.; Fischer, C.; Bierl, R. Fast sound field characterization of beamforming capable capacitive micromachined ultrasonic transducer (CMUT) arrays by refracto-vibrometry. In Proceedings of the 2023 IEEE International Ultrasonics Symposium (IUS), Montreal, QC, Canada, 3–8 September 2023; pp. 1–3.
23. Georgiev, M.; Bregović, R.; Gotchev, A. Fixed-Pattern Noise Modeling and Removal in Time-of-Flight Sensing. *IEEE Trans. Instrum. Meas.* **2016**, *65*, 808–820. [CrossRef]
24. Qiu, Z.; Lu, Y.; Qiu, Z. Review of Ultrasonic Ranging Methods and Their Current Challenges. *Micromachines* **2022**, *13*, 520 [CrossRef]
25. Zhang, K.; Wu, T.; Meng, Q.; Meng, Q. Ultrasonic measurement of oil film thickness using piezoelectric element. *Int. J. Adv. Manuf. Technol.* **2018**, *94*, 3209–3215. [CrossRef]
26. Bühling, B.; Küttenbaum, S.; Maack, S.; Strangfeld, C. Development of an Accurate and Robust Air-Coupled Ultrasonic Time-of-Flight Measurement Technique. *Sensors* **2022**, *22*, 2135 [CrossRef] [PubMed]
27. Jia, L.; Shi, L.; Liu, C.; Yao, Y.; Sun, C.; Wu, G. Design and Characterization of an Aluminum Nitride-Based MEMS Hydrophone With Biologically Honeycomb Architecture. *IEEE Trans. Electron Devices* **2021**, *68*, 4656–4663. [CrossRef]
28. Huang, X.; Wang, H.; Yu, L. Investigation on Design Theory and Performance Analysis of Vacuum Capacitive Micromachined Ultrasonic Transducer. *Micromachines* **2021**, *12*, 1127 [CrossRef]
29. Zhang, W.; Zhang, H.; Jin, S.; Zeng, Z. A Two-Dimensional CMUT Linear Array for Underwater Applications: Directivity Analysis and Design Optimization. *J. Sensors* **2016**, *2016*, 5298197. [CrossRef]
30. Gong, Y.; Zhang, M.; Sun, S.; Guo, W.; Sun, C.; Pang, W. Piezoelectric Micromachined Ultrasonic Transducers With Superior Frequency Control. *J. Microelectromechanical Syst.* **2023**, *32*, 513–515. [CrossRef]
31. Erguri, A.S.; Yongli, H.; Xuefeng, Z.; Oralkan, O.; Yarahoglu, G.G.; Khuri-Yakub, B.T. Capacitive micromachined ultrasonic transducers: Fabrication technology. *IEEE Trans. Ultrason. Ferroelectr. Freq. Control* **2005**, *52*, 2242–2258. [CrossRef]
32. Zou, J.; Guo, X.; Wu, J.; Xu, D.; Xu, K.; Chen, P.; Ta, D.; Peng, H. A weavable and wearable polymer ultrasonic transducer with a large bandwidth. *Sci. China Mater.* **2024**, *67*, 2653–2660. [CrossRef]
33. Oevermann, J.; Weber, P.; Tretbar, S.H. Encapsulation of Capacitive Micromachined Ultrasonic Transducers (CMUTs) for the Acoustic Communication between Medical Implants. *Sensors* **2021**, *67*, 2653–2660. [CrossRef] [PubMed]
34. Gao, Y.; Chen, M.; Wu, Z.; Yao, L.; Tong, Z.; Zhang, S.; Gu, Y.A.; Lou, L. A miniaturized transit-time ultrasonic flowmeter based on ScAlN piezoelectric micromachined ultrasonic transducers for small-diameter applications. *Microsystems Nanoeng.* **2023**, *9*, 49. [CrossRef]
35. Pyun, J.Y.; Kim, Y.H.; Park, K.K. Design of Piezoelectric Acoustic Transducers for Underwater Applications. *Sensors* **2023**, *23*, 1821 [CrossRef]
36. Mohamed, I.; David, H.; Lee, D.; Stefano, L.; Marco, R. Ultrasonic NDE of thick polyurethane flexible riser stiffener material. *Nondestruct. Test. Eval.* **2017**, *32*, 343–362. [CrossRef]
37. Kurth, J.C.; Krauss, P.D.; Foster, S.W. Corrosion Management of Maritime Infrastructure. *Transp. Res. Rec.* **2019**, *2673*, 2–14. [CrossRef]
38. *GB/T 23900-2009*; Nondestructive Testing—Method for Measuring Ultrasonic Velocity in Materials. China Standards Press: Beijing, China, 2009.
39. Vishnu, C.V.; Rajeswari, R.; Manoj, N.R.; Unnikrishnan, K.C.; Mathiazhagan, A.; Shamsudeen, R.K. Underwater acoustic performance of a planar PVDF hydrophone: Simulation and experimental validation. *Measurement* **2025**, *245*, 116597. [CrossRef]
40. Boubenia, R.; Moal, P.L.; Bourbon, G.; Ramasso, E.; Joseph, E. CMUT-Based Sensor for Acoustic Emission Application: Experimental and Theoretical Contributions to Sensitivity Optimization. *Sensors* **2021**, *21*, 2042. [CrossRef]

micromachines

MDPI

Article

Novel Debris Material Identification Method Based on Impedance Microsensor

Haotian Shi [1], Yucai Xie [2] and Hongpeng Zhang [2,*]

[1] College of Safety Science and Engineering, Nanjing University of Science and Technology, Nanjing 210094, China; dmu6hao@163.com
[2] Marine Engineering College, Dalian Maritime University, Dalian 116026, China
* Correspondence: zhppeter@dlmu.edu.cn

Abstract: Oil condition monitoring can ensure the safe operation of mechanical equipment. Metal debris is full of friction information, and the identification of debris material helps to locate wear of parts. A method based on impedance analysis is proposed to identify debris material in this article. The differences in permeability and conductivity result in the nonlinear variation trend of inductance–resistance amplitude with debris volume. By establishing a database of amplitude–size curves, debris information (material and size) can be obtained through impedance analysis. Based on experimental and simulation results, iron, stainless steel, aluminum, copper, and brass particles are effectively distinguished. This method is not affected by oil's light transmittance, other impurities, and debris surface dirt and can be used to distinguish metals with similar colors. This work provides a novel solution for debris material identification, which is expected to promote the development of fault diagnosis.

Keywords: oil condition monitoring; impedance debris sensor; material identification; coil-based impedance detection

1. Introduction

Currently, mechanical equipment is developing toward high-precision, high-efficiency, and intelligence; thus, its reliability requirements are higher. As an advanced technology for ensuring the safety of mechanical equipment, condition monitoring and fault early warning under big data are developing rapidly. In addition to the external environment and human factors, system failures mainly stem from abnormal coordination caused by wear [1,2]. Condition monitoring technology mainly includes vibration monitoring, oil analysis, sound feedback, and thermal imaging. Vibration monitoring, sound feedback, and thermal imaging are less sensitive to the health conditions of mechanical systems due to environmental noise and temperature hysteresis. It is difficult to find abnormal wear in the early stage [3,4]. Lubrication systems can inhibit wear, and debris generated by mechanical friction and external intrusions is carried in lubricating oil. Oil condition monitoring can identify and predict mechanical faults [5]. By detecting and analyzing the contaminants in oil, the lubrication environment and wear condition of parts can be directly reflected.

Oil condition monitoring can be divided into oil property analysis and oil contamination monitoring [6]. Oil property analysis reveals its lubricating performance through the degree of oil deterioration. Oil contamination monitoring focuses on measuring different content to reflect wear trends, such as water, air, and debris [7]. In particular, metal debris is full of friction information, which can be used as an indicator of mechanical health. The size and concentration of metal debris represent the wear level, while the wear mechanism can be judged based on debris morphology. Therefore, oil debris monitoring (ODM) can provide important information for intelligent operation. ODM includes offline, inline, online, and onsite forms. Offline detection in the laboratory can obtain accurate, detailed, and comprehensive information. However, due to the sample delivery link, offline detection results lag, which cannot meet the requirements of condition-based maintenance. Real-time information enables operators to actively respond to sudden failures and take emergency remedial measures. Therefore, the current demand for inline, online, and onsite oil detection devices with stronger timeliness is more urgent. The core components of these devices are sensors based on various principles, mainly including optics [8,9], acoustics [10,11], impedance [12–14], and image methods [15,16].

Metal materials of mechanical parts are different, and the debris material helps to semi-qualitatively determine the wear location. Material identification has always been a critical problem in ODM. At present, the distinction of debris material is mainly based on spectral and image recognition methods [17,18], but their detection effectiveness is greatly affected by oil's light transmittance. It is of great significance to develop new debris material identification methods to provide more detailed information for fault diagnosis. It is worth noting that inductance methods based on electromagnetic induction can distinguish between ferromagnetic debris and non-ferromagnetic debris [19,20]. Recently, various inductance debris detection devices were studied and developed, mainly including self-inductance single-coil sensors [21], mutual-inductance double-coil sensors [22], balanced triple-coil sensors [23], etc. But material identification is also a weakness in inductive debris monitoring [24]. The metal debris information obtained by inductance detection methods needs to be extended from magnetic property differentiation to material identification. Our group has conducted research on debris material identification based on inductive sensors. Shi et al. [25] achieved metal debris distinction based on the differences in sensing characteristics between sensing units and analyzed the influence of coil turns and magnetic media on sensitivity. However, the sensor structure mentioned above is complex, and the distinguishing accuracy of debris materials still needs to be improved.

The electromagnetic field generated by metal debris will change the impedance parameters of the coil. Traditional inductive sensors present the impedance change in the form of voltage, ignoring that impedance is the combined result of inductance and alternating current (AC) resistance. The material information of the debris is contained in impedance signals. Permeability and conductivity depend on the metal material, and the characteristic curves between signal amplitude and particle size for various metal particles have different trends. By combining the debris information judged by inductance and resistance amplitude–size curves, the material and size can be determined. Therefore, this article proposes a debris material identification method based on impedance analysis to enrich the obtained debris information. And this method is not affected by oil's light transmittance, other impurities, and debris surface dirt.

2. The Coil-Based Impedance Detection Method

2.1. The Impedance Debris Microsensor

In this work, an impedance debris microsensor with Coulter Counting was selected, as shown in Figure 1. The sensor includes a sensor substrate, an oil inlet, an oil outlet, a solenoid coil, and a channel with a diameter of 900 μm. The coil parameters are as follows: 200 turns, 0.9 mm inner diameter, 3.7 mm outer diameter, and 0.7 mm height. The channel is perpendicular to the coil inner hole.

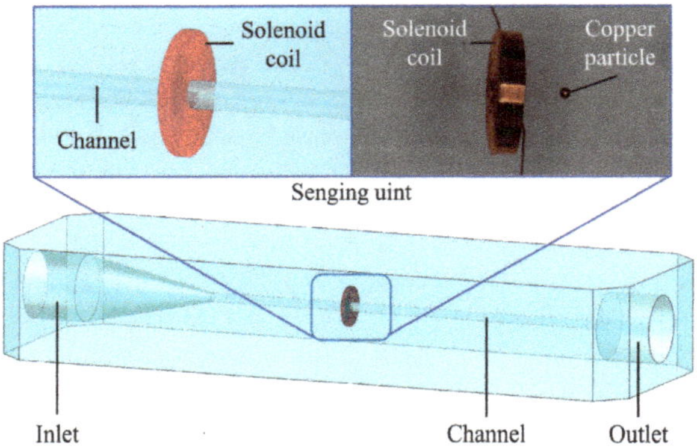

Figure 1. The impedance debris microsensor with Coulter Counting.

2.2. The Impedance Detection Principle Based on Coil

With the magnetization effect enhancing and eddy current weakening the magnetic field, the metal debris will change the magnetic flux and break the balance of the skin effect and the proximity effect, thereby causing changes in coil inductance and AC resistance, as shown in Figure 2. By measuring the impedance change of the coil, metal debris can be identified, sized, and counted.

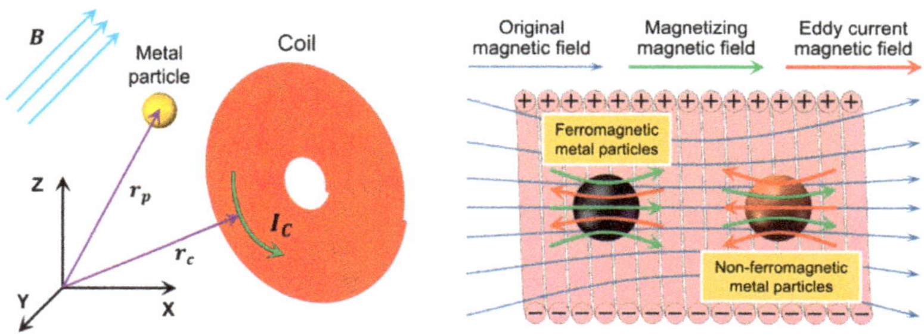

Figure 2. Detection principle based on alternating magnetic field.

Based on previous research [26], the impedance change Z_m caused by a metal particle in the self-excited coil is given by:

$$Z_m(\boldsymbol{r}_m) = j\mu_v V \chi_m |\boldsymbol{h}(\boldsymbol{r}_m)|^2 \tag{1}$$

where $\boldsymbol{r}_m$ is the position vector of the particle center, and μ_v is the permeability of vacuum, V is the volume of the metal particle. $\boldsymbol{h}$ is the sensitivity field of the coil:

$$\boldsymbol{h}(\boldsymbol{r}_m) = \frac{1}{4\pi}\iiint_{coil} \frac{n\boldsymbol{dl} \times (\boldsymbol{r}_R - \boldsymbol{r}_E)}{|\boldsymbol{r}_R - \boldsymbol{r}_E|^3} ds \tag{2}$$

where n is the turn density of the coil, $\boldsymbol{dl}$ is the current direction of the coil element, $\boldsymbol{r}_R$ is the position vector of the receiver current element, and $\boldsymbol{r}_E$ is the position vector of the excitation current element. χ_m is the magnetic susceptibility of the metal particle:

$$\chi_m = \frac{3}{2}\frac{\left(2\mu_m + 1 - a^2k^2\right)\sin ak + ak(2\mu_m + 1)\cos ak}{\left(\mu_m - 1 + a^2k^2\right)\sin ak - ak(\mu_m - 1)\cos ak} \tag{3}$$

where μ_m is the relative permeability of the metal, a is the radius of the metal particle. k is given by:

$$k = \sqrt{-j\omega\mu_v\mu_m\sigma_m} \tag{4}$$

Here, ω is the angular frequency of the alternating magnetic field and σ_m is the conductivity of the metal.

The impedance change of coil includes inductance and AC resistance:

$$L_m(\mathbf{r}_m) = \mu_v V|\mathbf{h}(\mathbf{r}_m)|^2 \text{Real}(\chi_m) \tag{5}$$

$$R_m(\mathbf{r}_m) = -\mu_v V|\mathbf{h}(\mathbf{r}_m)|^2 \text{Imaginary}(\chi_m) \tag{6}$$

As concluded from the above equations, it can be found that the impedance signal mainly depends on the sensor structure (coil parameters and channel position), excitation parameters (voltage and frequency), and metal debris (volume and material). The material represents magnetic permeability and conductivity, which affect the intensity of magnetization and the eddy current. This provides a theoretical basis for the material identification of metal debris.

2.3. The Simulation Analysis of Impedance Detection

To verify the theoretical analysis, simulations of impedance detection were conducted. As shown in Figure 3, a simulation model was built according to the coil structure used in the experiment, and the characteristics of magnetization and the eddy current with 500 μm metal particles were obtained. After applying a 2.0 V, 2.0 MHz excitation current, the distribution cloud diagram of the magnetic flux density originating from the coil is shown in Figure 3a. The magnetic flux density distributed in the coil inner hole ranged between 300 μT (inner hole center) and 350 μT (inner hole edge). The magnetic gradient established by the micro-coil was small, which reduced the signal fluctuation induced by the particle position. This was conducive to improving the detection stability.

In engineering, the materials of various friction pairs are different. The debris material helps to achieve wear localization. This article selected several typical metals for research, such as iron, stainless steel, aluminum, copper, brass, etc. The simulation results for a 500 μm Q235 steel particle, 304 stainless steel particle, 1060 aluminum particle, copper particle, and H62 brass particle are shown in Figure 3b. The magnetization field and eddy current field produced by the metal particles cancelled each other. Owing to the differences in permeability and conductivity, the ratio of magnetization to the eddy current of metal debris was nonidentical. For ferromagnetic particles, the magnetization field was dominant, which would enhance the inductance. For non-ferromagnetic particles, the eddy current field was dominant, which would weaken the inductance. Under the combined action of the two fields, AC resistance would increase. The magnetic permeability of the Q235 steel particle was higher, thus its magnetization effect was stronger than that of the 304 stainless steel particle. That is, σ_m of copper > σ_m of 1060 aluminum > σ_m of H62 brass, which made the eddy current in the copper particle the most intense, while the eddy current in the H62 brass particle was the weakest. The magnetization and eddy current characteristics could not intuitively reflect the changes in coil inductance and resistance. Therefore, the impedance signals caused by metal particles were obtained through parameterized scanning, as shown in Figure 4. The simulation data indicated that the inductance–resistance signal characteristics of various metals were different. The inductance pulses from the Q235 steel particle and 304 stainless steel were extremely close (the pulse amplitude of Q235 steel was 7.14 nH larger than that of 304 stainless steel), while the difference in resistance pulses was obvious (the pulse amplitude of Q235 steel was 2.449 Ω larger than that of 304 stainless steel). In addition, there were significant differences in the inductance and resistance pulses between the selected non-ferromagnetic metal particles. The amplitude–size curves for various metal particles had different trends. This was consistent with the theoretical analysis. Therefore, we propose a debris material identification method based on impedance analysis.

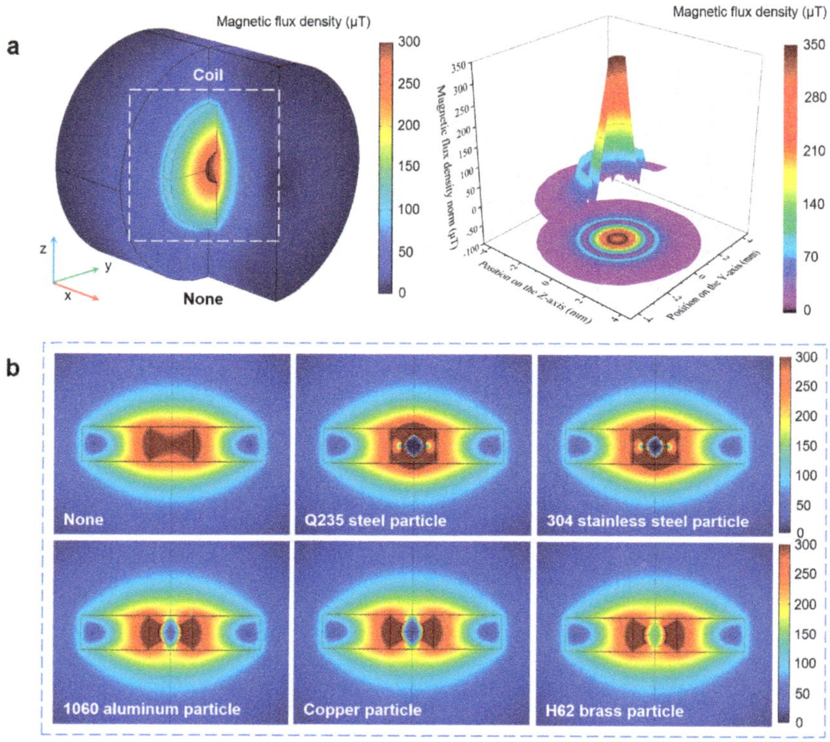

Figure 3. Simulation result of impedance sensor by COMSOL 5.4 (2.0 V, 2.0 MHz). (**a**) The distribution cloud diagram of the magnetic field around the coil. (**b**) The characteristics of magnetization and eddy current in various metal particles.

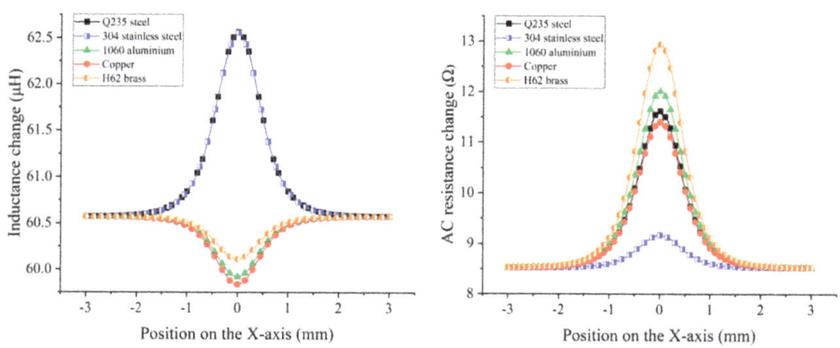

Figure 4. Inductance and AC resistance changes caused by 500 μm particles of different materials.

3. Experiments and Discussion

To verify the feasibility of the coil-based impedance detection method to distinguish debris material, the debris microsensor was fabricated based on the casting method [22].

The fabricating process included the following steps: coil winding, basic mold making, substrate curing, and channel forming. An Agilent E4980A meter (Agilent Technologies Inc., Bayan Lepas, Malaysia) was used to provide a 2.0 V, 2.0 MHz excitation for the sensor and to detect the impedance parameters. The 500 μm metal particle was added from the inlet and driven into the channel by a LSP02-1B syringe pump (Keysight E4980A, Baoding Qili Constant-flow-pump Ltd., Baoding, China). The high-frequency noise in the impedance signal was filtered out through a low-pass filter, and interpolation fitting method was used to achieve baseline zeroing. The measured data were processed to obtain the type, size, and number of debris.

The Q235 steel particle, 304 stainless steel particle, 1060 aluminum particle, copper particle, and H62 brass particle were repeatedly driven through the sensing region for detection. The photographs of metal particles under the microscope are shown in Figure 5a. The detection signals of 500 μm metal particles of different materials are shown in Figure 5b. Q235 steel and 304 stainless steel are paramagnetic materials, which produced positive inductance pulses. 1060 aluminum, copper, and H62 brass are diamagnetic materials, which produced negative inductance pulses. And all metal particles produced positive resistance pulses. Compared with the simulation results, there were errors caused by sensor manufacturing, circuit loss, and metal particles themselves. For instance, the 500 μm 1060 aluminum particle produced a −0.403 μH inductance pulse and a 4.241 Ω resistance pulse, which was less than the simulation result. The error source was that the aluminum particle was affected by oxidation, and the effective size was small. Therefore, for metal debris with obvious oxidation characteristic, the oxidation volume should be considered in actual size judgment.

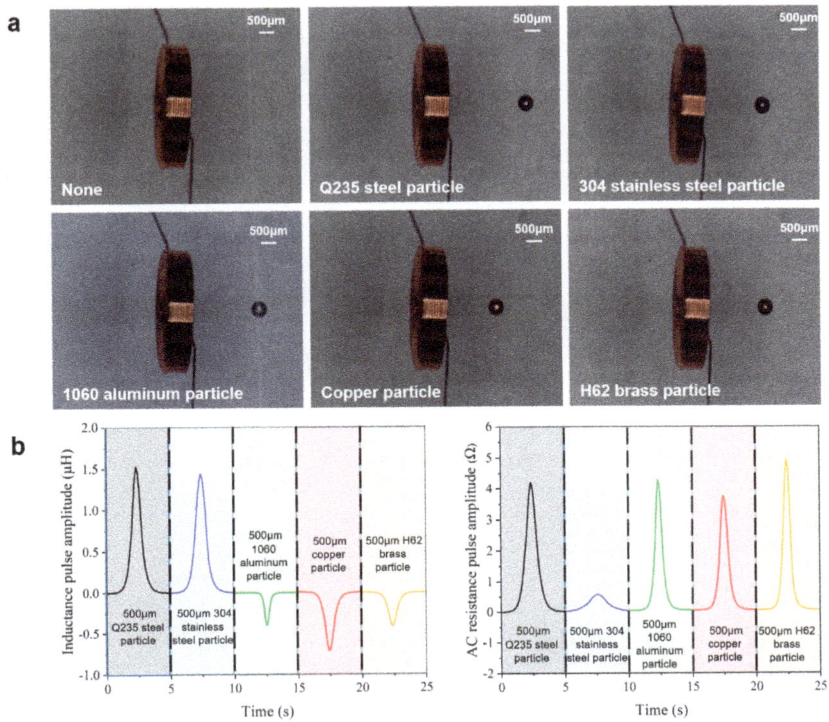

Figure 5. Research on the coil-based impedance detection method for identifying debris material. (**a**) The photographs of metal particles under the microscope. (**b**) The detection signals of 500 μm metal particles.

The coil-based impedance detection method was used to identify debris material by database comparison. The characteristic curves of the inductance–resistance signals for various metals were obtained in simulations, as shown in Figure 6. The results showed that the inductance–resistance pulse amplitudes grew nonlinearly with particle diameter. Eddy current and magnetization effects increased sharply with the particle volume, and the disturbances to the coil magnetic flux and the current distribution in the wire were more significant. With the detected impedance signal, metal debris could be analyzed according to the amplitude–size curves. A single inductance pulse amplitude or resistance pulse amplitude corresponded to multiple metal debris. By combining the debris information judged by inductance and resistance amplitude–size curves, the material and size could be determined. For example, debris A generated a 1.50 μH inductance pulse and a 2.00 Ω resistance pulse, while debris B generated a −1.50 μH inductance pulse and a 4.65 Ω resistance pulse. Selecting the simulation characteristic curves as the standard database, the debris information is analyzed in Figure 6. According to the direction (positive or negative) of the inductance pulses, particles could be determined to be ferromagnetic or non-ferromagnetic. Based on the inductance amplitude–size curves, debris A was judged as a 455 μm Q235 steel particle or 460 μm 304 stainless steel particle, and debris B was judged as a 620 μm copper particle, 640 μm 1060 aluminum particle, or 670 μm H62 brass particle. Based on the resistance amplitude–size curves, debris A was judged as a 455 μm Q235 steel particle or 660 μm 304 stainless steel particle, and debris B was judged as a 620 μm copper particle, 560 μm 1060 aluminum particle, or 510 μm H62 brass particle. Based on the comprehensive analysis of the two results, debris A was identified as 455 μm Q235 steel, and debris B was identified as 620 μm copper.

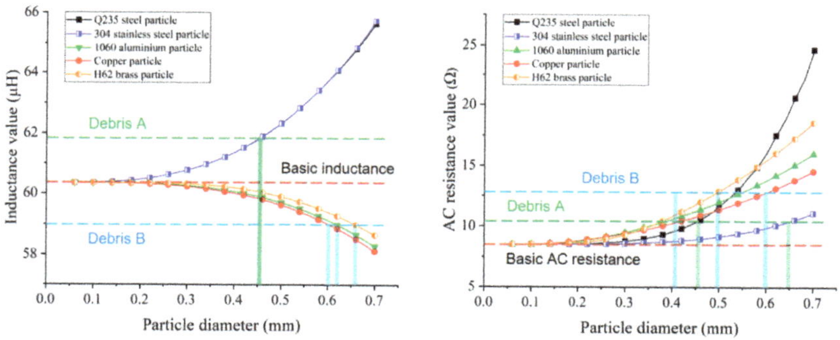

Figure 6. The characteristic curves of the inductance–resistance signals for various metals.

This debris material identification method relies on the predefined database of amplitude–size curves. In practical applications, particles that cannot be identified by a predefined database will be classified as unknown particles. Based on the inductance changes, these particles can be distinguished as ferromagnetic and non-ferromagnetic, and their size can be roughly estimated.

The debris material identification method based on impedance analysis is not affected by oil's light transmittance, other impurities, and debris surface dirt. And this method can be used to distinguish metals with similar colors. The greater difference between the amplitude–size curves, the more accurate the identified information. Therefore, the identification accuracy of debris material can be elevated by enhancing sensor performance. Furthermore, the pulse of small metal particles will be submerged in signal noise. The research and development of high-sensitivity impedance debris sensors is of great significance for the collection of debris information. In addition, this method can be combined with another method proposed in previous work [25], and impedance signal analysis based on multi-sensing units can improve the accuracy of debris material identification.

4. Conclusions

In conclusion, a coil-based impedance detection method is proposed to identify debris material. Firstly, the theoretical foundation of this method is established. The magnetization and eddy current differences between various particles are proven by electromagnetic simulations. The impedance microsensor fabricated using the casting method is used to study the inductance–resistance signal characteristics of metal debris. As each material has its unique permeability and conductivity, the nonlinear change trend of inductance and resistance pulse amplitude with debris volume is illustrated. Through database comparison, the method can effectively identify debris material. The identification accuracy of debris material can be improved by increasing the difference between the amplitude–size curves. This work provides a novel solution for debris material identification, enriching the detection information of debris sensors for mechanical fault diagnosis.

Author Contributions: Conceptualization, H.S. and H.Z.; methodology, H.S. and H.Z.; software, H.S. and Y.X.; validation, H.S. and Y.X.; formal analysis, H.S. and Y.X.; investigation, H.S.; writing—original draft preparation, H.S.; writing—review and editing, H.S.; funding acquisition, H.S. and H.Z. All authors have read and agreed to the published version of the manuscript.

Funding: This work was supported by the Natural Science Foundation of China (Grant No. 52405116, 52271303), Fundamental Research Funds for the Central Universities (Grant No. 30925010322), the Natural Science Foundation of Jiangsu Province (Grant No. BK20241474), and the Jiangsu Funding Program for Excellent Postdoctoral Talent (Grant No. 2024ZB147).

Data Availability Statement: Dataset available on request from the authors.

Conflicts of Interest: The authors declare no conflict of interest.

References

1. Lu, P.; Powrie, H.; Wood, R.; Harvey, T.; Harris, N. Early wear detection and its significance for condition monitoring. *Tribol. Int.* **2021**, *159*, 106946. [CrossRef]
2. Shanbhag, V.; Meyer, T.; Caspers, L.; Schlanbusch, R. Failure monitoring and predictive maintenance of hydraulic cylinder-state-of-the-art review. *IEEE/ASME Trans. Mechatron.* **2021**, *26*, 3087–3103. [CrossRef]
3. Poddar, S.; Tandon, N. Detection of particle contamination in journal bearing using acoustic emission and vibration monitoring techniques. *Tribol. Int.* **2019**, *134*, 154–164. [CrossRef]
4. Janssens, O.; Loccufier, M.; Hoecke, S. Thermal imaging and vibration based multi-sensor fault detection for rotating machinery. *IEEE Trans. Ind. Inform.* **2019**, *15*, 434–444. [CrossRef]
5. Wakiru, J.; Pintelon, L.; Muchiri, P.; Peter, K. A review on lubricant condition monitoring information analysis for maintenance decision support. *Mech. Syst. Signal Process.* **2018**, *118*, 108–132. [CrossRef]
6. Zhu, X.; Zhong, C.; Zhe, J. Lubricating oil conditioning sensors for online machine health monitoring—A review. *Tribol. Int.* **2017**, *109*, 473–484. [CrossRef]
7. Cao, W.; Dong, G.; Xie, Y.; Peng, Z. Prediction of wear trend of engines via on-line wear debris monitoring. *Tribol. Int.* **2018**, *120*, 510–519. [CrossRef]
8. Krogsøe, K.; Henneberg, M.; Eriksen, R. Model of a light extinction sensor for assessing wear particle distribution in a lubricated oil system. *Sensors* **2018**, *18*, 4091. [CrossRef]
9. Sun, J.; Wang, L.; Li, J.; Li, F.; Fang, Y. An on-line imaging sensor based on magnetic deposition and flowing dispersion for wear debris feature monitoring. *Mech. Syst. Signal Process.* **2024**, *212*, 111321. [CrossRef]
10. Xu, C.; Zhang, P.; Wang, H.; Li, Y.; Lv, C. Ultrasonic echo waveshape features extraction based on QPSO-matching pursuit for online wear debris discrimination. *Mech. Syst. Signal Process.* **2015**, *60–61*, 301–315. [CrossRef]
11. Reitz, T.; Fritzen, C. A novel baseline-free approach for acousto-ultrasonic crack monitoring of rotating axles. *Struct. Health Monit.* **2021**, *20*, 990–1003. [CrossRef]
12. Birkin, P.; Linfield, S.; Denuault, G.; Jones, R.; Youngs, J.; Wain, E. An analytical differential resistance pulse system relying on a time shift signal analysis–applications in Coulter Counting. *ACS Sens.* **2019**, *4*, 2190–2195. [CrossRef] [PubMed]

13. Murali, S.; Xia, X.; Jagtiani, A.; Carletta, J.; Zhe, J. Capacitive Coulter counting: Detection of metal wear particles in lubricant using a microfluidic device. *Smart Mater. Struct.* **2009**, *18*, 037001. [CrossRef]
14. Qian, M.; Ren, Y.; Zhao, G.; Feng, Z. Ultrasensitive inductive debris sensor with a two-stage autoasymmetrical compensation circuit. *IEEE Trans. Ind. Electron.* **2021**, *68*, 8885–8893. [CrossRef]
15. Li, B.; Wu, W.; Zhou, M.; Xi, Y.; Wei, H.; Mao, J. A full field-of-view online visual ferrograph debris detector based on reflected light microscopic imaging. *IEEE Sens. J.* **2021**, *21*, 16584–16597. [CrossRef]
16. Janssens, O.; Walle, R.; Loccufier, M.; Hoecke, S. Deep learning for infrared thermal image based machine health monitoring. *IEEE/ASME Trans. Mechatron.* **2018**, *23*, 151–159. [CrossRef]
17. Abdul-Munaim, A.; Reuter, M.; Abdulmunem, O.; Balzer, J.; Koch, M.; Watson, D. Using terahertz time-domain spectroscopy to discriminate among water contamination levels in diesel engine oil. *Trans. Asabe* **2016**, *59*, 795–801.
18. Wu, H.; Li, R.; Kwok, N.; Peng, Y.; Wu, T.; Peng, Z. Restoration of low-informative image for robust debris shape measurement in on-line wear debris monitoring. *Mech. Syst. Signal Process.* **2019**, *114*, 539–555. [CrossRef]
19. Yokouchi, T.; Kagawa, F.; Hirschberger, M.; Otani, Y.; Tokura, Y. Emergent electromagnetic induction in a helical-spin magnet. *Nature* **2020**, *586*, 232–236. [CrossRef]
20. Jia, B.; Ding, Z. High-precision induction coil sensor with high magnetic field homogeneity for online debris monitoring. *IEEE Sens. J.* **2025**, *25*, 18236–18248. [CrossRef]
21. Du, L.; Zhe, J. A high throughput inductive pulse sensor for online oil debris monitoring. *Tribol. Int.* **2011**, *44*, 175–179. [CrossRef]
22. Shi, H.; Zhang, H.; Ma, L.; Rogers, F.; Zhao, X.; Zeng, L. An impedance debris sensor based on a high-gradient magnetic field for high sensitivity and high throughput. *IEEE Trans. Ind. Electron.* **2021**, *68*, 5376–5384. [CrossRef]
23. Li, Y.; Wu, J.; Guo, Q. Electromagnetic sensor for detecting wear debris in lubricating oil. *IEEE Trans. Instrum. Meas.* **2020**, *69*, 2533–2541. [CrossRef]
24. Sun, J.; Wang, L.; Li, J.; Li, F.; Li, J.; Lu, H. Online oil debris monitoring of rotating machinery: A detailed review of more than three decades. *Mech. Syst. Signal Process.* **2021**, *149*, 107341. [CrossRef]
25. Shi, H.; Xie, Y.; Wang, C.; Liu, Y.; Zhang, H.; Li, W.; Zhang, S.; He, R.; Sun, Y.; Li, G.; et al. A novel distinction method of metal debris material based on inductive sensor with multi-sensing units. *Tribol. Int.* **2023**, *177*, 107948. [CrossRef]
26. Shi, H.; Xie, Y.; Zhang, H. Sensing model of metal particles in nonuniform time-harmonic magnetic field for high-sensitivity debris sensor. *IEEE Trans. Instrum. Meas.* **2025**, *74*, 3514609. [CrossRef]

micromachines

MDPI

Article

A Rotary Piezoelectric Electromagnetic Hybrid Energy Harvester

Zhiyang Yao and Chong Li *

School of Mechanical Engineering, Jiangsu University of Science and Technology, Zhenjiang 212100, China; 222241802734@stu.just.edu.cn
* Correspondence: lichong@just.edu.cn

Abstract: To collect the energy generated by rotational motion in the natural environment, a piezoelectric electromagnetic hybrid energy harvester (HEH) based on a planetary gear system is proposed. The harvester combines piezoelectric and electromagnetic effects and is mainly used for collecting low-frequency rotational energy. The HEH has a compact structure and contains four sets of piezoelectric energy harvesters (PEHs) and electromagnetic energy harvesters (EMHs) inside. The working principle of the energy harvester is analyzed, its theoretical model is established, and a simulation analysis is conducted. To verify the effectiveness of the design, an experimental device is constructed. The results indicate that the HEH can generate an average output power of 250 mW under eight magnets and an external excitation frequency of 7 Hz. In actual power supply testing, the HEH can light up 60 LEDs and provide stable power supply for the temperature–humidity meter.

Keywords: energy harvester; rotary motion; output power; planetary gear

1. Introduction

The development of microelectromechanical technology (MENS) and wireless sensor networks (WSNs) has greatly promoted the popularization of the Internet of Things (IoT) and it has been widely applied in fields such as environmental monitoring, smart cities, and industrial equipment health management [1–3]. The external power supply problem of sensors has become an obstacle to the sustainable development of IoT technology. However, self-powered technology can effectively solve the above problems [4,5].

There is abundant low-frequency mechanical vibration energy in the natural environment, such as wind energy [6–8], wave energy in the ocean [9,10], human motion energy in daily life [11–13], and so on, with most of the energy frequencies below 100 Hz. The current mainstream mechanisms for converting other forms of energy into electrical energy include piezoelectric [14–16], electromagnetic [17–19], magnetic energy, triboelectricity [20–22], and electrostatic mechanisms. Among them, the piezoelectric energy harvester utilizes the piezoelectric effect to convert mechanical vibration energy into electrical energy, and has the characteristics of a high energy density, no electromagnetic interference, and compact structure. The electromagnetic type is based on Faraday's law of electromagnetic induction, which uses a coil to generate induced current in a changing magnetic field, especially suitable for low-frequency and large-displacement-vibration environments. These two forms of energy harvesting are considered more suitable for harvesting energy in mechanical vibrations due to their high conversion efficiency and resistance to environmental influences.

Liao et al. [23] proposed a centrifugal spring mechanism that could achieve adaptive magnetic force enhancement at different wind speeds. This structure has the characteristics of an ultra-low start-up wind speed (1.5 m/s), high output power (5.97 mW at 4.75 m/s), and a wide bandwidth (1.5–5.5 m/s), while demonstrating the dual functions of powering microelectronics and implementing intelligent wind sensing through a CNN algorithm. To achieve the kinetic energy recovery of complex ship movements, Li et al. [24] developed a multiple-degree-of-freedom energy harvesting system based on a club mechanism. The device achieved a maximum output power of 173.1 μW in actual ship testing and demonstrated efficient capacitor charging capability, providing a self-powered solution for low-power equipment on board ships. Yin et al. [25] proposed a rolling-swing electromagnetic energy harvester that utilized the reverse rotation between a magnet and a coil to amplify magnetic field changes under ultra-low frequency vibrations (0.1–3 Hz). The energy harvester adopted a unique alternating magnetic pole arrangement to double the output voltage, achieving a milliwatt-level power output, and successfully demonstrating the ability to power multiple low-power devices (such as thermometers and calculators) under laboratory and actual walking conditions, suitable for self-powered IoT applications.

Due to the randomness and multi-directional dispersion of vibrations in natural environments, a single energy harvesting mechanism often can only collect energy in a specific direction, and its output form is relatively limited [26–28]. Taking piezoelectric and electromagnetic mechanisms as examples, the former can output a high voltage but has a limited output current, while the latter outputs a low voltage but a high current. Therefore, some researchers have designed composite-vibration energy harvesting devices by integrating two or more forms of energy harvesting [29–31] to improve overall energy harvesting efficiency. Zhao et al. [32] proposed an intelligent mechanical wave energy harvester. That design adaptively converted low-frequency, irregular wave motion into reciprocating rolling and high-speed unidirectional rotation through a ratchet and magnetic coupling. The prototype could operate effectively at ultra-low frequencies (0.1 Hz), charging a 0.47 F capacitor to 5 V within 6 min under simulated waves, achieving a self-powered wireless marine environment monitoring system for sensing light, temperature, and water quality. Bai et al. [33] proposed a friction piezoelectric electromagnetic hybrid wind energy collector based on a bouncing bistable mechanism. This design integrated three energy conversion mechanisms at the maximum strain, contact area, and displacement point, achieving a low wind speed (3 m/s), high power density, and continuous stable operation. It successfully lit up 1000 LED lights and demonstrated self-powered wireless environmental monitoring (such as temperature and wind speed). In addition, Zhao et al. [34] designed a high-performance frictional electromagnetic hybrid nanogenerator based on a NdFeB/EC composite film for collecting wind energy. At a wind speed of 15.5 m/s, their nanogenerator unit could generate an open-circuit voltage of 55 V, a short-circuit current of 5.4 μA, and a maximum power of 99.2 μW. The electromagnetic generator unit generated an open-circuit voltage and a short-circuit current of approximately 0.16 V and 5.93 μA, respectively, with an output power of 2.5 μW.

This article proposes a piezoelectric electromagnetic hybrid energy harvester (HEH) based on a planetary gear system design, aimed at efficiently capturing low-frequency mechanical energy in the environment. The innovation of this device lies in the gear meshing transmission through a planetary gear system (transmission ratio 2:1), which amplifies the external excitation frequency twofold to achieve a higher energy output. The device has a compact structure and contains four sets of piezoelectric energy harvesting units (PEHs) and four sets of electromagnetic energy harvesting units (EMHs) inside. In addition, the device also adopts a center crank slider mechanism to convert the external rotational excitation into a linear motion of magnets. The movement of the magnet drives the cantilever beam to bend and deform, thereby exciting the PZT to output electrical energy. On the other hand, the change in magnetic flux inside the coil generates an induced electromotive force, which outputs the corresponding induced current. This magnetic coupling excitation design improves the service life and reliability of the HEH.

2. Design and Operating Principle

The hybrid energy harvester (HEH) proposed in this article consists of four circularly distributed piezoelectric energy harvesters (PEHs) and four circularly distributed electromagnetic energy harvesters (EMHs), as shown in Figure 1.

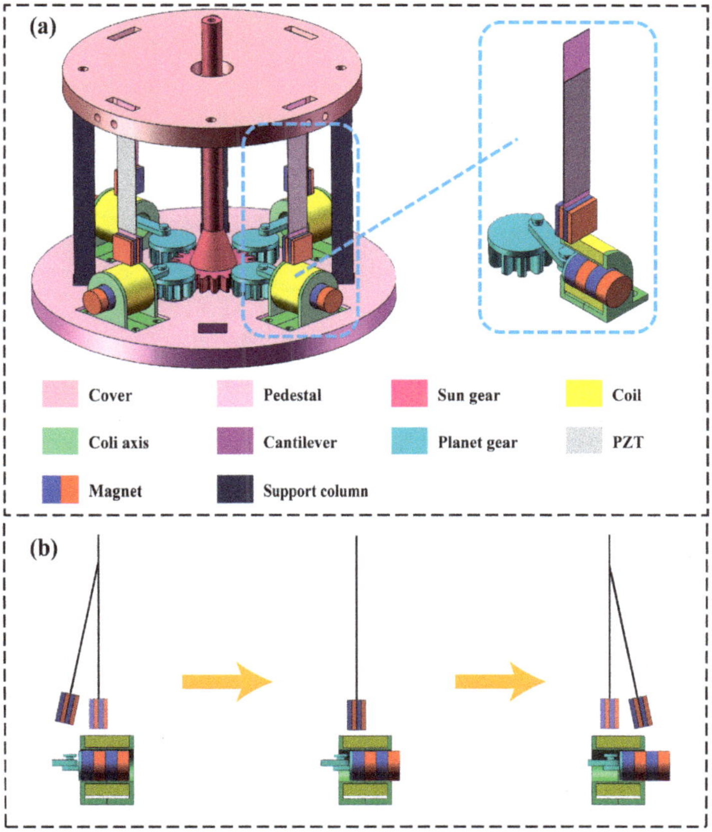

Figure 1. The piezoelectric electromagnetic hybrid energy harvester. (**a**) Schematic diagram of the prototype structure; (**b**) the changing process of the cantilever during work.

The EMH is mainly composed of circular magnets, coils, and coil tubes. By the reciprocating motion of a circular magnet inside the tube, a constantly changing magnetic field is generated around the coil, resulting in induced current inside the coil. The PEH is mainly composed of a PZT material, cantilever beam, and rectangular magnet attached to the bottom of the cantilever beam. Due to the constantly changing gravitational force between the circular magnet and the square magnet, the cantilever beam deforms, causing voltage to be generated inside the PZT material. The HEH utilizes magnetic coupling mechanisms to achieve the conversion of mechanical energy into electrical energy. Tables 1 and 2 present in detail the structural dimensions of the HEH and the technical parameters of its main components. The size of the entire HEH is Φ140 × 95 mm. Due to its compact size and low installation space requirements, the device has good application prospects.

Table 1. Dimension parameters of the HEH.

Description	Value
Dimensions of the PZT	$40 \times 10 \times 0.2$ mm^3
Dimensions of the cantilever	$60 \times 10 \times 0.2$ mm^3
Dimensions of the square magnet	$10 \times 10 \times 3$ mm^3
Dimensions of the circular magnet	$5^2\pi \times 2$ mm^3
Dimensions of the cover	$57^2\pi \times 8$ mm^3
Dimensions of the pedestal	$70^2\pi \times 8$ mm^3
Inner, outer, and height of the coil	12 mm, 20 mm, 15 mm
Wire diameters of the coil	0.35 mm
Turns of the coil	432
Impedance of the coil	4Ω
Module of the sun gear and planet gear	1.5
Number of teeth of the sun gear	20

Table 2. Material parameters of the HEH.

Description	Material
PZT	PZT-5H
Cantilever	301 stainless steel
Square magnet	NdFeB 35
Circular magnet	NdFeB 52
Coil	Copper

When the sun gear of the HEH is excited by external rotational energy, it drives the rotation of four planet gears through gear meshing. At the same time, a centering crank slider mechanism is attached to the planet gear, with one end of the connecting rod connected to a circular magnet, converting the rotational motion of the planet gear into the linear reciprocating motion of the magnet. Figure 1b illustrates the working principles of the EMH and PEH. On the one hand, the linear reciprocating motion of the magnet causes the magnetic induction intensity around the coil to constantly change, thereby continuously generating the current induced in the coil. On the other hand, circular magnets attract square magnets to move synchronously during their motion, causing deformation of the cantilever beam and generating a polarization voltage inside the PZT material, achieving the output of piezoelectric energy. Throughout the entire motion process, due to the constant existence of the magnetic coupling, the HEH can achieve a stable and efficient energy output. The characteristic of this article is that by adjusting the gear ratio of the sun gear and the planet gear, the external low-frequency energy excitation can be amplified, thereby effectively achieving the collection of low-frequency energy. The transmission ratio of this HEH is 2:1, which can amplify the external excitation applied to the sun gear twofold and transmit it to the EMH and PEH, thereby achieving a greater energy output.

3. Modelling and Analysis

3.1. Theoretical Modeling

The HEH designed in this article utilizes the centric slider crank mechanism to convert the rotational motion of the planet gear into the linear motion of a permanent magnet. The motion structure is shown in Figure 2.

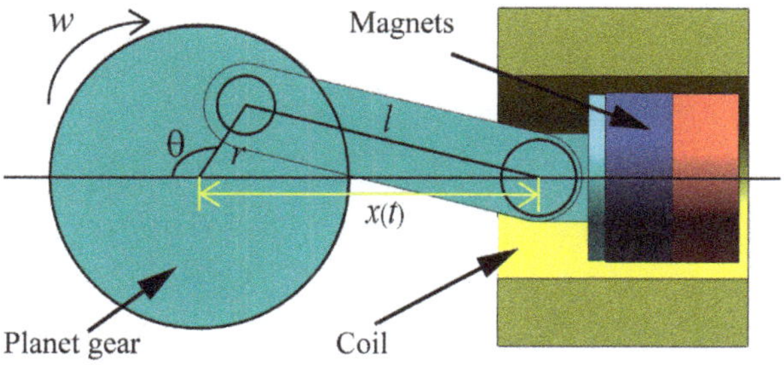

Figure 2. Schematic diagram of the movement of the magnets.

Assuming that the angular velocity of the planet gear during rotation is ω, the time is t, and the initial state is $\theta = 0°$, the displacement $x(t)$ of the permanent magnet is:

$$x(t) = rcos(\omega t) + \sqrt{l^2 - r^2 sin^2(\omega t)}, \tag{1}$$

where r is the distance radius from the planet gear to the connecting rod, and l is the length of the connecting rod.

Meanwhile, by taking the derivative of $x(t)$, the moving speed $v(t)$ of the permanent magnet can be calculated as follows:

$$\begin{aligned} v(t) &= \frac{dx}{dt} = -r\omega sin(\omega t) - \frac{r^2\omega sin(\omega t)cos(\omega t)}{\sqrt{l^2 - r^2 sin^2(\omega t)}} \\ &= -r\omega sin(\omega t)\left[1 + \frac{rcos(\omega t)}{\sqrt{l^2 - r^2 sin^2(\omega t)}}\right] \end{aligned}. \tag{2}$$

The induced electromotive force generated by the law of electromagnetic induction is divided into an induced electromotive force and a dynamic electromotive force. In the EMH system designed in this article, the magnetic field around the coil constantly changes due to the motion of the permanent magnet, so the electromotive force generated inside the coil is an induced electromotive force. The electromotive force can be expressed as:

$$\begin{aligned} E_1 &= -\frac{d\Phi}{dt} \\ &= NAv(t)\frac{dB}{d\xi} \end{aligned}, \tag{3}$$

where N and A are the number of turns and cross-sectional area of the coil; B is the magnetic induction strength of the magnet, and ξ is the instantaneous distance between the center $x1$ of the coil and the center $x(t)$ of the permanent magnet, $\xi = x_1 - x(t)$.

The relationship between the magnetic induction intensity generated by a magnet and the instantaneous distance between the coil and the magnet is as follows:

$$B(\xi) = \frac{B_r}{2}\left[\frac{\xi + \frac{L}{2}}{\sqrt{\left(\xi + \frac{L}{2}\right)^2 + R^2}} - \frac{\xi - \frac{L}{2}}{\sqrt{\left(\xi - \frac{L}{2}\right)^2 + R^2}}\right], \tag{4}$$

where B_r is the residual magnetic flux density of the permanent magnet itself, and R and L are the radius and length of the cylindrical magnet, respectively.

Due to the constant motion of permanent magnets, the magnetic field strength around the coil is also constantly changing, and the magnetic field gradient can be used to represent the rate of change in magnetic induction intensity around the coil with the instantaneous distance, expressed as:

$$\frac{dB}{d\xi} = \frac{B_r R^2}{2}\left(\frac{1}{\left[\left(\xi + \frac{L}{2}\right)^2 + R^2\right]^{3/2}} - \frac{1}{\left[\left(\xi - \frac{L}{2}\right)^2 + R^2\right]^{3/2}}\right). \quad (5)$$

Therefore, the induced electromotive force generated by the final EMH can be expressed as:

$$E_1 = -NA\frac{B_r R^2}{2} r\omega sin(\omega t)\left[1 + \frac{rcos(\omega t)}{\sqrt{l^2 - r^2 sin^2(\omega t)}}\right] \times \left(\frac{1}{\left[\left((x_c - x_m(t)) + \frac{L}{2}\right)^2 + R^2\right]^{3/2}} - \frac{1}{\left[\left((x_c - x_m(t)) - \frac{L}{2}\right)^2 + R^2\right]^{3/2}}\right). \quad (6)$$

To better understand the working principle of the HEH structure, an equivalent dynamic model was established, as shown in Figure 3. The equivalent models of the PEH and EMH are shown in Figure 3a and 3b, respectively.

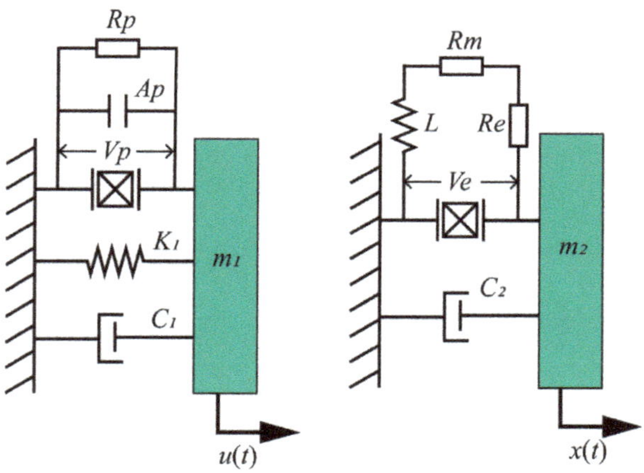

Figure 3. Equivalent model of the energy harvester. (**a**) Piezoelectric module; (**b**) electromagnetic module.

In the equivalent model of the PEH shown in Figure 3, R_p and A_p represent the load resistance and equivalent capacitance in the piezoelectric system circuit; V_p represents the output voltage of the piezoelectric system; K_1 is the equivalent stiffness of the cantilever beam; C_1 is the mechanical piezoelectric equivalent damping coefficient, consisting of a piezoelectric damping coefficient and a mechanical damping coefficient; M_1 is the equivalent mass of the cantilever beam and its end magnets; $U(t)$ represents the external motivation it receives. For the equivalent model of the EMH, R_m and R_e represent the load resistance and internal resistance in the electromagnetic system circuit, and L represents the equivalent inductance of the electromagnetic system; V_e represents the output voltage of the electromagnetic system; C_2 is its equivalent damping coefficient, mainly derived from air resistance; M_2 is the equivalent mass of the magnet inside the coil tube; and $x(t)$ represents the external excitation received by the electromagnetic system.

According to the equivalent dynamic model of the device, the motion's differential equation of the piezoelectric energy harvesting part can be established as follows:

$$m_1\ddot{u}(t) + C_1\dot{u}(t) + K_1 u(t) = F_m - \gamma_1 V_0(t), \quad (7)$$

where γ_1 is the electromechanical coupling coefficient of the PEH, and F_m is the magnetic force between magnets, which can be expressed as:

$$F_m = \frac{\mu_0 v_1 v_2 J_1 J_2}{4\pi}\left(\frac{5p}{(p^2+R_0^2)^{\frac{5}{2}}} - \frac{5p^3}{(p^2+R_0^2)^{\frac{7}{2}}}\right), \quad (8)$$

where μ_0 is the magnetic permeability of the vacuum medium, ε_1 is the volume of the magnet inside the coil tube, ε_2 is the volume of the response magnet at the end of the piezoelectric cantilever beam, J_1 is the magnetization intensity of the magnet inside the coil tube, J_2 is the magnetization intensity of the response magnet at the end of the piezoelectric cantilever beam, R_0 is the distance between the center of the driving magnet at the end of the piezoelectric cantilever beam and the magnet inside the coil tube, and p is the distance between the magnet inside the coil tube and the coil tube.

From Equation (8), it can be inferred that the magnetic force Fm is not only related to the distance between magnets but also to the number of magnets and the properties of the magnets themselves.

The PEH in the device can be equivalent to a parallel connection of voltage source V_p, capacitor A_p, and resistor R_p. According to the equivalent power model and Kirchhoff's law, the following can be obtained:

$$A_p\dot{V}_p(t) + V_p(t)R_p^{-1} - \gamma_1\dot{u}(t) = 0. \quad (9)$$

When the piezoelectric cantilever beam is excited by a magnetic force, according to the analysis of the positive piezoelectric effect, the open-circuit voltage output of the PEH can be obtained as:

$$V_p = \frac{3d_{31}t_h E t_b \sigma F_m}{8m_1 t_l^2 A_p \omega^2}, \quad (10)$$

where t_b is the thickness of the piezoelectric sheet, t_h is the height of the piezoelectric cantilever beam, t_l is the length of the piezoelectric cantilever beam, σ is the average surface stress of the piezoelectric cantilever, E_p is the Young's modulus of the piezoelectric cantilever beam, and ω is the excited vibration frequency.

Furthermore, the output power of the PEH can be derived as:

$$P_1 = \frac{V_p^2}{R_p} = \frac{9}{64}\left[\frac{(d_{31}t_h E_p t_b \sigma F_m)^2}{R_p m_1^2 t_l^4 C_{pe}^2 \omega^4}\right]. \quad (11)$$

According to the established equivalent model, the motion's differential equation of the electromagnetic energy harvesting part can be established, and its expression is:

$$m_2\ddot{x}(t) + C_2\dot{x}(t) = F_m - \rho_1 I_0(t), \quad (12)$$

where $I_0(t)$ is the voltage across the external load resistor of the electromagnetic part, and ρ_1 is the electromechanical coupling coefficient of the electromagnetic energy harvesting part. When the magnet inside the coil tube moves in a straight line, the coil cuts the magnetic induction line, and a corresponding induced electromotive force is generated inside, which can be expressed as:

$$V_e = \frac{d\phi}{dt} = nB_0L_0\dot{x}(t), \tag{13}$$

where n is the number of coil turns, B_0 is the magnetic induction intensity, and L_0 is the effective length of the coil cutting the magnetic induction line.

In the equivalent circuit of the electromagnetic part, since the inductance of the coil is very small and can be ignored, only the load's internal resistance R_m and the internal resistance R_e of the coil are considered. Therefore, the output power of the coil part can be expressed as:

$$P_2 = \frac{1}{T}\int_0^T \frac{V_e^2}{R_m + R_e}dt = \frac{1}{2}\frac{(nB_0L_0)^2}{R_m + R_e}\omega^2 z^2, \tag{14}$$

where ω represents the excitation frequency, and z represents the displacement of the magnet relative to the coil.

By combining Equations (11) and (14), the total output power of the entire device can be obtained as follows:

$$P = P_1 + P_2 = \frac{9}{64}\left[\frac{(d_{31}t_hE_pt_b\sigma F_m)^2}{R_pm_1^2t_l^4C_{pe}^2\omega^4}\right] + \frac{1}{2}\frac{(nB_0L_0)^2}{R_m + R_e}\omega^2 z^2 \tag{15}$$

3.2. Simulation Analysis

To verify the output characteristics and force conditions of the HEH, this paper used COMSOL 6.2 software and ANSYS 2022R1 software to simulate the HEH and verify the effectiveness of the model device and mathematical model.

The EMH module consisted of N52 permanent magnets and induction coils. Due to the symmetrical structure of the EMH, a two-dimensional plane was chosen for modeling and simulation. In the parameter setting section, the number of turns of the coil in the model was 432, the number of magnetic oscillators was 8, the magnetic induction intensity was 1.48 T, the peak displacement was set to 5 mm, and the excitation method was simplified to a sine wave. This model was mainly used to solve the electrical output characteristics of coils, so the field integration method was used for the calculation.

Figure 4 shows the voltage changes of the EMH under four different excitation frequencies. From the results, as the excitation frequency increased, the motion period gradually shortened, while the peak voltage output produced by the EMH gradually increased, allowing for more energy to be obtained per unit time. In addition to the voltage output characteristics of the EMH, similar conclusions could also be drawn regarding changes in output power.

Figure 5 shows the output power of the EMH module under different excitation frequencies. As shown in the figure, the output power of the coil increased with the increase in frequency, and the motion period gradually decreased, enabling the EMH to achieve a larger and faster power output per unit time. When the driving frequency was 8 Hz, the maximum output power of the electromagnetic module reached 58 mW.

Due to the non-contact magnetic excitation used for the vibration excitation of piezoelectric cantilever beams, it is particularly important to analyze the magnetic field and force of the interaction between magnets. Based on the relevant parameters in Tables 1 and 2, a magnet simulation model was established in COMSOL 6.2 software. The circular magnet was set to a sinusoidal linear reciprocating motion with a peak displacement of 5 mm. Under these conditions, the magnetic field and magnetic force were solved. The magnetic flux density mode between magnets is shown in Figure 6a, with the maximum magnetic field strength near the magnets.

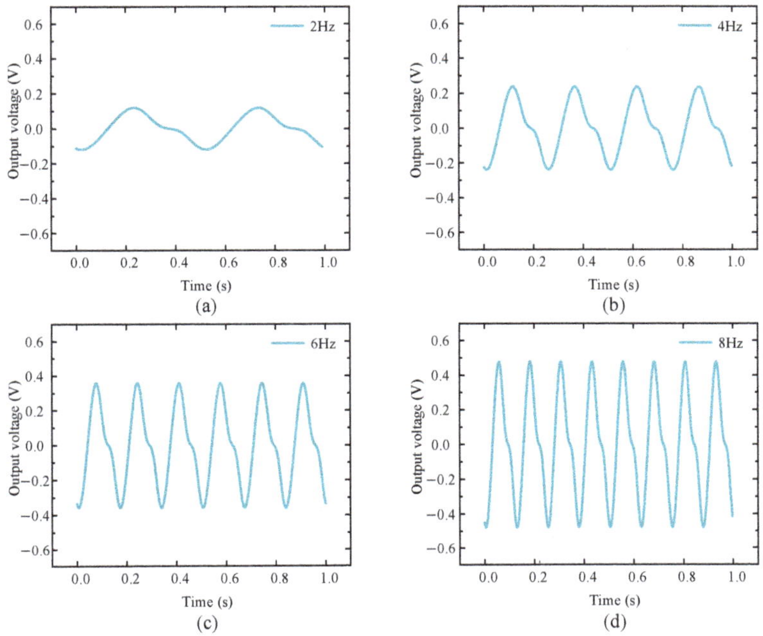

Figure 4. Output voltage of the electromagnetic module under different frequencies: (**a**) 2 Hz; (**b**) 4 Hz; (**c**) 6 Hz; (**d**) 8 Hz.

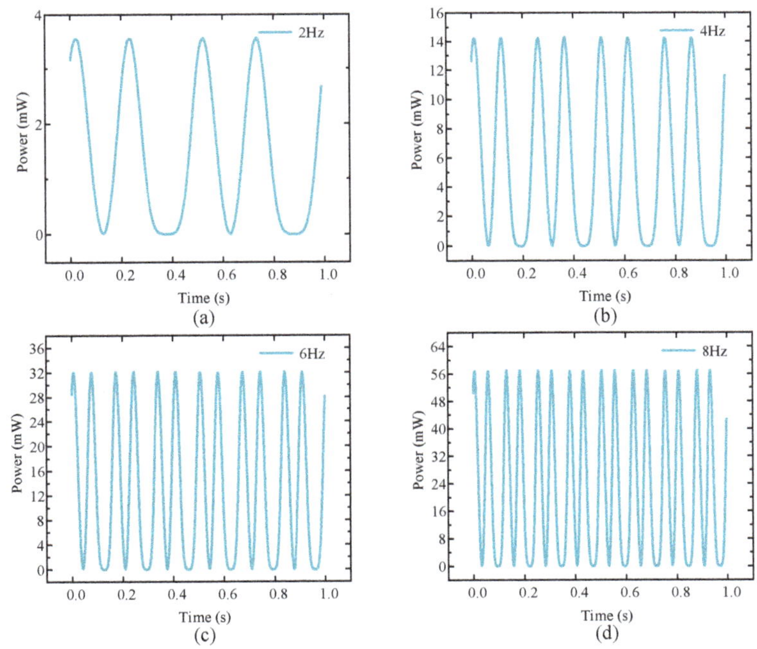

Figure 5. Output power of the electromagnetic module under different frequencies: (**a**) 2 Hz; (**b**) 4 Hz; (**c**) 6 Hz; (**d**) 8 Hz.

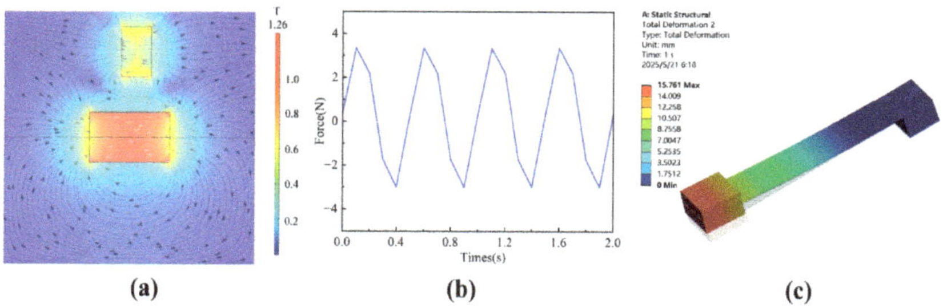

Figure 6. Simulation analysis of the piezoelectric cantilever beam under the action of a magnetic field. (**a**) Magnetic flux density mode; (**b**) force between magnets with time; (**c**) simulated displacement.

Figure 6b shows the variation in the interaction force between magnets over time. Due to the sinusoidal motion of the circular magnet, its force was approximately expressed as a sine wave, with a maximum value of about 3.3 N and an effective value of 2.3 N. Subsequently, a simulation model of a piezoelectric cantilever beam was constructed in ANSYS 2022R1 software, with one end of the cantilever beam set as a fixed constraint and a pressure of 2.3 N applied to the surface of the bottom magnet to simulate the force situation of the cantilever beam during the motion process. The simulation results showed that the maximum deformation of the piezoelectric cantilever beam under that condition reached 15.761 mm, and this deformation was mainly distributed in the end magnets, which would not cause damage to the PZT. This result indicates that magnetic excitation can effectively deform piezoelectric cantilever beams.

4. Experimental Results and Discussion

4.1. Output Characteristics Analysis

To further verify the energy output characteristics of the HEH structure under rotational excitation conditions, a prototype of the HEH was manufactured using 3D printing technology, and an experimental setup was established, as shown in Figure 7. We fixed the experimental device on the experimental platform with bolts and fixed the sun gear and planet gear on the pedestal with bearing bolts. We used a servo motor (ASDA-B2) and its driver to drive the sun gear to simulate rotational excitation conditions, the servo motor was connected to the sun gear through a coupling. We used an oscilloscope to display and record the electrical output characteristics of the HEH. In the experiment, the PZT was attached to a cantilever beam using epoxy resin, while the circular magnet was composed of several N52 permanent magnets with a diameter of 10 mm and a height of 2 mm. Considering the low-frequency operating characteristics of the energy harvester, the operating frequency of the servo motor was set to 1–10 Hz by setting parameters on the PC host to control the driver, and the excitation provided by the servo motor was sinusoidal.

To verify the amplification effect of the planetary gear system on an external low-frequency energy excitation, the output voltages of the EMH and PEH in the experimental device were collected, and the results are shown in Figure 8. The frequencies shown in the figure are the excitation frequencies applied by the servo motor to the experimental device. The number of N52 permanent magnets used was eight, and the overall height of the magnets was 16 mm, which was consistent with the parameters of the coil simulation. The waveform of the PEH indicated that 7 Hz was the resonant frequency of the cantilever beam at which the maximum voltage was collected.

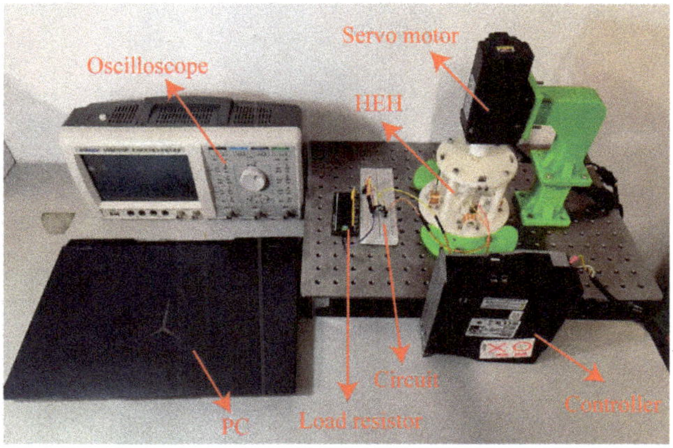

Figure 7. The experimental setup of the HEH.

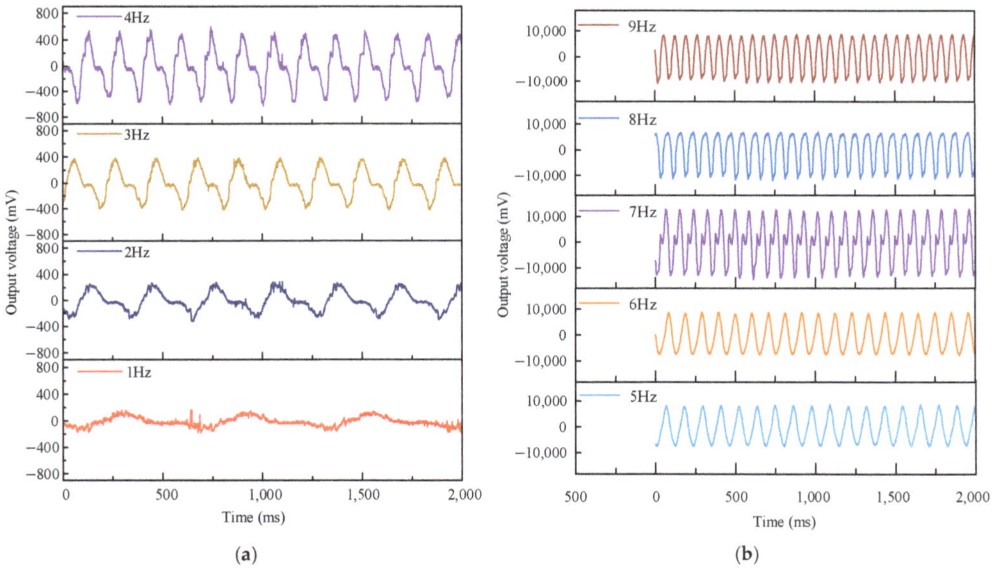

Figure 8. Output voltage of the EMH and PEH under different frequencies. (**a**) Output voltage of the EMH; (**b**) output voltage of the PEH.

Results showed that the experimental waveforms, peak voltages, and periods at each frequency were similar to the simulation results corresponding to twice the frequency in Figure 4. For example, in the experiment, the peak voltage at the 4 Hz excitation frequency was 550 mV with a period of 150 ms, which was consistent with the 500 mV peak voltage and 125 ms period at 8 Hz in the simulation. These results effectively demonstrate that planetary gear systems have a certain amplification effect on external low-frequency energy excitation. However, there was still a deviation between experimental data and simulation data, which may have been mainly due to the friction between the magnet and the coil axis during its motion. This friction hindered the speed at which the coil cuts the magnetic induction line, resulting in a delay in the experimental period compared to the simulation period.

The number of magnets at the center of the coil also affects the performance of the proposed EMH. Therefore, experimental tests were conducted on the EMH under different quantities of permanent magnets. The effective values of its output voltage were collected using an oscilloscope, and the output frequency of the EMH was calculated. The results are shown in Figure 9.

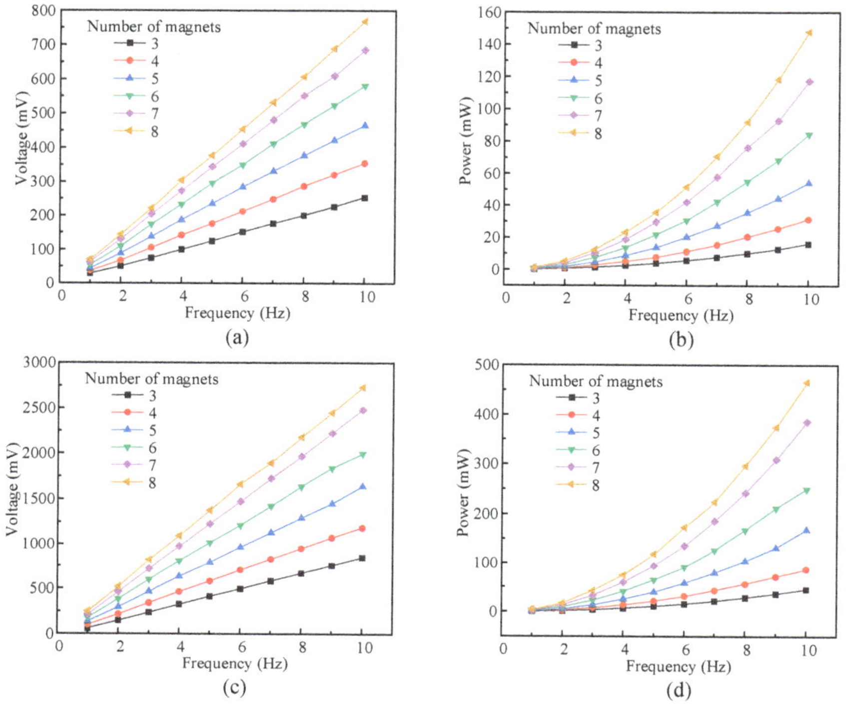

Figure 9. Experimental output results of the EMH. (**a**) Output voltage of a single EMH module; (**b**) output power of a single EMH module; (**c**) output voltage of a multiple-EMH module; (**d**) Output power of a multiple-EMH module.

Figure 9a shows the variation in output voltage of a single EMH group with an excitation frequency for different numbers of permanent magnets. As the excitation frequency and number of magnets increase, the output voltage of a single EMH group also increases accordingly. This is because as the excitation frequency and the number of magnets increase, the magnetic induction intensity around the coil increases, and the speed at which the coil cuts the magnetic induction line accelerates, resulting in an increase in the rate of change in the magnetic flux inside the coil. The combined effect of these factors increases the induced electromotive force generated. Correspondingly, the output power of the EMH also increases, as shown in Figure 9b.

To measure the output characteristics of multiple sets of EMHs in the HEH, four sets of EMHs were connected in series to collect data on their output voltage and power. The results are shown in Figure 9c,d. Under the excitation frequency conditions of eight magnets and 10 Hz, the output voltage and power of the entire EMH could reach 2.728 V and 465 mW. However, the output voltage of the multiple-EMH module was not exactly equal to four times the output voltage of a single EMH. The reason for this may be that the speed or movement path of the four sets of permanent magnets was unstable during the motion process, and there were differences in the rate of change in magnetic flux generated by each set of coils, resulting in phase differences in the induced electromotive force generated by each coil, and the total voltage may have experienced vector superposition due to phase differences.

Figure 10 shows the output voltage of the EMH when the excitation frequency fluctuates. The results show that when the frequency fluctuates, the instantaneous output power changes. In order to achieve a stable output of energy, the collected energy needs to be stabilized, processed, and stored in a battery so that it can supply power to low-power appliances.

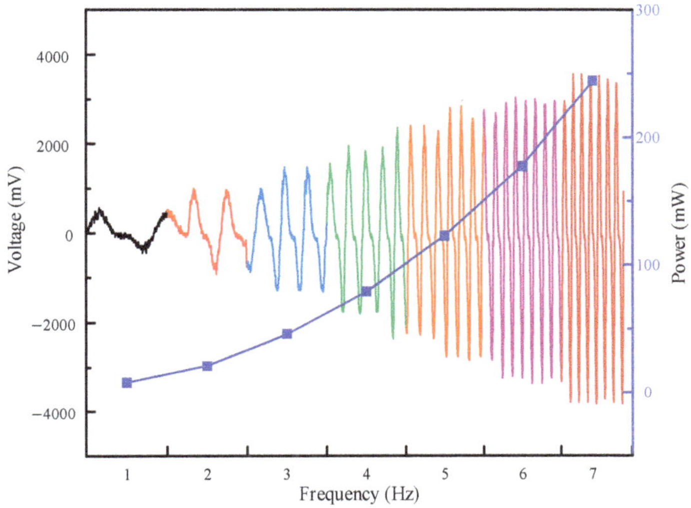

Figure 10. Voltage changes in the EMH during frequency fluctuations.

For the PEH modules, their electrical output characteristics largely depend on the resonance of the device structure, and the vibration frequency of the cantilever beam is a key external factor affecting the performance of the PEH. This article tested the variation in open circuit voltage with the excitation frequency of the PEH under different numbers of permanent magnets on a self-built experimental device, as shown in Figure 11.

When the number of permanent magnets at the center of the coil was fixed, as the excitation frequency increased from 1 Hz to 10 Hz, the measured output voltage of the PEH showed a trend of first increasing and then decreasing. When the rotational excitation frequency was 7 Hz, the output voltage reached its peak, indicating that the cantilever beam in the PEH was in a resonant state. Since the HEH designed in this article could amplify the external excitation frequency twofold through the planetary gear, when the external excitation frequency reached 7 Hz, the vibration frequency of the cantilever beam in the PEH was 14 Hz, which was highly consistent with the finite element simulation results. The experimental results showed that when configuring eight permanent magnets and using 7 Hz as the excitation frequency, the output voltage of a single PEH module could reach 9.2 V.

Due to the capacitive properties exhibited by the piezoelectric effect, the internal impedance of the PZT piezoelectric ceramic sheet in the PEH structure is not constant but varies with the external excitation, which directly affects the output voltage and power of the PEH. Therefore, it is crucial to find the best load impedance for the PZT in the operating range. Considering that 7 Hz was the optimum excitation frequency for the PEH structure and the internal impedance of the PZT was not affected by the variation in the structural device's own parameters, in this paper, the optimal impedance of the PZT was chosen to be estimated at a 7 Hz excitation frequency. A variable load resistor was connected to the output port of the PEH, and an oscilloscope was used to record the output voltage and power of the PEH for different numbers of permanent magnets; the results are shown in Figure 12.

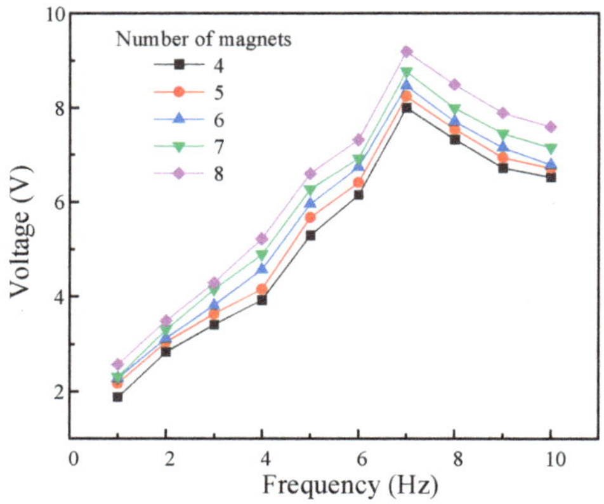

Figure 11. Output voltage of PEH under different numbers of magnets.

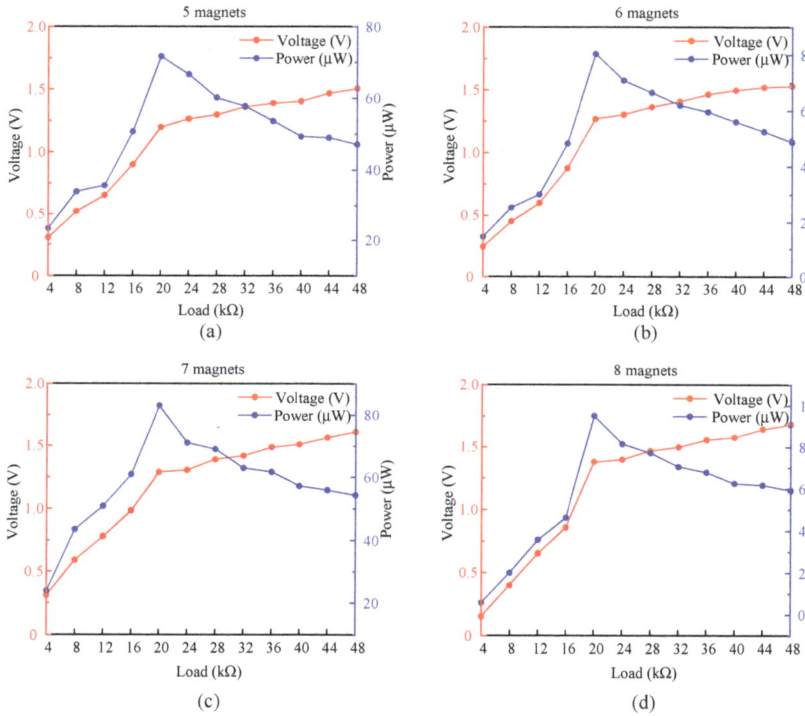

Figure 12. Output voltage and power of the PEH with various loads under different numbers of magnets: (**a**) 5 magnets; (**b**) 6 magnets; (**c**) 7 magnets; (**d**) 8 magnets.

In Figure 12, with the same number of permanent magnets, the output voltage gradually increases with the increase in load resistance, while the output power shows a trend of increasing and then decreasing. Through a comparative analysis, it was found that PEH reached its maximum output power at a 20 kΩ load for all different configurations of the number of permanent magnets. Therefore, 20 kΩ could be considered as the optimum load impedance in this frequency band. Under this load condition, the maximum output power of the PEH module was 95 μW.

Under the external excitation of eight magnets at 7 Hz, the output voltage obtained by experimentally measuring the four EMHs in series was 1.9 V, and the actual output power was calculated to be about 230 mW, so the output efficiency of the actual EMHs was calculated to be 64%.

Similarly, with eight magnets, an external excitation frequency of 7 Hz, and an external load of 20 kΩ, the actual measured output power of the four PEHs was about 180 μW, and the calculated output efficiency of the PEHs was about 47%.

In summary, it can be deduced that the total output power of the HEH reached 250 mW with an overall efficiency of 67% under eight magnets and a 7 Hz external excitation frequency. The loss of efficiency was due to the fact that during the movement of the magnet, its speed or path was not consistent, resulting in a slight difference in the amount of flux generated by each group of coils, which caused a phase difference between the induced electromotive force generated by each coil. This phase difference caused the total output power to be reduced due to vector superposition.

4.2. Application Analysis

The HEH proposed in this paper is essentially a piezoelectric–electromagnetic hybrid energy harvester. Due to the significant differences in the output electrical characteristics of the EMH and PEH, two independent channels of energy harvesting were used to fully utilize their respective advantages. The electrical energy generated by an EMH has low-voltage and high-current characteristics, so a voltage doubler rectifier circuit, which is a circuit that stores energy through capacitors to realize multiple voltage rectification, was used, as shown in Figure 13a. For the PEH, the electrical energy generated is an alternating current (AC), characterized by a high voltage and low current, which makes it impossible to supply power to the load continuously. For this reason, an AC–DC bridge rectifier circuit was used to efficiently collect the energy generated by the PEH, which consisted of an AC–DC rectifier module and a filter capacitor, as shown in Figure 13b.

By connecting the two rectifier circuits in series, the overall energy harvesting of the HEH was realized, and thus continuous power supply to the load was achieved. Through experiments as well as calculations, under the condition of eight magnets and an external excitation frequency of 7 Hz, the output voltage of the HEH could reach 10.4 V, and the average output power was 250 mW. The results show that the HEH was good at low-frequency energy harvesting.

To verify the performance of the designed HEH in supplying power to low-power electronic devices, electronic load experiments were conducted. Under eight permanent magnets and a 7 Hz excitation frequency, the HEH successfully lit 60 LEDs and provided a stable power supply for the temperature–humidity meter, as shown in Figure 14. This experimental result indicates that the HEH designed in this paper has the potential to provide reliable energy for some low-power electronic products in low-frequency environments.

In addition to the experiments mentioned above, further applications for the HEH are possible. In practical applications, the HEH can be combined with wind cups to convert the driving mode from electric motors to wind power, thereby converting wind energy in the natural environment into electrical energy, as shown in Figure 15. At the same time, the rectification circuit of the HEH can be connected to an energy storage circuit for storing the collected electrical energy, which can be used for some small electronic devices such as thermometers, indicator lights, street lighting, etc.

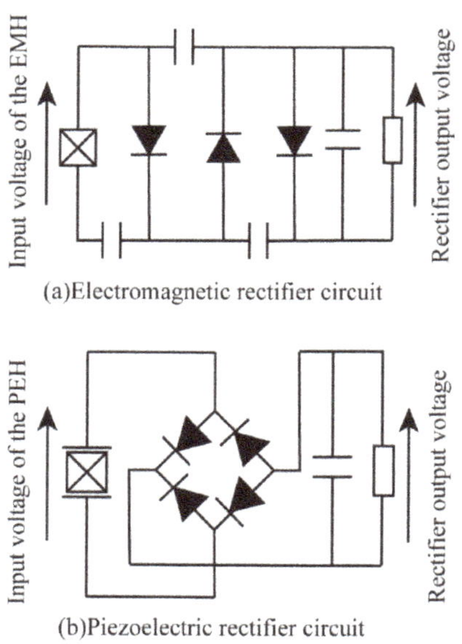

Figure 13. Rectifier circuit of the HEH. (**a**) Electromagnetic rectifier circuit; (**b**) piezoelectric rectifier circuit.

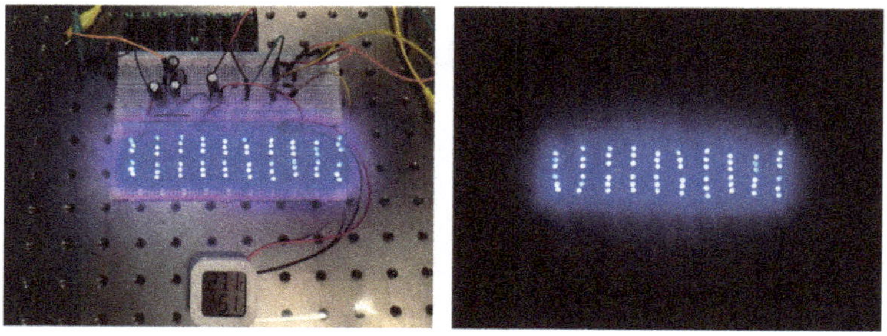

Figure 14. Practical application of the HEH.

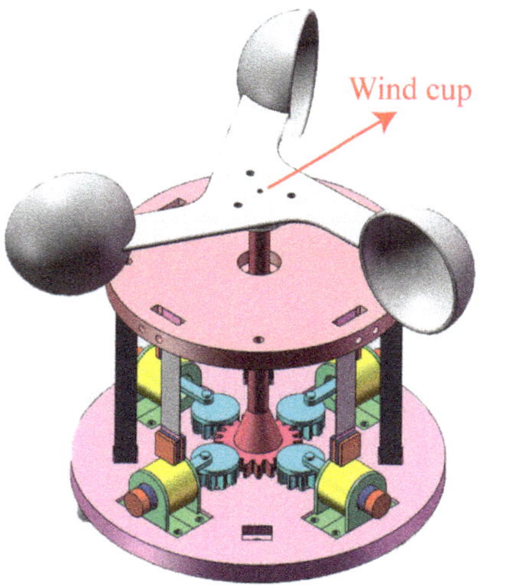

Figure 15. Further practical application of the HEH.

5. Conclusions

This article proposed a piezoelectric electromagnetic hybrid energy harvester (HEH) based on a planetary gear system, which combined piezoelectric and electromagnetic effects and was mainly used for collecting low-frequency rotational energy. The HEH had a compact structure with dimensions of 140 × 95 mm and contained four sets of piezoelectric energy harvesters (PEH) and electromagnetic energy harvesters (EMH) inside. Through the planetary gear system inside the device, the device could amplify the external excitation frequency, thereby improving energy output efficiency. To verify the effectiveness of the design, an experimental research device was constructed, and external rotational excitation conditions were simulated using motors. The influence of different numbers of magnets and external excitation frequencies on the output performance of the HEH was studied. The results indicated that the HEH could generate an output voltage of 10.4 V and an average output power of 250 mW under eight magnets and an external excitation frequency of 7 Hz. In actual power supply testing, the HEH successfully lit up 60 LEDs, fully verifying its potential for application in low-frequency environments.

Author Contributions: Investigation, data curation, Z.Y.; writing—original draft preparation, writing—review and editing, C.L. All authors have read and agreed to the published version of the manuscript.

Funding: This research was funded by Qinglan Project of Jiangsu Province of China, Key Research and Development Program of Jiangsu Province (No. BE2022062-3).

Data Availability Statement: The original contributions presented in the study are included in the article.

Conflicts of Interest: The authors declare no conflicts of interest.

References

1. Wang, L.; Fei, Z.; Duan, C.; Han, X.; Li, M.; Gao, W.; Xia, Y.; Jia, C.; Lin, Q.; Zhao, Y. Self-sustained and self-wakeup wireless vibration sensors by electromagnetic-piezoelectric-triboelectric hybrid energy harvesting. *Appl. Energy* **2024**, *355*, 122207. [CrossRef]
2. Babayo, A.A.; Anisi, M.H.; Ali, I. A review on energy management schemes in energy harvesting wireless sensor networks. *Renew. Sustain. Energy Rev.* **2017**, *76*, 1176–1184. [CrossRef]
3. Liu, H.; Fu, H.; Sun, L.; Lee, C.; Yeatman, E.M. Hybrid energy harvesting technology: From materials, structural design, system integration to applications. *Renew. Sustain. Energy Rev.* **2021**, *137*, 110473. [CrossRef]
4. Mrozik, W.; Rajaeifar, M.A.; Heidrich, O.; Christensen, P. Environmental impacts, pollution sources and pathways of spent lithium-ion batteries. *Energy Environ. Sci.* **2021**, *14*, 6099–6121. [CrossRef]
5. Hossain, M.I.; Zahid, M.S.; Chowdhury, M.A.; Hossain, M.M.M.; Hossain, N. MEMS-based energy harvesting devices for low-power applications—A review. *Results Eng.* **2023**, *19*, 101264. [CrossRef]
6. He, L.; Han, Y.; Liu, R.; Hu, R.; Yu, G.; Cheng, G. Design and performance study of a rotating piezoelectric wind energy harvesting device with wind turbine structure. *Energy* **2022**, *256*, 124675. [CrossRef]
7. Wang, G.; Song, R.; Luo, L.; Yu, P.; Yang, X.; Zhang, L. Multi-piezoelectric energy harvesters array based on wind-induced vibration: Design, simulation, and experimental evaluation. *Energy* **2024**, *300*, 131509. [CrossRef]
8. Zhou, Z.; Cao, D.; Huang, H.; Qin, W.; Du, W.; Zhu, P. Biomimetic swallowtail V-shaped attachments for enhanced low-speed wind energy harvesting by a galloping piezoelectric energy harvester. *Energy* **2024**, *304*, 132063. [CrossRef]
9. He, L.; Liu, R.; Liu, X.; Zheng, X.; Zhang, L.; Lin, J. A piezoelectric-electromagnetic hybrid energy harvester for low-frequency wave motion and self-sensing wave environment monitoring. *Energy Convers. Manag.* **2024**, *300*, 117920. [CrossRef]
10. Du, X.; Li, P.; Li, Z.; Liu, X.; Wang, W.; Feng, Q.; Du, L.; Yu, H.; Wang, J.; Xie, X. Multi-pillar piezoelectric stack harvests ocean wave energy with oscillating float buoy. *Energy* **2024**, *298*, 131347. [CrossRef]
11. Fan, K.; Liu, Z.; Liu, H.; Wang, L.; Zhu, Y.; Yu, B. Scavenging energy from human walking through a shoe-mounted piezoelectric harvester. *Appl. Phys. Lett.* **2017**, *110*, 143902. [CrossRef]
12. Chen, Z.; Xu, M.; Zhou, C.; Hu, Z.; Du, Z.; Fu, X.; Song, Y.; Jia, Y.; Wen, X.; Wang, J. Phase transformation enabled textile triboelectric nanogenerators for wearable energy harvesting and personal thermoregulation. *Nano Energy* **2024**, *132*, 110361. [CrossRef]

13. Hao, Y.; Yang, J.; Niu, Z.; Wang, M.; Liu, H.; Qin, Y.; Zhang, C.; Li, X. High-output triboelectric nanogenerator based on L-cystine/nylon composite nanofiber for human bio-mechanical energy harvesting. *Nano Energy* **2023**, *118*, 108964. [CrossRef]
14. Kuang, Z.; Zhang, Z.; Liao, W.; Lin, S.; Wang, K.; Zhang, J.; Kan, J. Magnetic transfer piezoelectric wind energy harvester with dual vibration mode conversion. *Energy* **2024**, *308*, 133020. [CrossRef]
15. Shen, J.; Wan, S.; Fu, J.; Li, S.; Lv, D.; Dekemele, K. A magnetic plucking frequency up-conversion piezoelectric energy harvester with nonlinear energy sink structure. *Appl. Energy* **2024**, *376*, 124326. [CrossRef]
16. Li, S.; Feng, Z.; He, X.; Ye, Y.; Li, J. An in-plane omnidirectional flutter piezoelectric wind energy harvester. *Mech. Syst. Signal Process.* **2023**, *200*, 110637. [CrossRef]
17. Wang, S.; Miao, G.; Zhou, S.; Yang, Z.; Yurchenko, D. A novel electromagnetic energy harvester based on the bending of the sole. *Appl. Energy* **2022**, *314*, 119000. [CrossRef]
18. Miao, G.; Fang, S.; Wang, S.; Zhou, S. A low-frequency rotational electromagnetic energy harvester using a magnetic plucking mechanism. *Appl. Energy* **2022**, *305*, 117838. [CrossRef]
19. Zhang, T.; Kong, L.; Zhu, Z.; Wu, X.; Li, H.; Zhang, Z.; Yan, J. An electromagnetic vibration energy harvesting system based on series coupling input mechanism for freight railroads. *Appl. Energy* **2024**, *353*, 122047. [CrossRef]
20. Zhao, C.; Hu, G.; Li, X.; Liu, Z.; Yuan, W.; Yang, Y. Wide-bandwidth triboelectric energy harvester combining impact nonlinearity and multi-resonance method. *Appl. Energy* **2023**, *348*, 121530. [CrossRef]
21. Pang, Y.; Zhu, X.; Jin, Y.; Yang, Z.; Liu, S.; Shen, L.; Li, X.; Lee, C. Textile-inspired triboelectric nanogenerator as intelligent pavement energy harvester and self-powered skid resistance sensor. *Appl. Energy* **2023**, *348*, 121515. [CrossRef]
22. Li, W.; Wan, L.; Lin, Y.; Liu, G.; Qu, H.; Wen, H.; Ding, J.; Ning, H.; Yao, H. Synchronous nanogenerator with intermittent sliding friction self-excitation for water wave energy harvesting. *Nano Energy* **2022**, *95*, 106994. [CrossRef]
23. Liao, W.; Su, X.; Fang, F. A centrifugal spring mechanism empowers self-adjusting in piezoelectric wind energy harvesting. *Nano Energy* **2025**, *133*, 110462. [CrossRef]
24. Li, Y.; Zhao, Z.; Fan, Z.; Fan, D.; Jiang, F.; Hu, X. A novel multi-degree of freedom kinetic energy harvester for self-powered low-power applications in ships. *Energy Convers. Manag.* **2024**, *302*, 118096. [CrossRef]
25. Yin, P.; Tang, L.; Li, Z.; Xia, C.; Li, Z.; Aw, K.C. Harnessing ultra-low-frequency vibration energy by a rolling-swing electromagnetic energy harvester with counter-rotations. *Appl. Energy* **2025**, *377*, 124507. [CrossRef]
26. Chen, W.; He, Z.; Zhao, J.; Mo, J.; Ouyang, H. Hybrid triboelectric-piezoelectric energy harvesting via a bistable swing-impact structure with a tuneable potential barrier and frequency-up conversion effects. *Appl. Energy* **2024**, *375*, 124123. [CrossRef]
27. Du, X.; Han, Y.; Guo, D.; Wang, W.; Liu, X.; Wang, S.; Yu, H.; Wang, J.; Tang, L.; Aw, K.C. Hybrid piezo-triboelectric wind energy harvesting mechanism with flag-dragging the cantilever beam vibration. *Nano Energy* **2024**, *131*, 110274. [CrossRef]
28. Du, X.; Chen, H.; Li, C.; Li, Z.; Wang, W.; Guo, D.; Yu, H.; Wang, J.; Tang, L. Wake galloping piezoelectric-electromagnetic hybrid ocean wave energy harvesting with oscillating water column. *Appl. Energy* **2024**, *353*, 122081. [CrossRef]
29. Peng, W.; Ni, Q.; Zhu, R.; Fu, X.; Zhu, X.; Zhang, C.; Liao, L. Triboelectric-electromagnetic hybrid wind energy harvesting and multifunctional sensing device for self-powered smart agricultural monitoring. *Nano Energy* **2024**, *131*, 110272. [CrossRef]
30. Qu, Z.; Wang, X.P.; Huang, M.K.; Chen, C.X.; An, Y.; Yin, W.; Li, X. An eccentric-structured hybrid triboelectric-electromagnetic nanogenerator for low-frequency mechanical energy harvesting. *Nano Energy* **2023**, *107*, 108094. [CrossRef]
31. Wang, C.; Chai, H.; Li, G.; Wang, W.; Tian, R.; Wen, G.L.; Wang, C.H.; Lai, S.K. Boosting biomechanical and wave energy harvesting efficiency through a novel triple hybridization of piezoelectric, electromagnetic, and triboelectric generators. *Appl. Energy* **2024**, *374*, 123876. [CrossRef]
32. Zhao, L.C.; Zou, H.X.; Xie, X.; Guo, D.H.; Gao, Q.H.; Wu, Z.Y.; Yan, G.; Wei, K.X.; Zhang, W.M. Mechanical intelligent wave energy harvesting and self-powered marine environment monitoring. *Nano Energy* **2023**, *108*, 108222. [CrossRef]
33. Bai, Q.; Gan, C.Z.; Zhou, T.; Du, Z.C.; Wang, J.H.; Wang, Q.; Wei, K.X.; Zou, H.X. A triboelectric-piezoelectric-electromagnetic hybrid wind energy harvester based on a snap-through bistable mechanism. *Energy Convers. Manag.* **2024**, *306*, 118323. [CrossRef]
34. Zhao, K.; Song, Z.; Gao, Z.; Gao, W.; Liu, M.J.; Gu, B.; Guo, J.; Chueh, Y.L. High-performance triboelectric-electromagnetic hybrid nanogenerator using dual-functional flexible neodymium iron boron/ethyl cellulose (NdFeB/EC) composite films for wind energy scavenging. *Chem. Eng. J.* **2024**, *489*, 150763. [CrossRef]

micromachines

MDPI

Article

Study on the Driving Performance and Influencing Factors of Multi-Electrothermal Co-Actuation Devices Considering Application Environments

Yujuan Tang [1,*], Zihao Guo [1], Yujiao Ding [2] and Xinjie Wang [2]

[1] School of Intelligent Science and Control Engineering, Jinling Institute of Technology, Nanjing 211169, China; zhguo020528@163.com

[2] School of Mechanical Engineering, Nanjing University of Science and Technology, Nanjing 210094, China; yujiaoding@njust.edu.cn (Y.D.); xjwang@njust.edu.cn (X.W.)

* Correspondence: yjtang@jit.edu.cn

Abstract: Electrothermal actuators, with their simple structure, small size, strong anti-interference ability, and easy integration, have emerged as a promising solution for micro-drive technology. However, deploying them in extreme environments, such as the fuze systems—which demand exceptional reliability under high mechanical overloads. In this study, a device based on multi-electrothermal co-actuation is designed for the fuze system of loitering munition. The overall structure and work principle of the multi-electrothermal co-actuation device is discussed. Considering application environments, the effect factors of V-beam numbers, air gap, type of contact surface, external load force, periodic voltage and gas damping on the output performance of the multi-electrothermal co-actuation device are systematically addressed via simulation and experimental method. Furthermore, the high overload resistance performance of the co-actuation device applied in loitering munition is studied. The results show that the proposed multi-electrothermal co-actuation device could operate stably under a high overload (12,000 g/73.79 μs) environment, fully meeting the demanding requirements of fuze system for loitering munition. In addition, this study identifies laser processing-induced thermal gradients and mechanical stresses as critical fabrication challenges. This study provides significant insights into the design and optimization of multi-electrothermal actuation systems for next-generation fuze applications, establishing a valuable framework for future development in this field.

Keywords: electrothermal actuator; multi-electrothermal; co-actuation device; influencing factors; drive performance; simulation analysis

1. Introduction

Micro-actuators convert other forms of energy into mechanical energy and are the main devices in the mechanism that accomplish the drive and enable the output of force or displacement [1]. In general, micro-actuators can be classified into electrostatic [2], piezoelectric [3,4], electromagnetic [5], electrothermal [6], and so on. Compared with electromagnetic, electrostatic, shape memory alloy, and piezoelectric actuators, electrothermal actuators have become one of the most important driving methods in the field of micro-actuation technology [7] and neurotechnology [8] due to the advantages of the simpler structure, smaller size, no interference from electrostatic or magnetic conditions, and better compatibility with integrated circuits.

In order to realize the function optimization from small displacements into larger ones, the electrothermal actuators are usually combined with the lever, pawl flywheel, and other displacement amplification mechanisms, which greatly enhances the drive performance and applicability of the device. Many research studies have explored various novel configurations of electrothermal actuators and the analysis or optimization of their driving performance. A novel V-shaped electrothermal actuator proposed by Shen et al. [9] using surface silicon technology achieved significant displacement at lower voltages within a shorter response time. Nguyen et al. [10] developed an electrothermal V-shaped actuator using a surface sputtering process. Under a voltage of 16 V, the displacement of the sputtering V-shaped actuator was twice that of the non-sputtering, enhancing the displacement output capability of electrothermal actuators under the same voltage conditions. Tecpoyotl et al. [11] designed a novel asymmetric Z-shaped electrothermal actuator. At a voltage of 2 V, the output force increased by 370.48%, while the displacement decreased by 29.8%, making it suitable for applications where a significant driving force is required, but displacement output is not a critical factor. Vargas et al. [12] studied an electrothermal actuator based on a rotary system, whose claw displacement of the actuator reached 21.2 μm, with a driving force of approximately 34.2 μN. Although this actuator offers significant advantages in miniaturization, the array-based layout of its driving beams results in a complex structure. A new in-plane parallel electrothermal actuator proposed by Yen et al. [13] can drive the lever to produce in-plane displacement of about 177 μm under 6 V and 115.32 mW.

In recent years, electrothermal actuators have been applied in the Safety and Arming Device (SAD) of the fuze system. In modern fuze safety systems, functional requirements not only mandate that SADs incorporate recoverable functionality but also demand that the miniaturization of fuze isolation mechanisms be achieved while maintaining a certain explosion isolation distance to ensure reliable blast containment. However, in the existing SADs, the isolation mechanisms based on micro-actuators still face limitations in terms of size, displacement, and recoverable function. Researchers have begun to explore new structures and drive technologies to achieve more significant outputs. Ostrow et al. [14] designed a pawl-driven flywheel MEMS safe and arming device. Under 250 Hz periodic voltage, the slider moves from the initial (safety state) to the specified (un-safety state) position in only 0.325 s, with an output displacement of 785 μm. A high-speed linear electrothermal actuator for SAD systems designed by Li et al. [15] consists of four V-shaped micro electrothermal actuators, four lever amplification structures, a slider, and a spring. The slider's maximum displacement and output force were under 23 V and 5.27 mN. An environmental force sensing combined with a circuit-controlled MEMS fuze was developed by Wang et al. [16]. The environmental force-sensing inertial switch controls the state of the first fuze according to the impact load; the electrothermal actuator is connected to the slider and acts to limit the displacement of the slider. Hu et al. [17] designed a bidirectional large-displacement MEMS platform consisting of multiple V-shaped and U-shaped electrothermal actuators. However, this platform still has the disadvantages of a complex structure and excessive size.

In our previous published paper [18], we presented the co-actuation device based on multi-electrothermal actuators and established the mathematical model. In this paper, we focus on investigating the output performance and influencing factors of multi-electrothermal co-actuation devices in applied environmental SADs. Firstly, the electrical–thermal–mechanical coupled simulation model of the multi-electrothermal co-actuation device is established. Subsequently, the high overload resistance performance of the multi-electrothermal co-actuation device is investigated. In addition, the effect factors of V-beams umbers, air gap, type of contact surface, external load force, periodic voltage, and gas damping on the output performance of the multi-electrothermal co-actuation device are analyzed in detail via simulation and experimental method.

2. Overall Structure of Multi-Electrothermal Co-Actuation Device

The overall structure of the multi-electrothermal co-actuation device is shown in Figure 1, which mainly includes the following: main-actuate unit (i.e., the two V-shaped electrothermal actuators responsible for the output displacement), sub-actuate unit (i.e., the two V-shaped electrothermal actuators responsible for the position-holding function), the substrate, the flameproof slider, the metal limit frame, and the upper and lower package covers. The V-shaped electrothermal actuator and the metal limit frame are fixed to the substrate through the positioning holes and fix screws.

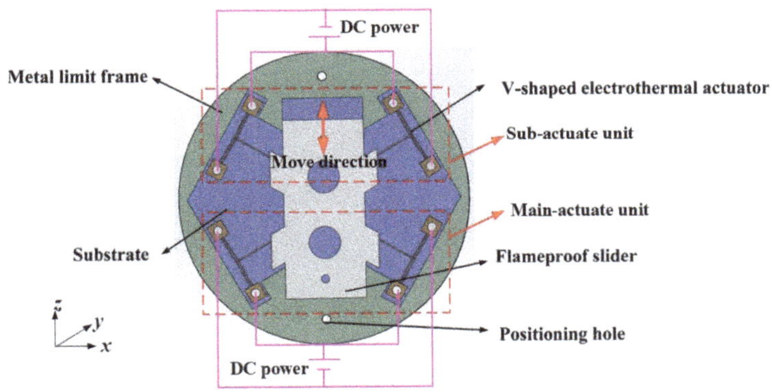

Figure 1. The structure diagram of the multi-electrothermal co-actuation device.

The V-shaped electrothermal actuator [19] consists of anchor points at both ends and a V-beam (cantilever beam), the structure of which is shown in Figure 2. The two ends of the V-beam are fixed to the substrate through anchor points to form a stable support.

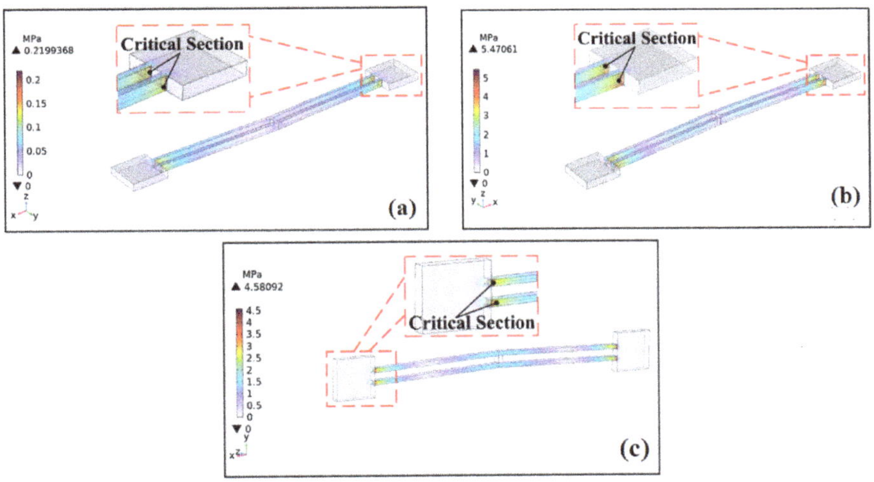

Figure 2. Stress distribution and critical sections under multi-directional impact loads: (**a**) $\pm x$ direction; (**b**) $\pm y$ direction; (**c**) $\pm z$ direction.

When a voltage is applied across the anchor points, there is an electric current flowing through the V-beam and generating Joule heating within the beam structure. This heating causes thermal expansion of the electrothermal actuator. Due to the inclined angle of the V-beam, a non-uniform thermal stress field is induced during thermal expansion, resulting in bending deformation at the V-beam tips. This deformation is manifested as a linear output displacement and actuating force at the tip of the push lever, enabling the actuator to push or pull the target as required to meet the actuation demands. When the flameproof slider is required to output displacement in the +z direction, it is mainly actuated by the main-actuate unit. Conversely, when the slider needs to output displacement in the $-z$ direction, it is mainly actuated by sub-actuate unit.

According to the miniaturization requirements of the fuze system for loitering munition, the final geometric parameters of the single V-shaped electrothermal actuator are provided in Table 1.

Table 1. The parameters of the single V-shaped electrothermal actuator.

Parameter Name	Parameters	Physical Unit	Specific Value
Length of the actuator	L_z	μm	7000
Thickness of the actuator	d	μm	200
Width of the actuator	a	μm	100
Tilt angle of the V beam	θ	°	2
Length of the lever	L_p	μm	5000

3. Analysis of High Overload Resistance Performance of the Multi-Electrothermal Co-Actuation Device

The multi-electrothermal co-actuation device studied in this paper is primarily designed to drive the isolation device of the SAD. As a vital driving part, it plays an essential role in initiating and controlling the fuze. It is required not only to output the target displacement to drive the flameproof slider but also to possess sufficient resistance to high overload conditions. The impact overload for the V-shaped electrothermal is primarily based on the acceleration generated by ammunition drops during the service phase, which is approximated as a half-sine function [20], given as follows:

$$a(t) = \begin{cases} G\sin\left(\frac{\pi}{\Delta t}t\right) & t \leq \Delta t \\ 0 & t > \Delta t \end{cases} \quad (1)$$

where G is the overload acceleration amplitude, and Δt is the pulse width.

With an acceleration set to 10,000 g and a pulse width of 75 μs, the stress distribution of the actuator under impact loading applied in six directions—$\pm x$, $\pm y$, and $\pm z$—is shown in Figure 2.

The electrothermal actuator is made of stainless steel 304. If the maximum stress afforded by the electrothermal actuator is below the allowable stress of the material, $[\sigma]$ = 386 MPa, the actuator can maintain regular operation. In all six directions, the maximum stress (i.e., at the critical section) occurs at the connection between the V-beam and the anchor point. Depending on the direction of the applied impact load, the critical section appears on different surfaces of the connection. The maximum stress at the critical section is about 0.0164 MPa, significantly lower than the allowable stress of the material.

Since the V-shaped electrothermal actuator is designed with a push lever, a structural transition exists at the connection between the push lever and the V-beam. In addition, the push lever is relatively long. Therefore, it is necessary to analyze whether the push lever deforms when subjected to high overload in the direction of the electrothermal actuator's deformation (x direction).

Figure 3 compares the maximum stress at the actuator and the push lever under an impact load with a pulse width of 100 μs and an amplitude of 10,000 g. Under high-overload impact conditions, the maximum stress at the push lever is about 0.01048 MPa, significantly lower than the allowable stress of the material. Therefore, the push lever does not affect the location of the actuator's critical section, which remains at the connection between the anchor and the V-beam.

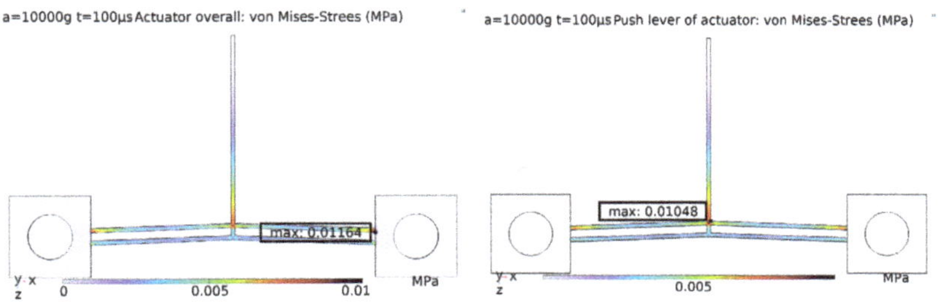

Figure 3. Stress distribution of the actuator under impact load in the x direction: (**a**) overall actuator; (**b**) push lever of the actuator.

Figure 4 shows the deformation of the push lever under impact loads in the x direction with varying pulse widths and amplitudes. Under an impact load with a pulse width of 70 μs and an amplitude of 10,000 g, the deformation of the push lever is minimal. When subjected to an impact load with a pulse width of 80 μs and an amplitude of 30,000 g, the push lever exhibits slight deformation.

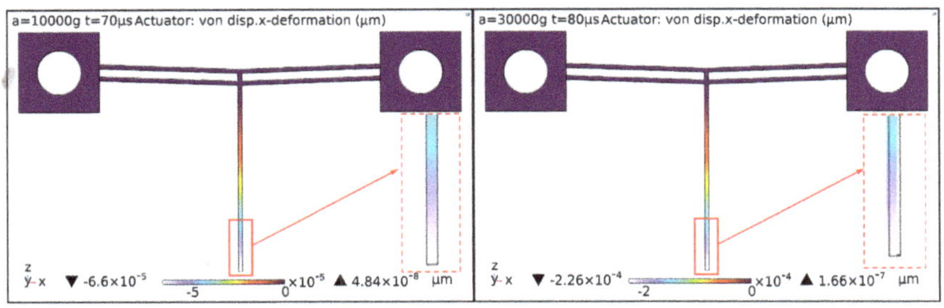

Figure 4. Deformation of the push lever under impact loads in the x direction: (**a**) 70 μs/10,000 g impact load; (**b**) 80 μs/30,000 g impact load.

The analysis results indicate that the metal-based V-shaped electrothermal actuator with a push lever can withstand high-overload impact loads with pulse widths ranging from 50 μs to 100 μs and amplitudes reaching tens of thousands of g. It can resist various impacts encountered by loitering munitions during the service phase without fracture or significant deformation, fully meeting the high-overload resistance requirements of the SAD of the loitering munition. In addition, simulation results demonstrate that the metal-based electrothermal actuator offers superior overload resistance performance compared to silicon-based electrothermal actuators [21].

4. Simulation Analysis of Performance-Influencing Factors of the Multi-Electrothermal Co-Actuation Device

4.1. Structural Influencing Factors

4.1.1. Number of the Actuating V-Beams

The V-beam is critical in actuation performance as the core component responsible for the electrothermal actuator's output displacement. The number of cascaded V-beams directly affects both the output performance and structural stability of the V-shaped electrothermal actuator. Figure 5 shows the displacement and driving force variation under various voltages with different numbers of V-beams.

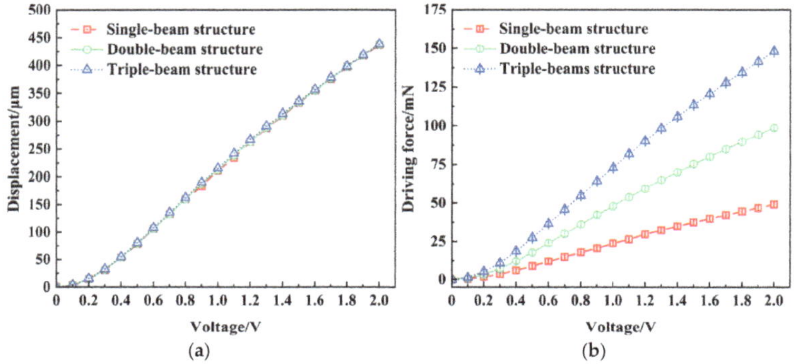

Figure 5. Influence of cascaded V-beams number on electrothermal actuator's performance: (**a**) displacement variation; (**b**) driving force variation.

The results indicate that the number of V-beams has a relatively minor effect on the output displacement. However, under the same voltage, the driving force of the electrothermal actuator greatly increases with the increase in the number of V-beams.

Therefore, the number of V-beams can be selected when designing V-shaped electrothermal actuators based on specific application requirements. A single-beam or double-beam structure with the advantage of a simple structure is preferred for the applications where a high driving force is not essential. In contrast, a multi-beam cascaded structure can be adopted for applications requiring a higher driving force. The double-beam structure of the V-shaped electrothermal actuator is adopted in the following section of this paper.

4.1.2. Air Gap

As a thermal insulation layer, the air gap can effectively regulate the heat conduction efficiency between the electrothermal actuator and its surrounding environment. Herein, the air gap refers explicitly to the gap between the bottom of the electrothermal actuator and the substrate. The temperature and displacement of the electrothermal actuator vary with the thickness of this gap, shown in Figure 6.

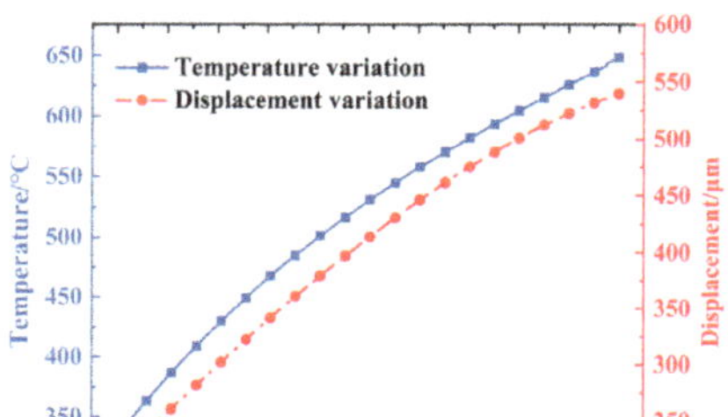

Figure 6. Influence of air gap thickness on V-shaped electrothermal actuator under 1.0 V.

At the same voltage, the electrothermal actuator's temperature and displacement increase with the air gap's enlargement. The reason is that the thermal conduction between the bottom of the actuator and the substrate is weakened as the air gap increases, which reduces heat dissipation through this path [22]. Consequently, more heat is retained within the electrothermal actuator, increasing internal temperature and a more significant output displacement. This effect has become more pronounced with larger air gap thicknesses.

In practical applications in SAD, the thickness of the air gap should be reasonably selected based on specific operating conditions. This ensures that the displacement is minimally affected by the friction between the bottom of the actuator and the substrate and helps to enhance the displacement under specific voltage or temperature constraints. Since the electrothermal actuator has a thickness of 200 μm and the flameproof slider has a thickness of 300 μm, an airgap of 100 μm is selected in the following section of this paper.

4.1.3. Type of Contact Surface

During the actuation process, the push lever of the actuator exerts an interaction force on the flameproof slider. Different types of contact surfaces result in different force distributions on the slider, thereby affecting its displacement. Finite element models were established for three contact surface types to analyze this effect. Figure 7 shows the simulation results of the slider's displacement under each contact condition.

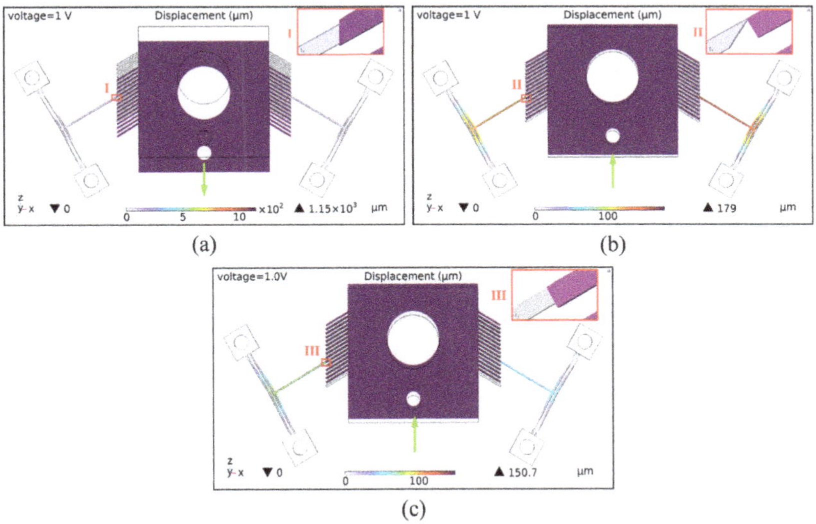

Figure 7. Displacement of the slider under different types of contact surfaces: (**a**) contact type I: vertical surface contact; (**b**) contact type II: point contact; (**c**) contact type III: inclined surface contact.

Figure 7a illustrates a "vertical surface contact" type, where the contact surface lies in the same plane as the slider's displacement direction. Upon actuation, the actuator deforms within the x-o-z plane, but due to the minimal friction force along the z direction at the contact surface, it is insufficient to drive the slider upward, resulting in displacement of 0 μm.

Figure 7b shows a "point contact" type, in which the slanted tip of the push lever (with an inclined surface) contacts an oblique ratchet tooth (whose tooth surface forms an angle with the z-axis). Under a 1 V voltage, the slider achieves an upward displacement of about 179 μm along the +z direction. Although the contact surface is very small, the force transfer path is more direct in this contact type.

Figure 7c presents an "inclined surface contact" type, where the straight tip of the push lever (with a horizontal surface) contacts an oblique ratchet tooth (whose tooth surface forms an angle with the z-axis). Under the same voltage, the slider achieves a displacement of about 150.7 μm. However, the direction of force transfer in this contact surface may be easily shifted.

In summary, the contact type II configuration enables the slider to perform better displacement output. This will be further validated through experimental studies, fabricating and testing prototypes with different contact configurations.

4.2. External Influencing Factors

4.2.1. External Load Force

During the electrothermal actuator's operation as a slider's driving source, external load forces directly affect the actuator's displacement performance. When the external load exceeds the maximum load capacity of the actuator, it may lead to irreversible deformation, malfunctioning, and reduced service life. Figure 8 shows the variation in the output displacement and the maximum thermal stress in the electrothermal actuator's y direction (displacement direction) at a steady state under different external load forces.

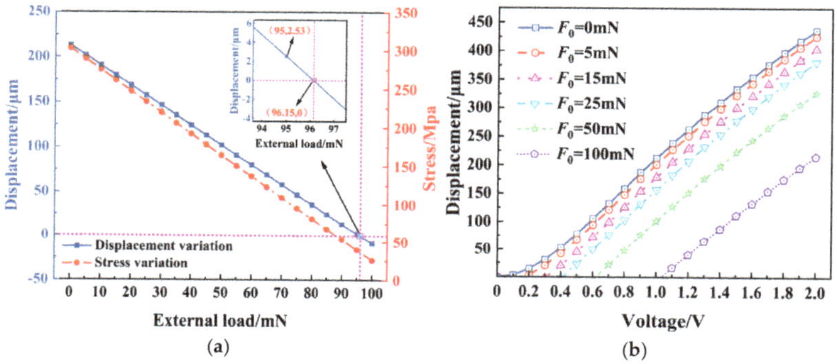

Figure 8. Influence of external loading on electrothermal actuator performance: (**a**) under 1 V; (**b**) under different voltages.

As shown in Figure 8a, the maximum load that the V-shaped electrothermal actuator can withstand is approximately 96.15 mN. The actuator's displacement and thermal stress exhibit a linear decreasing trend with the increase in external load. When the external load has become excessively large (though still within the material's allowable stress capacity), the direction of the displacement and the thermal stress in the y direction may reverse, indicating that the electrothermal actuator has entered a failure state.

According to Figure 8b, the displacement of the actuator under a constant voltage gradually decreases with an increasing external load force. Under smaller loads, the displacement is primarily determined by the voltage, while at larger loads, the load force plays a dominant role.

Overall, in the design process of electrothermal actuators, it is essential to rigorously control the loading conditions and ensure that the applied load remains within the actuator's maximum load capacity. For working environments where high external loads are unavoidable, the actuator's load tolerance can be improved by increasing the driving voltage or optimizing geometric parameters—such as increasing the number of cascaded V-beams.

4.2.2. Periodic Voltage

From the perspective of energy flow [23], there is an electrical–thermal–mechanical energy conversion throughout the drive's operation. Therefore, for the same actuator, the actuation performance ultimately depends on the amplitude of the voltage. However, for electrothermal actuators controlled by the periodic voltage to realize co-actuation driving, the frequency and phase of the periodic voltage may also have a different influence on the output performance. Figure 9 shows the variation curves of temperature and displacement with voltage amplitude for the electrothermal actuator in a steady state.

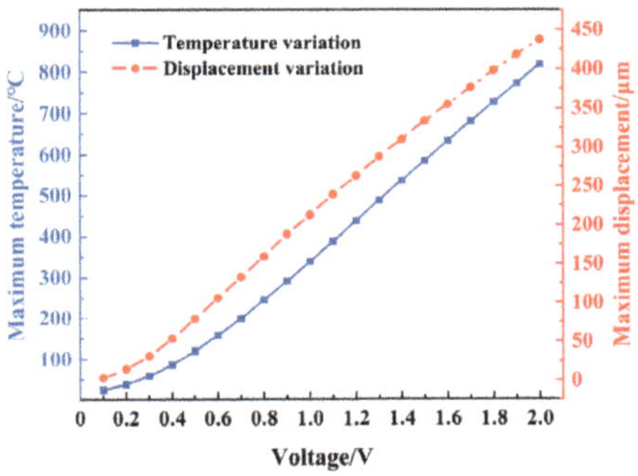

Figure 9. Variation in temperature and displacement with voltage amplitude.

According to Figure 9, once the dimensions of electrothermal actuator are determined, the temperature and displacement with voltage in a steady state are synchronized with a nonlinear increasing trend as the voltage increases. The temperature should be within the safe range, and the voltage amplitude can be appropriately increased to obtain a more significant displacement.

The total duration of the periodic voltage and simulation calculation is given as follows:

$$\begin{cases} V = \sin(2\pi f t + \varphi) \\ T_0 = \frac{T}{4} \end{cases} \tag{2}$$

where φ and f are the phase and the frequency of the periodic voltage, respectively, and T_0 is the total simulation time.

When voltage signals applied to the same actuate unit vary in frequency (from 10 Hz to 100 Hz) and phase (from 0 to π), the temperature and displacement of the electrothermal actuator are shown in Figure 10. According to Figure 10a, the temperature and displacement of the actuator will decrease with an increase in the frequency. It can be seen from Figure 10b that both the displacement and temperature of the actuator exhibit sinusoidal variations with the phase.

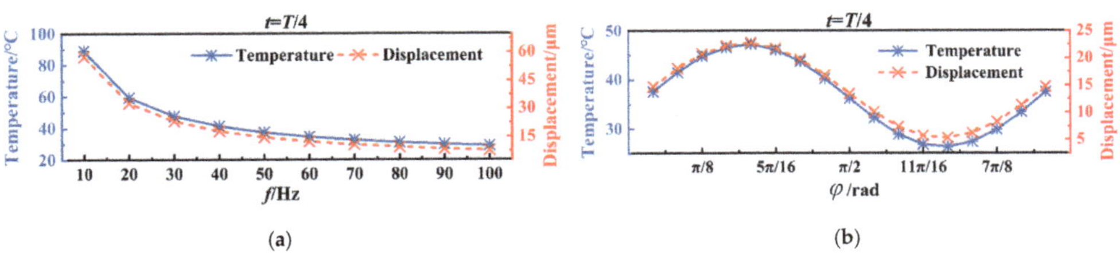

Figure 10. Influence of periodic voltage parameters on electrothermal actuator performance: (**a**) frequency variation, when $\varphi = \pi/2$; (**b**) phase variation, when f = 100 Hz.

As mentioned above, both the main-actuate unit and sub-actuate unit consist of one pair of V-shaped electrothermal actuators. When periodic voltages are applied to the two electrothermal actuators of the main-actuate unit or sub-actuate unit, the frequencies of the periodic voltages should be identical and the phases of the periodic voltages should be the same or have a phase difference of $\pi/2$, which ensures actuation synchronization among the actuate units and maintains the overall co-actuation of the device.

4.2.3. Heat Transfer

Heat transfer is an important process during the operation of the multi-electrothermal actuator co-actuation device that directly affects the performance and reliability of the fuze system. An electrothermal actuator essentially functions as a resistive heating element, and in normal operation, the heat comes mainly from the joule heat generated internally.

The heat conduction coefficient between the bottom of the actuator and the substrate is defined as S_t. In contrast, the heat convection coefficient between the surface of the actuator and the surrounding air is denoted as h_c. If the voltage of 1.0 V is applied to the electrothermal actuator, the temperature and displacement vary with the heat conduction coefficient and heat convection coefficient, shown in Figure 11.

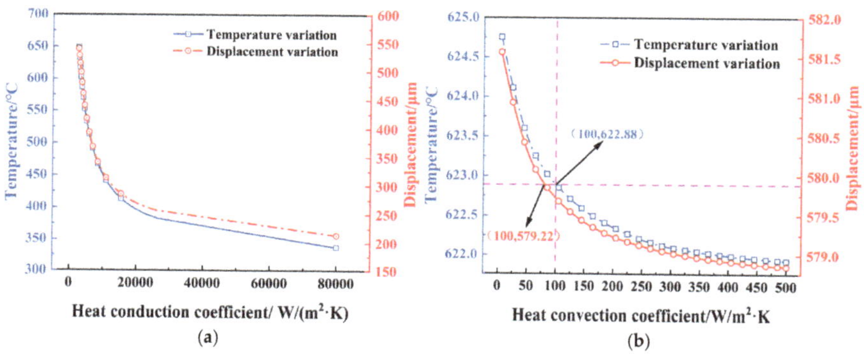

Figure 11. Influence of heat transfer on electrothermal actuator performance under 1.0 V: (**a**) heat conduction between the bottom of the actuator and the substrate; (**b**) heat convection between the surface of the actuator and the surrounding air.

As shown in Figure 11a, when S_t decreases, the heat dissipated through this part of the electrothermal actuator during the working process is reduced accordingly. As a result, the heat is retained more within the actuator, leading to a more significant displacement by the thermal expansion effect. Figure 11b shows that when h_c increases, both the temperature and displacement of the actuator change within a small range, indicating that the effect caused by heat convection is small.

Under typical working conditions of electrothermal actuator applied in SAD, the absolute temperature ΔT of the actuator's surface is generally around 300 °C~400 °C, and its thermal radiation power is much lower than that of high-temperature equipment (such as furnace walls, etc.). Assuming the absolute surface temperature of 400 °C, the heat radiation power is calculated as follows:

$$E_\varepsilon = \varepsilon\sigma\Delta T^4 = 145.152\ \mathrm{W/m^2} \tag{3}$$

where σ is the Stefan–Boltzmann constant, i.e., $\sigma = 5.67 \times 10^{-8}\ \mathrm{W{\cdot}m^{-2}{\cdot}K^{-4}}$, ε is the thermal emissivity, and the thermal emissivity of stainless-steel materials typically ranges from 0.1 to 0.3 [24].

Under the same conditions, when the heat convection coefficient h_c is 70 $\mathrm{W/m^2{\cdot}K}$, the heat dissipated through this part can be calculated as follows:

$$q_c = h_c\Delta T = 28{,}000\ \mathrm{W/m^2} \tag{4}$$

In summary, heat loss due to heat radiation accounts for less than 0.5% compared to that caused by heat convection. Moreover, the influence of heat convection is much smaller than that of heat conduction. Therefore, heat conduction is the primary method of heat transfer in the operation of electrothermal actuators, which is consistent with the results in reference [25]. Therefore, the heat conduction paths of the V-shaped electrothermal actuator are illustrated in Figure 12.

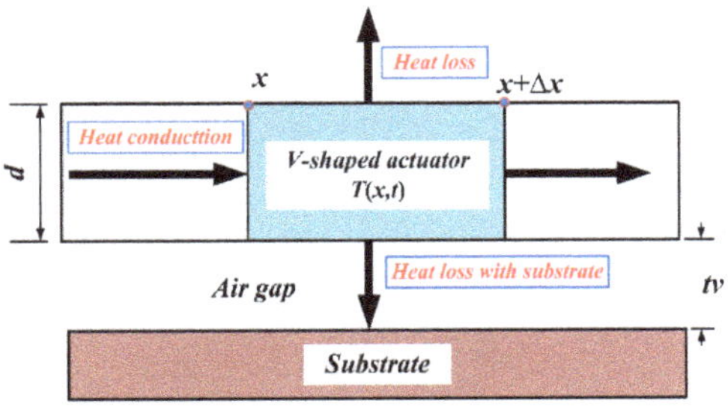

Figure 12. Schematic diagram of the heat conduction paths in the V-shaped electrothermal actuator.

4.2.4. Gas Damping

The multi-electrothermal co-actuation device for SAD proposed in this paper is primarily influenced by sliding film damping during operation, while the effect of squeezing film damping is negligible. The film thickness is set to t_v = 100 μm, and the total duration is defined as $T_1 = 10\ T = 20\pi/\omega$. A sinusoidal vibration velocity is applied to the actuator, expressed as

$$u(t) = sin(\omega t) \quad . (5)$$

Figure 13 shows the damping force and output displacement variation curves for the vibration of angular frequency. The damping force exhibits a clear nonlinear increasing trend with the increase in the angular frequency. However, the displacement remains stable from 200 μm to 210 μm. Even in high-frequency vibration conditions, the actuator maintains a minimum displacement of 200 μm. Therefore, the influence of damping on output performance can be considered negligible.

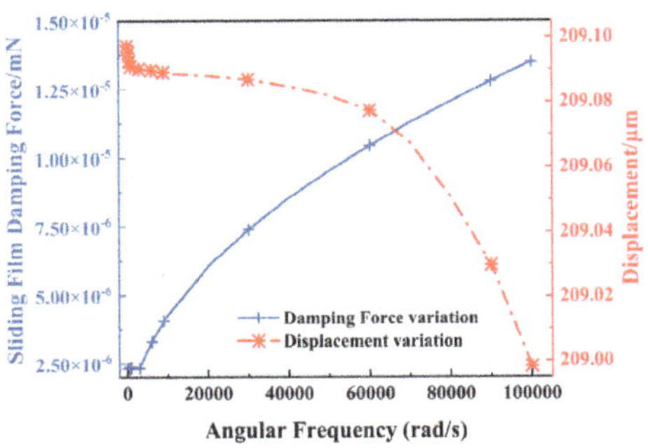

Figure 13. Influence of sliding film damping on electrothermal actuator performance.

In addition, damping can be further optimized to ensure the stability of the fuze system at low frequencies and generate a sufficient response in a specific high-frequency range to prevent accidental triggering, thus improving the overall reliability of the fuze system.

5. Performance Experiments of the Multi-Electrothermal Co-Actuation Device

This section processes the prototype to verify the simulation analysis results, and the performance of the multi-electrothermal co-actuation device under the effect of different influencing factors is experimentally investigated.

5.1. Experiments of Processing Technology Influence on the Actuator Performance

As illustrated in Figure 14a, the experiment platform comprises a microscope, a DC power supply, an ammeter, a high-speed camera with a monitor, and dedicated measurement software. The V-shaped electrothermal actuator, fabricated using laser processing technology, is anchored to the substrate at both ends, as shown in Figure 14b.

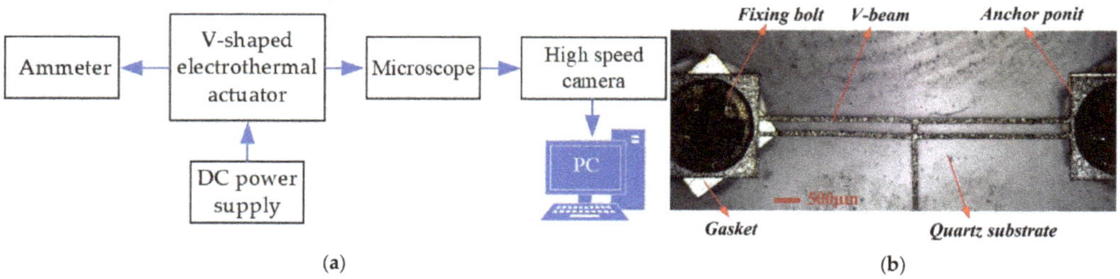

Figure 14. The experimental preparation: (**a**) the schematic block diagrams of experimental platform; (**b**) the prototype of the V-shaped electrothermal actuator.

In general, factors such as manufacturing techniques, processes, and tolerances affect the fabrication of the prototypes of the multi-electrothermal co-actuation device. In the case of laser processing technology, deviations in geometric parameters due to processing errors and various external factors during processing (such as temperature, external forces, etc.) may also influence the performance of the electrothermal actuator. This subsection analyzes the effects of temperature and external forces on the actuator's performance during the laser fabrication process.

5.1.1. Influence of Material Characteristics Changes on Actuator Performance

The impact of excessive temperature during laser processing—caused by insufficient coolant or other factors—on the performance of the electrothermal actuator is investigated here. Such overheating may change the metal's material properties or even oxidation. Figure 15 compares the actuator prototypes before and after oxidation.

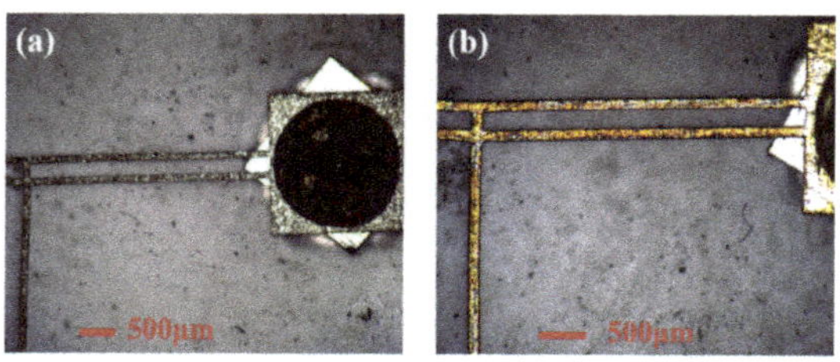

Figure 15. The prototypes of V-shaped electrothermal actuators: (**a**) before oxidation; (**b**) after oxidation.

Figure 16 shows the experimental results of the oxidized prototype in Figure 15b under different voltages. Figure 17 shows the displacement comparison between the unoxidized and oxidized prototype. The displacements of the actuator in the non-oxidized case at 0.5 V, 1.0 V, and 1.2 V are 38.834 μm, 133.524 μm, and 174.768 μm, respectively, which are significantly lower than those in the non-oxidized case under the same voltage. Oxidation of the material induced by the high temperature can change its coefficient of thermal expansion, which can adversely affect the actuator's output performance.

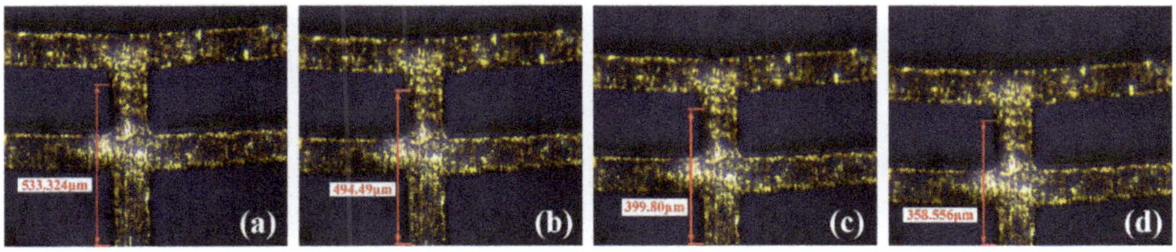

Figure 16. The deformation of the oxidized actuator under different voltages: (**a**) initial state; (**b**) under 0.5 V voltage; (**c**) under 1.0 V voltage; (**d**) under 1.2 V voltage.

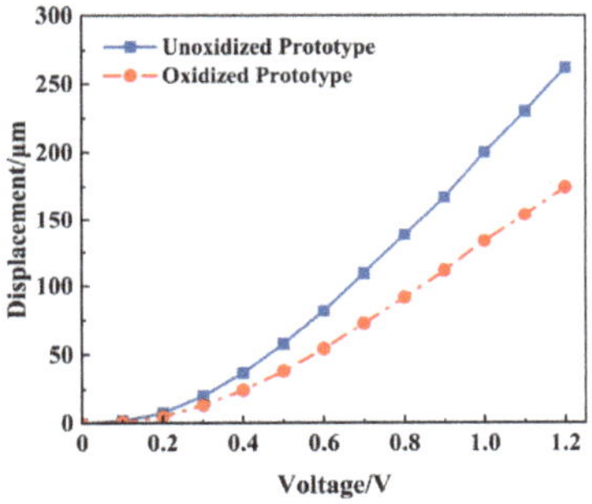

Figure 17. The experiment displacements of unoxidized and oxidized actuator prototypes.

5.1.2. Influence of Deformation Caused by External Forces on Actuator Performance

The prototype could be irreversibly deformed on the V-beam due to external force before use, as shown in Figure 18. The displacements of this deformed prototype under different voltages are tested and shown in Figure 19. The corresponding displacements of the deformed actuator under 0.5 V, 1.0 V, and 1.2 V are 29.79 μm, 114.88 μm, and 159.11 μm, respectively. Furthermore, the displacement comparison between the undeformed and deformed V-shaped beams is shown in Figure 20. According to Figures 18–20, under the same voltage, the displacement of the deformed prototype is reduced by at least 50% compared to the undeformed one. The deformation that occurs on the V-beam structure adversely affects its output, leading to a reduction in its output displacement.

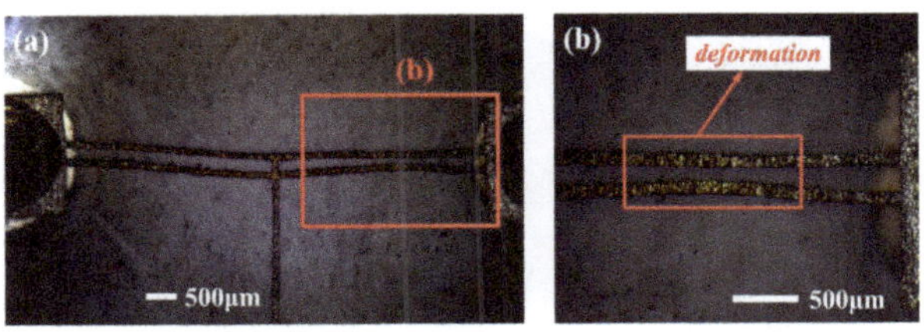

Figure 18. The prototypes of deformed V-shaped electrothermal actuators: (**a**) overall schematic; (**b**) detailed view.

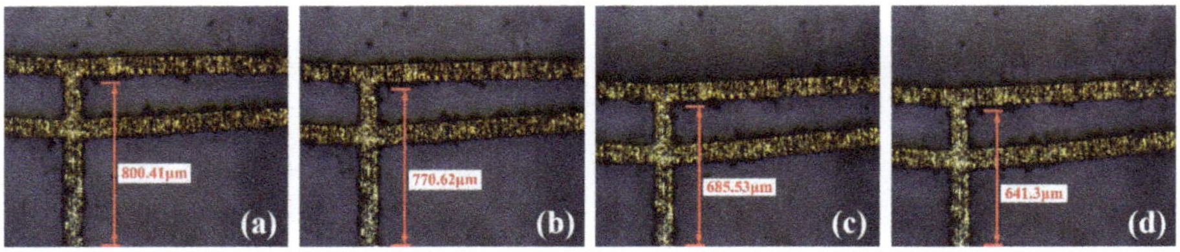

Figure 19. The displacements of the deformed actuator under different voltages: (**a**) initial state; (**b**) under 0.5 V voltage; (**c**) under 1.0 V voltage; (**d**) under 1.2 V voltage.

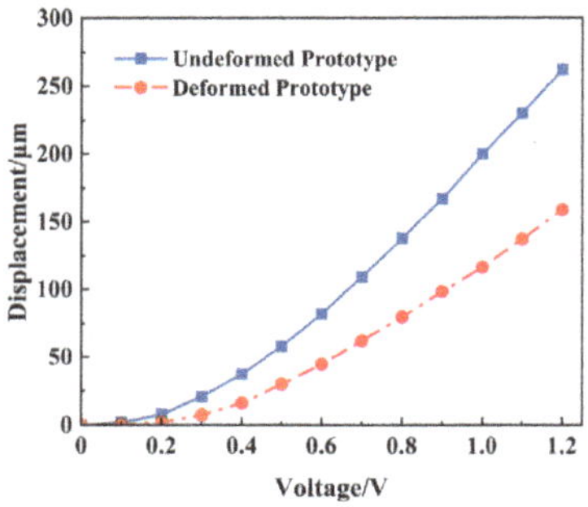

Figure 20. The experiment displacements of undeformed and deformed actuator prototypes.

From the above experimental results, both high temperature and external force will inhibit the output performance of the electrothermal actuator. Therefore, such influencing factors should be avoided as much as possible in the fabrication process to restore the ideal model and improve the reliability of the experimental results.

5.2. Experiment of Contact Surface Type Influence on the Slider Output Performance

The three prototypes corresponding to the simulation analysis in Section 4.1.3 were fabricated and numbered prototypes 1, 2, and 3, respectively, representing contact types I, II, and III.

Figure 21 shows the slider's displacement of different contact surface types when 1 V voltage is simultaneously applied to the main-actuate unit.

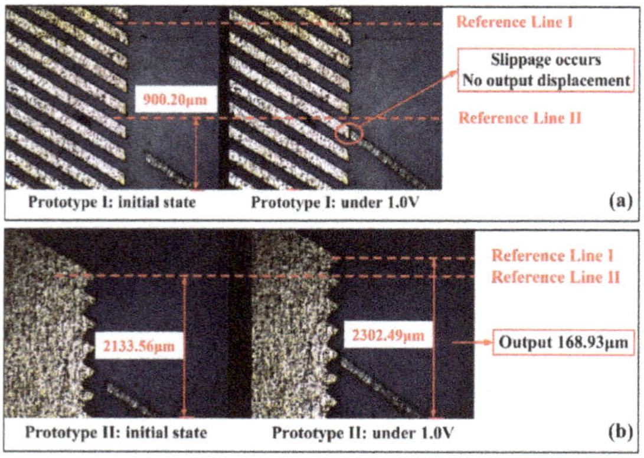

Figure 21. *Cont.*

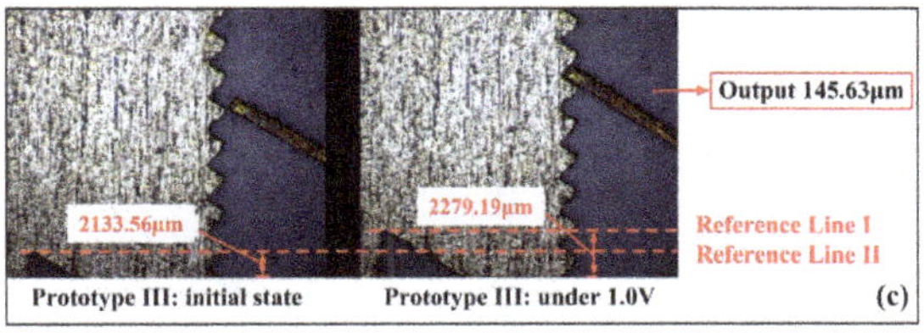

Figure 21. The slider displacement under different types of contact surfaces: (**a**) contact type I: vertical surface contact; (**b**) contact type II: point contact; (**c**) contact type III: inclined surface contact.

For prototype I, the displacement of the slider is almost 0 μm. Under the condition of prototype II, the slider outputs a displacement of about 168.93 μm when applying the voltage of 1.0 V, and under the same voltage, the slider in prototype III outputs a displacement of about 145.63 μm. This is consistent with the simulation results in Section 4.1.3, which verify the accuracy of the simulation analysis and the reasonability of choosing type II as the final contact surface of the device.

5.3. Experiment of High Overload Resistance Performance

The high overload resistance test is mainly accomplished with the help of the Marshall hammer impact test platform, which mainly consists of Marshall hammerheads, a piezoelectric accelerometer (CA-YD-111), a charge amplifier (YE5854A), and an oscilloscope, as shown in Figure 22.

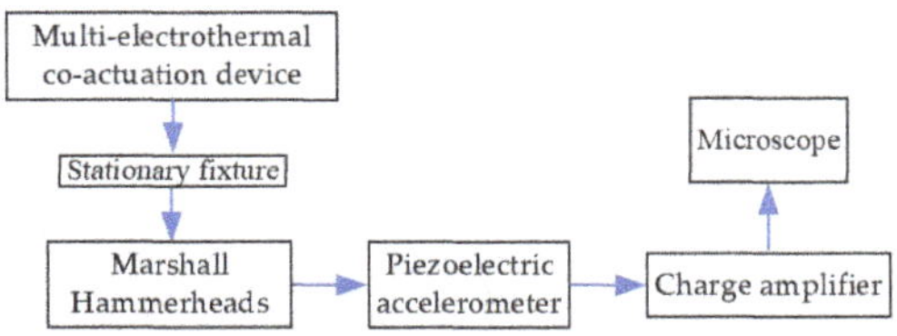

Figure 22. The Marshall hammer impact test platform.

Since loitering munition experiences relatively low launch overloads, the loitering munition is subjected to impact loads with pulse widths of 50 μs~500 μs and amplitudes of 5000 g~30,000 g, when it falls on hard targets such as concrete, steel plates, and cast steel during the loitering munition's service phase. Herein, the high overload resistance of the co-actuation device under the impact loads with pulse widths of 70 μs~80 μs and amplitudes of 6000 g~9000 g are tested in the next section.

5.3.1. Experiment Under Impact Load in Six Directions

Three impact loads are applied in six directions: $\pm x$, $\pm y$, $\pm z$; the conditions of the prototype after impact, focusing on whether fractures or other damages occur at the critical sections of the prototype, are observed. The specific data of each impact load and the state of the prototype after impact are recorded in Table 2. The maximum impact loads in the six directions of $\pm x$, $\pm y$, $\pm z$ are 10,600 g, 11,000 g, 12,000 g, 10,600 g, 12,200 g, and 12,600 g. It can withstand an impact load of more than 10,000 g without fracture.

Table 2. The experimental results under impact loading in six directions.

Prototype	Direction	Amplitude/g	Pulse Width/μs	Break or Not
Prototype 1	$+x$	8600/10,600/9200	71.79/91.79/73.29	Not
Prototype 2	$-x$	8400/9600/11,000	75.79/89.79/77.79	Not
Prototype 3	$+y$	7200/9600/12,000	79.79/75.79/73.79	Not
Prototype 4	$-y$	6400/10,600/9400	81.79/77.79/75.79	Not
Prototype 5	$+z$	8600/10,600/12,200	77.79/89.79/81.79	Not
Prototype 6	$-z$	7800/8400/12,600	71.79/71.79/75.79	Not

5.3.2. Experiment Under Impact Load at an Angle of 30° to the *y* Direction

The V-shaped electrothermal actuator in the co-actuation device is arranged at an inclined angle of 60° to the displacement direction (*y* direction) of the flameproof slider, while the deformation direction of the V-beam is at an angle of 30° to the displacement direction of the slider. Eight impact tests were carried out, and the deformation or fracture of the actuator was assessed when the prototype was subjected to an impact load at an angle of 30° to the y direction, as shown in Table 3.

Table 3. The experimental results under impact loading at an angle of 30° to the *y* direction.

Impact Load Number	Amplitude/g	Pulse Width/μs	Deformity or Not
Impact load 1	7800	71.79	Not
Impact load 2	8600	75.79	Not
Impact load 3	9200	73.29	Not
Impact load 4	10,600	77.79	Not
Impact load 5	11,000	77.79	Not
Impact load 6	12,000	73.79	Not
Impact load 7	12,200	69.79	Not
Impact load 8	13,800	67.79	A little

When the electrothermal actuator is subjected to an impact load with a pulse width of 73.79 μs and an amplitude of 12,000 g, it is hardly deformed, while at an impact load with a pulse width of 67.79 μs and an amplitude of 13,800 g, the push lever of the actuator undergoes a slight deformation, which can be observed through a microscope.

Figure 23 shows the state of the prototype before and after being subjected to impact load 8. The overall structure of the prototype remained intact after impact load 8, with no fracture of the slider, the V-beam, the anchor points, the push lever, or the connection between the anchor points and the V-beam (critical section). Some of the push levers of the electrothermal actuators showed a slight deformation, which could be observed through a microscope. Moreover, this deformation is within the tolerance of the material. The experimental results show that the device has excellent high overload resistance performance and strong bending resistance ability.

As shown in Figure 24, after being subjected to impact load 6, the slider achieved a displacement of 163.21 μm per step under 1.2 V. The prototype's actuation performance was almost unaffected by the impact load.

The above experimental results verified that the co-actuation device studied in this paper has excellent high overload resistance performance and bending deformation ability. It can resist various high overload impacts of loitering munition in the service phase and meets the requirements of high stability and reliability of the device under the application of the fuze system.

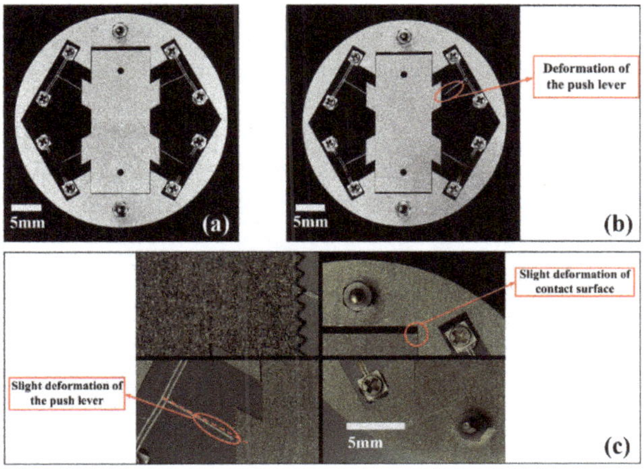

Figure 23. Comparison of the prototype before and after impact load 8 (13,800 g/67.79μs): (**a**) before impact load; (**b**) after impact load 8; (**c**) detailed view.

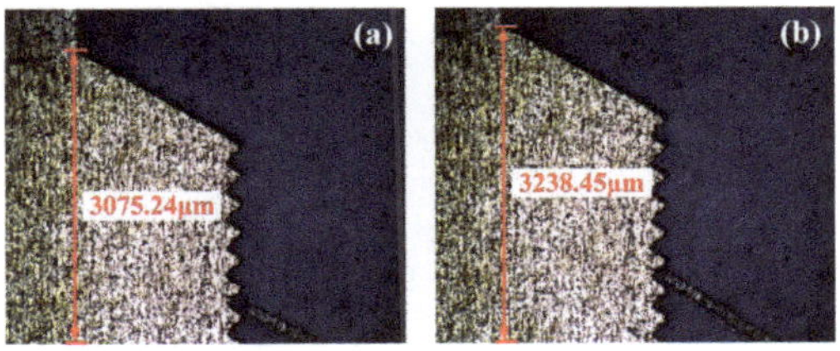

Figure 24. Displacement of the prototype after impact load 6 (12,000 g/73.79 μs): (**a**) initial state; (**b**) under 1.2 V voltage.

6. Conclusions

This study focuses on the overall structure, performance simulation analysis, and experimental verification of the multi-electrothermal co-actuation device, particularly emphasizing its application value in the fuze safety system of loitering munition. The main research conclusions of this paper are listed as follows:

1. Based on finite element simulation, the effects of influence factors such as the number of V-beams, air gaps, contact surface type, external load force, periodic voltage, heat transfer, and slip film damping on the multi-electrothermal co-actuation device's performance are investigated. The simulation results indicate that adequately optimizing structural parameters and driving conditions can significantly enhance the device's performance. In addition, the flameproof slider's contact surface II structure provides the best output displacement performance.
2. The multi-electrothermal co-actuation device has excellent high overload resistance performance. It maintains stable driving performance under high-impact load conditions, with only slight and recoverable deformation observed at the push lever, which meets the high overload resistance requirements of fuze applications.
3. Temperature changes and external forces during the laser fabrication process will influence the actuator's performance to different degrees. Experimental results show that changes in the material properties of the structural deformation lead to performance degradation, so strict control of influence factors in the fabrication process is necessary to ensure the stability of the driving performance.

Author Contributions: Conceptualization, Y.T. and X.W.; methodology, Y.T. and Y.D.; software, Z.G. and Y.D.; validation, Y.D. and Z.G.; formal analysis, Y.T. and Y.D.; investigation, Y.T. and X.W.; resources, X.W.; data curation, Y.D.; writing—original draft preparation, Y.D. and Z.G.; writing—review and editing, Y.T. and X.W.; visualization, Z.G.; supervision, Y.T. and X.W.; project administration, Y.T.; funding acquisition, Y.T. All authors have read and agreed to the published version of the manuscript.

Funding: This research was funded by the funding support from the Natural Science Foundation of Jiangsu Higher Education Institutions of China (No. 22KJA460005 and No. 24KJA460004).

Data Availability Statement: All data are presented within the paper.

Conflicts of Interest: The authors declare no conflicts of interest.

References

1. Ishihara, H.; Arai, F.; Fukuda, T. Micro mechatronics and microactuators. *IEEE/ASME Trans. Mechatron.* **1996**, *1*, 68–79. [CrossRef]
2. Algoos, Y.; Lenz, W.B.; Khan, F.; Younis, M.I. An electrostatically actuated MEMS microgripper with high amplification for precision manipulation. *Sens. Actuators A Phys.* **2024**, *378*, 115800. [CrossRef]
3. Duan, B.; Hu, H.; Lei, Y.; Chen, J. Design and experimental investigation of a novel stick-slip piezoelectric actuator based on V-shaped driving mechanism. *Mechatronics* **2023**, *95*, 103067. [CrossRef]
4. Nah, S.K.; Zhong, Z.W. A microgripper using piezoelectric actuation for micro-object manipulation. *Sens. Actuators A Phys.* **2007**, *133*, 218–224. [CrossRef]
5. Zhu, K.; Li, H.; Zhang, X.; Xu, T.; Li, S. A miniature electromagnetic linear actuator with 3D MEMS coil. *IEEE Electron Device Lett.* **2023**, *44*, 1732–1735. [CrossRef]
6. Sciberras, T.; Demicoli, M.; Grech, I.; Mallia, B.; Mollicone, P.; Sammut, N. Coupled finite element-finite volume multi-physics analysis of MEMS electrothermal actuators. *Micromachines* **2021**, *13*, 8. [CrossRef]
7. Potekhina, A.; Wang, C. Review of electrothermal actuators and applications. *Actuators* **2019**, *8*, 69. [CrossRef]
8. Jackson, N.; Sridharan, A.; Anand, S.; Baker, M.; Okandan, M.; Muthuswamy, J. Long-term neural recordings using MEMS based moveable microelectrodes in the brain. *Front. Neuroeng.* **2010**, *3*, 1302. [CrossRef]
9. Shen, Y.Y.; Song, R.C.; Tang, H.M.; Chen, Q.S.; Guan, P. The Dynamic Characteristics Research of MEMS V-Shaped SiO_2/Al Electrothermal Actuator. *Key Eng. Mater.* **2015**, *645*, 963–967. [CrossRef]
10. Nguyen, D.T.; Hoang, K.T.; Pham, P.H. Larger displacement of silicon electrothermal V-shaped actuator using surface sputtering process. *Microsyst. Technol.* **2021**, *27*, 1985–1991. [CrossRef]
11. Tecpoyotl-Torres, M.; Vargas-Chable, P.; Escobedo-Alatorre, J.; Cisneros-Villalobos, L.; Sarabia-Vergara, J. Z-Shaped Electrothermal Microgripper Based on Novel Asymmetric Actuator. *Micromachines* **2022**, *13*, 1460. [CrossRef] [PubMed]
12. Vargas-Chable, P.; Tecpoyotl-Torres, M.; Vera-Dimas, G.; Grimalsky, V.; Mireles García, J., Jr. Novel Electrothermal Microgrippers Based on a Rotary Actuator System. *Micromachines* **2022**, *13*, 2189. [CrossRef]
13. Ho, Y.N.; Michael, A.; Kwok, C.Y.; Pulikkaseril, C. Parallel in-plane electrothermal actuators. *J. Microelectromech. Syst.* **2024**, *33*, 305–307. [CrossRef]
14. Ostrow, S.A.; Lake, R.A.; Lombardi, J.P.; Coutu, R.A.; Starman, L.A. Fabrication process comparison and dynamics evaluation of electrothermal actuators for a prototype MEMS safe and arming devices. *Exp. Mech.* **2012**, *52*, 1229–1238. [CrossRef]
15. Li, X.; Zhao, Y.; Hu, T.; Bai, Y. Design of a high-speed electrothermal linear micromotor for microelectromechanical systems safety-and-arming devices. *Micro Nano Lett.* **2016**, *11*, 692–696. [CrossRef]
16. Wang, D.; Lou, W.; Feng, Y.; Wang, F. Design, Fabrication and Analysis of Threshold-Value Judging Mechanism in MEMS Safety and Arming Device. *Preprints* **2017**. [CrossRef]
17. Hu, T.; Fang, K.; Zhang, Z.; Jiang, X.; Zhao, Y. Design and research on large displacement bidirectional MEMS stage with interlock mechanism. *Sens. Actuators A Phys.* **2018**, *283*, 26–33. [CrossRef]
18. Tang, Y.; Guo, Z.; Ding, Y.; Wang, X. Design and Driving Characteristics of a Bidirectional Micro-Device Based on Multi-Electrothermal Co-Actuation. *Micromachines* **2025**, *16*, 487. [CrossRef]
19. Zhang, Z.; Yu, Y.; Liu, X.; Zhang, X. Dynamic modelling and analysis of V-and Z-shaped electrothermal microactuators. *Microsyst. Technol.* **2017**, *23*, 3775–3789. [CrossRef]
20. Qin, H.; Jiang, X. Low-g MEMS inertial switch with spiral spring for fuze. *J. Terahertz Sci. Electron. Inf. Technol.* **2024**, *22*, 459–466.
21. Chen, W.; Cao, Y.; Kong, X.; Xie, H.; Xi, Z. Research of Anti-High Overload Performance of U and V- Shaped Micro Electrothermal Actuators. *J. Shanghai Jiaotong Univ.* **2024**. [CrossRef]
22. Gwinn, J.P.; Webb, R.L. Performance and testing of thermal interface materials. *Microelectron. J.* **2003**, *34*, 215–222. [CrossRef]
23. Ding, Y.; Wang, X.; Wang, X. Design and analysis of large displacement bidirectional device based on multi-electrothermal co-actuation. *J. Phys. Conf. Ser.* **2024**, *2853*, 012033. [CrossRef]

24. Bityukov, V.K.; Khudak, Y.I.; Gusein-Zade, N.G. Analytical derivation of the Stefan–Boltzmann law for integral radiance from Planck's law for spectral radiance. *Bull. Lebedev Phys. Inst.* **2018**, *45*, 46–50. [CrossRef]
25. Zhang, Z.; Zhang, W.; Wu, Q.; Yu, Y.; Liu, X.; Zhang, X. A comprehensive analytical model and experimental validation of Z-shaped electrothermal microactuators. In *Recent Advances in Mechanism Design for Robotics, Proceedings of the 3rd IFToMM Symposium on Mechanism Design for Robotics, Aalborg, Denmark, 2–4 June 2015*; Springer: Cham, Switzerland, 2015; pp. 177–187.

micromachines

MDPI

Article

Enhancement of Quality Factors in a 6.5 GHz Resonator Using Mo/SiC Composite Microstructures

Binghui Lin [1], Yupeng Zheng [1], Haiyang Li [1], Yuqi Ren [1], Tingting Yang [1], Zekai Wang [1], Yao Cai [1], Qinwen Xu [1,*] and Chengliang Sun [1,2,*]

1 Hubei Key Laboratory of Electronic Manufacturing and Packaging Integration, The Institute of Technological Sciences, Wuhan University, Wuhan 430072, China; cyrus0108@whu.edu.cn (B.L.); zhengyupeng@whu.edu.cn (Y.Z.); lhy2022@whu.edu.cn (H.L.); renyuqi@whu.edu.cn (Y.R.); tingtingyang@whu.edu.cn (T.Y.); zekaiwang@whu.edu.cn (Z.W.); caiyao999@whu.edu.cn (Y.C.)

2 Wuhan Institute of Quantum Technology, Wuhan University, Wuhan 430072, China

* Correspondence: qinwen_xu@whu.edu.cn (Q.X.); sunc@whu.edu.cn (C.S.)

Abstract: This study addresses the critical challenge of lateral acoustic wave energy leakage in high-frequency film bulk acoustic resonators (FBARs) and elucidates the reflection mechanism of acoustic waves at acoustic reflection boundaries. Based on the theory of acoustic impedance mismatch, a novel Mo/SiC composite microstructure is designed to strategically establish multiple acoustic reflection boundaries along the lateral acoustic wave leakage paths. Finite element simulations reveal that SiC microstructures effectively suppress vibration amplitudes in non-resonant regions, thereby preventing acoustic wave leakage. By integrating Mo and SiC microstructures, the proposed composite structure significantly enhances the resonator's acoustic confinement and energy retention capabilities. A resonator incorporating this Mo/SiC composite microstructure is fabricated, achieving a series resonance frequency of 6.488 GHz and a remarkable quality factor (Q) of 310. This represents a substantial 51.2% improvement in Q compared to the basic FBAR, confirming the effectiveness of the proposed design in mitigating lateral acoustic wave leakage and enhancing resonator performance for high-frequency, low-loss applications. This work offers valuable insights into the design of next-generation RF resonators for advanced wireless communication systems.

Keywords: FBAR; SiCOI substrate; microstructure; quality factor

1. Introduction

With the rapid evolution of wireless communication technology, the global deployment of fifth-generation (5G) mobile networks is accelerating, while research efforts toward sixth-generation (6G) technology are intensifying [1,2]. The increasing demand for high-speed, low-latency communication necessitates the utilization of higher frequency bands, thereby driving the need for high-performance, low-loss RF devices [3,4]. In this context, FBARs play a crucial role in RF filtering applications due to their high electromechanical coupling efficiency and compact form factor. However, as operating frequencies continue to rise, conventional FBARs suffer from severe intrinsic losses in piezoelectric film and lateral acoustic wave leakage, leading to significant energy dissipation and reduced quality factors, which ultimately degrade device performance.

To address this challenge, researchers have focused on optimizing the quality of piezoelectric thin films and developing novel resonator architectures to suppress energy losses. However, the inherent limitations of piezoelectric film growth on non-single-crystalline substrates restrict the extent of quality improvement achievable through growth optimization and thermal treatment [5–11]. Moreover, while considerable efforts have been devoted to suppressing acoustic wave leakage along the longitudinal direction, relatively little attention has been given to mitigating lateral acoustic wave energy dissipation [12–19]. Few studies have reported the implementation of Bragg reflectors or frame structures in the lateral direction of resonators [20,21]. Their work provides a good guide for suppressing lateral acoustic wave energy leakage. However, these structures typically incorporate top electrode materials, which prevent the formation of a well-defined acoustic impedance mismatch boundary on the surface. The issue becomes increasingly critical at high frequencies, where reduced piezoelectric film thicknesses exacerbate intrinsic losses and shorter acoustic wavelengths enhance wave transmission, leading to substantial energy leakage from the resonator surface and edges.

In this work, we propose and investigate an innovative Mo/SiC composite microstructured FBAR that effectively suppresses lateral acoustic wave leakage by establishing multiple acoustic reflection boundaries. Through a combination of theoretical analysis, finite element simulations, and experimental validation, we demonstrate the ability of this microstructure to enhance acoustic confinement and improve resonator performance. Based on high-quality $Sc_{0.2}Al_{0.8}N$ grown on SiCOI substrates, the fabrication process is refined using an optimized thin-film bonding and transfer technique, ensuring high structural integrity and device reliability. The experimental results confirm that the Mo/SiC composite microstructured FBAR achieves a significant Q factor enhancement of 51.2%, paving the way for the next generation of high-frequency acoustic devices.

2. Structural Design and Simulation

FBARs typically consist of a sandwich structure comprising a top electrode layer, a piezoelectric layer, and a bottom electrode layer. When a voltage is applied across the electrodes, the d33 piezoelectric coefficient of the piezoelectric film excites the fundamental mode (thickness-extensional mode), while the d31 coefficient induces parasitic modes (lateral modes). Additionally, in practical applications, the metal routing paths of the top and bottom electrodes are not perfectly aligned along the z-axis. This misalignment at the resonator's edge results in lateral electric field components, which exacerbate lateral vibrations in the piezoelectric film. As illustrated in Figure 1a, these lateral vibrations generate acoustic waves that propagate along the x-axis within the film. While some waves are reflected at the interface between the top electrode layer and the passivation layer, a portion of the acoustic energy still escapes from the resonator region.

Acoustic waves undergo reflection at acoustic boundaries where impedance mismatches occur. By constructing multiple acoustic boundaries, it is possible to maximize acoustic energy retention. To explore the feasibility of local acoustic impedance modulation within the resonator structure, two candidate materials—molybdenum (Mo) and silicon carbide (SiC)—were evaluated based on their acoustic properties. Table 1 summarizes the density, longitudinal acoustic velocity, and calculated acoustic impedance of these materials. Mo exhibits a high acoustic impedance due to its high density and velocity, whereas SiC shows a lower impedance, resulting from its relatively lower density but much higher velocity. These significant differences in acoustic impedance compared to that of the surrounding materials suggest that both Mo and SiC can serve as effective localized reflectors or impedance modulation regions. Moreover, the superior mechanical strength of SiC and the ease of thin-film deposition for Mo render them highly suitable for the design of acoustic wave reflection structures. When introduced in the lateral plane of the resonator, they are expected to generate impedance mismatch boundaries that partially reflect the laterally propagating acoustic waves, thereby suppressing lateral energy leakage and improving Q performance.

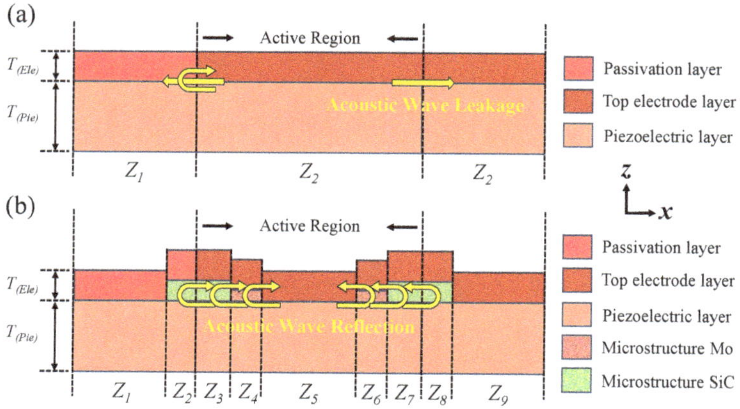

Figure 1. Schematic diagram of the acoustic boundary. (**a**) The basic FBAR has lateral energy leakage due to the lack of an acoustic reflection boundary. (**b**) The FBAR configured with Mo/SiC composite microstructures constructs an acoustic reflection boundary to suppress lateral energy leakage.

Table 1. Main material properties of Mo and SiC.

Material	Density (kg/m^3)	Velocity (m/s)	Acoustic Impedance (10^6 kg/m^2·s)	Mechanical Strength	Depositable
Mo	10,200	6260	63.9	Moderate	Yes
SiC	3200	12,000	38.4	High	Yes

This study proposes a Mo/SiC composite microstructure, as illustrated in Figure 1b, which depicts the acoustic boundary configuration of an FBAR incorporating Mo and SiC microstructures. Region 1, covered by the passivation layer, exhibits an effective acoustic impedance determined by the density and Young's modulus of both the passivation and piezoelectric layers. Regions 5 and 9, covered by the top electrode layer, have an effective acoustic impedance influenced by the density and Young's modulus of both the top electrode and piezoelectric layers. Regions 4 and 6, covered by both the top electrode and Mo microstructure, have an effective acoustic impedance determined by the density and Young's modulus of the top electrode, Mo microstructure, and piezoelectric layer. Similarly, regions 3, 7, and 8, covered by the top electrode and SiC microstructure, exhibit an effective acoustic impedance dependent on the top electrode, SiC microstructure, and piezoelectric layer. By introducing Mo and SiC microstructures within the active region, multiple acoustic reflection boundaries near the resonator's edges are created. This configuration facilitates multiple reflections of laterally leaked acoustic waves before they escape the resonator region, effectively suppressing lateral acoustic leakage. Furthermore, the high crystalline quality of single-crystal SiC prevents excessive energy dissipation within the material. Figure 2 shows the schematic top view of the basic FBAR and the FBAR configured with Mo/SiC composite microstructures. The Mo/SiC composite microstructures are distributed around the boundary of the resonance region.

This paper utilizes COMSOL® finite element simulation to optimize the design of high-frequency film bulk acoustic wave resonators. To validate this theoretical analysis, simulations compare the performance of two FBAR configurations: (1) a resonator without microstructures and (2) a resonator with microstructures. Two-dimensional simulation models are performed to investigate the resonance frequency and the propagation mode of the FBAR. The substrate, cavity, and sandwich stack (including the top electrode layer, piezoelectric layer, and bottom electrode layer) are constructed in the geometry module. The relative position relationship of each part and the width and thickness of each film layer are confirmed. The width of the two-dimensional model of the film bulk acoustic wave resonator is 200 μm, the thickness of the substrate is 10 μm, the width and depth of the cavity are 100 μm and 1.5 μm, respectively, the thicknesses of the top electrode layer and the bottom electrode layer are 65 nm each, and the thickness of the piezoelectric layer is 270 nm. The design parameters of the FBAR are shown in Table 2.

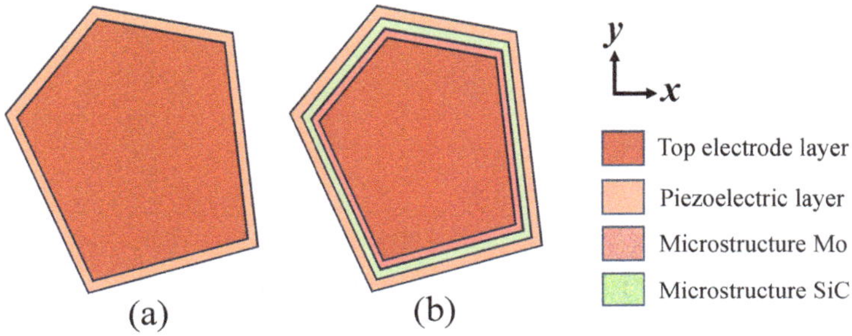

Figure 2. Schematic top view of (**a**) the basic FBAR and (**b**) the FBAR configured with Mo/SiC composite microstructures. The Mo/SiC composite microstructures are distributed around the boundary of the resonance region.

Table 2. Design parameters of the resonator.

Function	Material	Thickness
Passivation layer	AlN	50 nm
Top electrode	Mo	65 nm
Piezoelectric layer	$Sc_{0.2}Al_{0.8}N$	270 nm
Bottom electrode	Mo	65 nm
Insulation layer	SiO_2	100 nm
Microstructure 1	Mo	25 nm~200 nm
Microstructure 2	SiC	25 nm~200 nm

As shown in the results in Figure 3, at the resonant frequency, the vibration of the film is mainly concentrated in the resonant region above the cavity, forming a significant main resonant mode, and the acoustic wave energy is also concentrated in the resonant region. The basic FBAR shows that acoustic energy leaks continuously from the resonator to the non-resonant region, resulting in an average vibration displacement of approximately 3.98×10^{-5} μm. However, when 50-nanometer-thick SiC microstructures are introduced, the average vibration displacement in the non-resonant region is reduced to 1.51×10^{-5} μm, demonstrating significant suppression of lateral acoustic leakage.

The material properties and dimensions of the microstructures are critical for suppressing lateral acoustic energy leakage. The key acoustic properties (acoustic impedance) of the Mo and SiC materials used in the simulation are listed in Table 1. To effectively reflect and interfere with lateral acoustic waves, the width of the microstructure should be designed as an odd multiple of a quarter wavelength of the corresponding lateral wave mode at the resonance frequency [21]:

$$\text{width} = (2\text{m} + 1)\frac{\lambda}{4} \tag{1}$$

where m is an ideally positive integer.

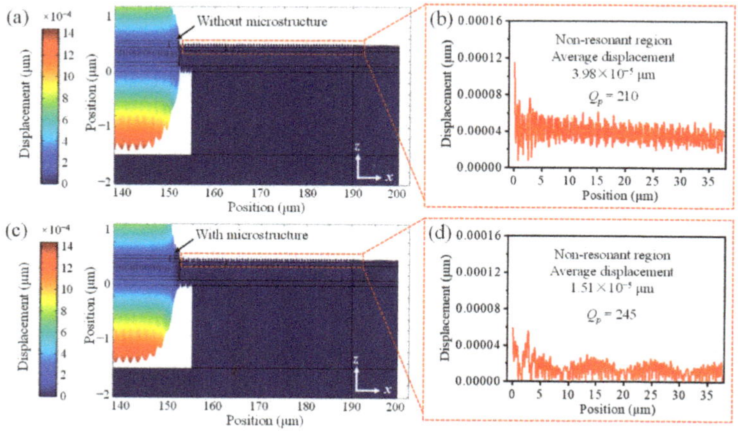

Figure 3. Mode shape and displacement curve of the resonator (**a**,**b**) without and (**c**,**d**) with SiC microstructures.

According to the simulated dispersion characteristics of the FBAR configuration, the wavelength of the lateral S_0 mode is 840 nm, and that of the A_0 mode is 810 nm. Therefore, the width of the microstructures should be designed as an odd multiple of 202.5 nm or 210 nm to achieve maximum constructive interference and reflection of lateral acoustic waves, thereby effectively suppressing energy leakage.

By incorporating Mo microstructure within the resonant region, local material thickness can be modified, thereby adjusting the effective density and Young's modulus to establish acoustic reflection boundaries. This section investigates the effects of Mo microstructure width and thickness on the performance of the resonator. As illustrated in Figure 4a, the quality factor (Qp) of the resonator is analyzed under various Mo microstructure configurations, with widths of 1 μm, 2 μm, 3 μm, and 4 μm, and thicknesses ranging from 25 nm to 200 nm in increments of 25 nm. The results indicate that the introduction of Mo microstructures leads to varying degrees of Qp enhancement. Overall, the Qp values of the resonator incorporating Mo microstructures range from 218 to 273, representing an improvement of 3.8% to 30% compared to the basic resonator (Qp = 210). This demonstrates the effectiveness of Mo microstructures in mitigating acoustic energy leakage in thin-film bulk acoustic resonators (FBARs). Furthermore, the Qp variation exhibits a nonlinear dependence on the width and thickness of the Mo microstructures. Excessively narrow Mo microstructures may lack sufficient mechanical strength, leading to deformation during resonance and ineffective reflection of high-frequency acoustic waves. Conversely, overly thick Mo microstructures can introduce modal conversion and stress concentration, exacerbating energy leakage. The most favorable enhancement in Qp is observed when the Mo microstructure width is within the range of 1.5 μm to 3.5 μm and the thickness is below 125 nm.

By integrating SiC microstructures within the resonant region, local material composition can be altered, allowing for modulation of the effective density and Young's modulus to create acoustic reflection boundaries. This section examines the impact of SiC microstructure width and thickness on resonator performance. As shown in Figure 4b, the Qp values of the resonator are evaluated under different SiC microstructure configurations, with widths of 1 μm, 2 μm, 3 μm, and 4 μm, and thicknesses varying from 25 nm to 200 nm in increments of 25 nm. The results reveal that the inclusion of SiC microstructures significantly enhances the Qp of the resonator. Specifically, the Qp values range from 245 to 297, corresponding to an improvement of 16.7% to 41.4% compared to the basic resonator

(Qp = 210). This confirms the effectiveness of SiC microstructures in reducing acoustic energy leakage in FBARs. The relationship between Qp and SiC microstructure dimensions follows a nonlinear trend. Due to its high mechanical strength, a sufficiently thick SiC microstructure minimizes structural deformation, thereby effectively suppressing lateral acoustic wave leakage and confining more acoustic energy within the resonant region. The most significant enhancement in Qp is achieved when the SiC microstructure width is between 2.5 μm and 4 μm, with a thickness exceeding 125 nm.

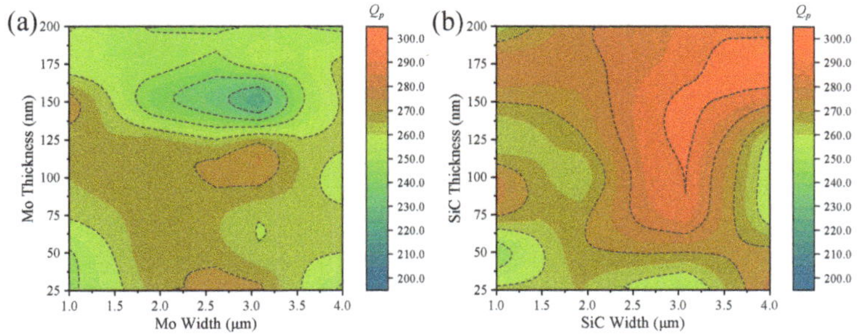

Figure 4. Trend in Qp variation in resonators with (**a**) Mo or (**b**) SiC microstructure dimensions.

However, a resonator incorporating microstructure exhibits significant spurious modes below the fundamental mode frequency. The frequency position of these spurious modes varies with the thickness of the microstructure. This phenomenon arises because when the microstructure is located within the resonant region, the layered thin-film structure in the thickness direction forms a local resonator. A portion of the electrical energy couples into localized standing waves, leading to resonance. Additionally, the increased thickness of the layered thin films in this region extends the acoustic wave propagation path, resulting in the appearance of spurious modes below the fundamental mode frequency. As the width of the microstructure increases, its proportion within the active region of the resonator grows, leading to an enhancement in the intensity of the spurious modes. Therefore, it is crucial to mitigate the impact of these spurious modes.

To minimize the impact of spurious modes on high-frequency communication bands, in addition to reducing the spurious mode frequency by increasing the film thickness, an alternative approach involves weakening the intensity of individual spurious modes through differential microstructures. As analyzed previously, the thickness of the microstructure affects the frequency position of the spurious modes, while its width influences their intensity. By dividing the microstructure into two sections with different thicknesses, it is possible to generate two spurious modes at distinct frequency positions, each with significantly reduced intensity. This design enhances the quality factor Qp of the resonator while ensuring that the spurious modes do not interfere with high-frequency communication bands.

According to previous analysis, a Mo microstructure with a thickness of 35 nm or a SiC microstructure with a thickness of approximately 150 nm exhibits a noticeable enhancement in Qp. Based on this, the present work proposes a Mo/SiC composite microstructure, where a 35-nanometer-thick Mo layer and a 150-nanometer-thick SiC layer are combined. The microstructure within the resonant region is divided into two sections along the width direction, and the width ratio of these sections is optimized to suppress spurious mode vibrations. The composite Mo/SiC microstructure within the active region configures with a fixed total width of 4 μm. As shown in Figure 5, eight design variations were considered by adjusting the relative widths of Mo and SiC: the width of Mo was set to 3.5 μm, 3 μm, 2.5 μm, 2 μm, 1.5 μm, 1 μm, 0.5 μm, and 0 μm, respectively, corresponding to SiC widths of 0.5 μm, 1 μm, 1.5 μm, 2 μm, 2.5 μm, 3 μm, 3.5 μm, and 4 μm. By evaluating the intensity of spurious modes across these eight configurations, the optimal structure was selected to achieve enhanced Q for the fundamental mode while minimizing spurious mode interference. When the Mo microstructure width was 3.5 μm, a prominent spurious mode appeared near 5.9 GHz. Conversely, when the SiC microstructure width was 4 μm, a noticeable spurious mode emerged around 4.5 GHz. However, when the Mo microstructure width was set to 1.5 μm and the SiC microstructure width to 2.5 μm, no significant spurious modes were observed near 5.9 GHz or 4.5 GHz. These results preliminarily validate the feasibility of mitigating spurious modes by segmenting the microstructure within the resonant region and optimizing the width ratio of the two sections.

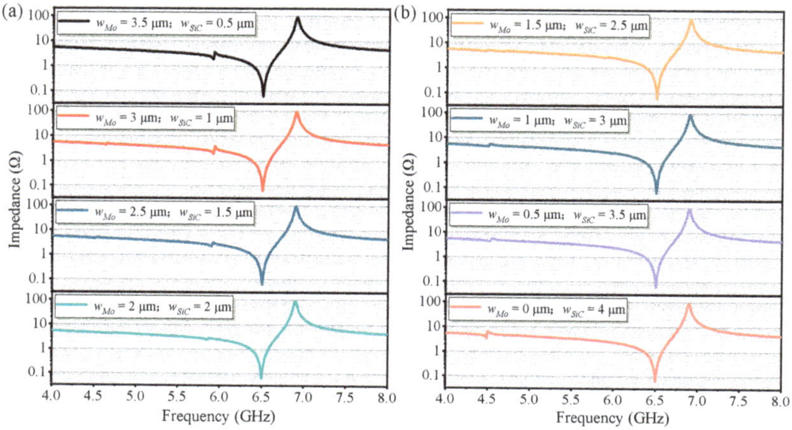

Figure 5. Simulated impedance curves of Mo/SiC composite microstructure resonators with different width ratios: (**a**) The width of Mo was set to 3.5 μm, 3 μm, 2.5 μm, and 2 μm, respectively. (**b**) The width of Mo was set to 1.5 μm, 1 μm, 0.5 μm, and 0 μm, respectively.

3. Thin Film Growth and Device Fabrication

Traditional FBAR fabrication relies on sacrificial SiO_2 layers embedded within the substrate. However, due to their amorphous nature and high surface roughness, these layers create suboptimal templates for piezoelectric film growth, leading to degraded film quality and increased energy losses. To overcome these limitations, we deposit a 270-nanometer-thick $Sc_{0.2}Al_{0.8}N$ film on a silicon-on-carbide (SiCOI) substrate.

As shown in Figure 6, to assess the surface morphology and crystalline orientation of the deposited films, scanning electron microscope (SEM) and X-ray diffraction (XRD) measurements, including rocking curve scans, were conducted. Benefiting from the SiC film, which provides an excellent template for the growth of piezoelectric films, the $Sc_{0.2}Al_{0.8}N$ film grown on the SiCOI substrate exhibits superior crystalline quality, reduced surface roughness, and a full-width at half maximum (FWHM) of 2.4° on the X-ray rocking curve, significantly lower than the 4.3° observed in conventional growth techniques. Moreover, no significant grain precipitation was observed on the film surface. This is very helpful in solving the severe intrinsic losses in piezoelectric film in high-frequency resonators.

The FBAR is fabricated using an optimized thin-film bonding and transfer process. The detailed fabrication process flow of the proposed FBAR is illustrated in Figure 7. Firstly, a 270-nanometer-thick $Sc_{0.2}Al_{0.8}N$ piezoelectric layer is deposited on the SiCOI substrate. A 65-nanometer-thick Mo is subsequently deposited and etched to form the bottom electrode. Then, a 100 nm-thick SiO_2 is deposited to cover the wafer by plasma-enhanced chemical vapor deposition. The 1.4-micrometer-thick poly-Si layer is deposited and patterned to form the sacrificial layer. Next, a thick SiO_2 support layer is deposited to fulfill the step height, and the wafer is flattened by using chemical mechanical polishing.

After that, wafer bonding was performed using gold as the bonding medium between the two substrates. The Si substrate and SiO_2 film are removed by wafer thinning and wet etching. In the next step, the SiC film is patterned and dry-etched to form a 150-nanometer-thick SiC microstructure. Subsequently, the piezoelectric layer is etched to expose the pad of the bottom electrode, followed by the deposition and etching of the 35-nanometer-thick Mo microstructure and the 65-nanometer-thick top electrode. After a 1-micrometer-thick Au film is deposited and patterned to define the contact pad, the 50-nanometer-thick passivation layer AlN covers the active region. At last, the release holes are etched, and the air cavity is released by xenon difluoride (XeF_2), thus finishing the fabrication of the proposed FBAR.

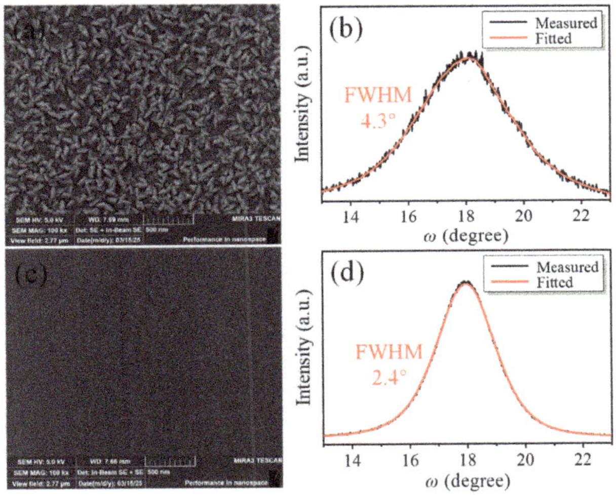

Figure 6. SEM images and rocking curves of 270-nanometer-thick $Sc_{0.2}Al_{0.8}N$ films grown on (**a**,**b**) SiO_2 film and (**c**,**d**) SiCOI substrate.

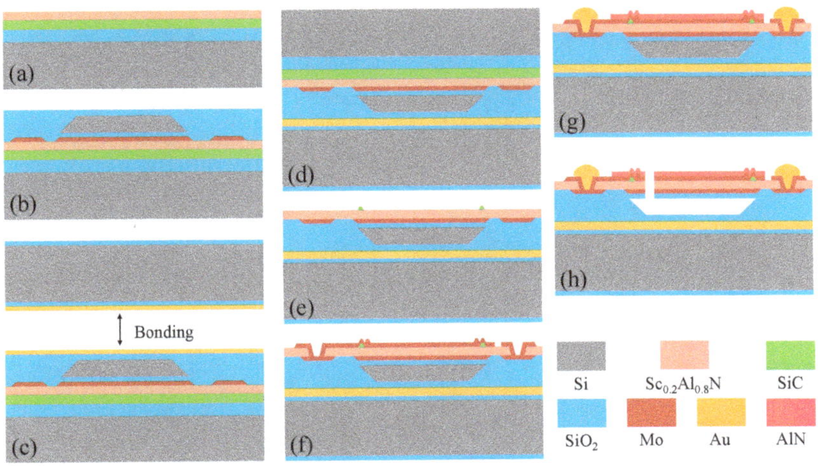

Figure 7. The main process step for the fabrication of FBAR. (**a**) Piezoelectric layer deposition. (**b**) Bottom electrode, sacrificial layer, and support layer fabrication. (**c**) Bonding material deposition. (**d**) Wafer bonding. (**e**) Si substrate and SiO_2 film removal, and SiC microstructure fabrication. (**f**) Mo microstructure and top electrode fabrication. (**g**) Contact pad fabrication. (**h**) Release holes are etched, and a cavity is formed.

4. Measurement and Results Analysis

Figure 8a,b show the SEM cross-sectional view and top view of the basic FBAR. In the top-view magnified observation of the active region edge, the boundaries of the cavity, bottom electrode, and top electrode can be clearly identified. Figure 8c,d show the SEM top view and the cross-sectional view of the optimized FBAR. The Mo/SiC composite microstructures are arranged along the active region edge, with the first-stage Mo structure measuring 1.5 μm in width and the second-stage SiC structure measuring 2.5 μm in width.

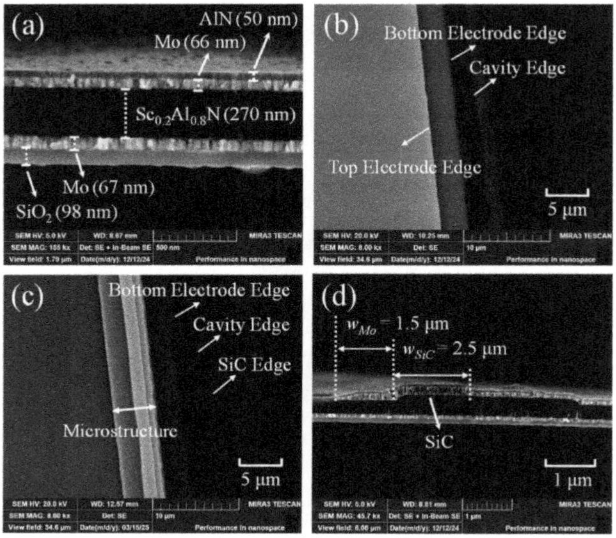

Figure 8. SEM images characterization: (**a**) The cross-sectional view of the basic FBAR. (**b**) Top view of the active region edge of the basic FBAR. (**c**) Top view of the Mo/SiC composite microstructure of the optimized FBAR. (**d**) The cross-sectional view of the Mo/SiC composite microstructure.

The fabricated FBARs were measured using a radio frequency (RF) S-parameter measurement system. The system primarily consists of a semi-automatic RF probe station, a vector network analyzer, and GSG probes. The effective electromechanical coupling coefficient (k^2_{eff}) and quality factor (Q) can be calculated by the following formulas [22,23]:

$$k^2_{eff} = \frac{\pi^2}{4} \cdot \left(\frac{f_s}{f_p}\right) \cdot \left(\frac{f_p - f_s}{f_p}\right) \tag{2}$$

$$Q(f) = \frac{2\pi f \tau(f) |S_{11}|}{1 - |S_{11}|^2} \tag{3}$$

where $\tau(f)$ is the group delay of S_{11}.

Figure 9 shows the measured impedance curves of high-frequency FBARs with Mo/SiC composite microstructures of varying width ratios within the active region, aiming to analyze the variation of spurious modes under different configurations. As the width of Mo in the composite microstructure decreases, the spurious mode near 5.9 GHz gradually becomes suppressed. Correspondingly, as the width of SiC increases, the spurious mode near 4.5 GHz becomes more prominent. These experimental results are consistent with the simulation conclusions. To avoid the generation of spurious modes in the resonator, a balance between the widths of Mo and SiC in the composite microstructure is required.

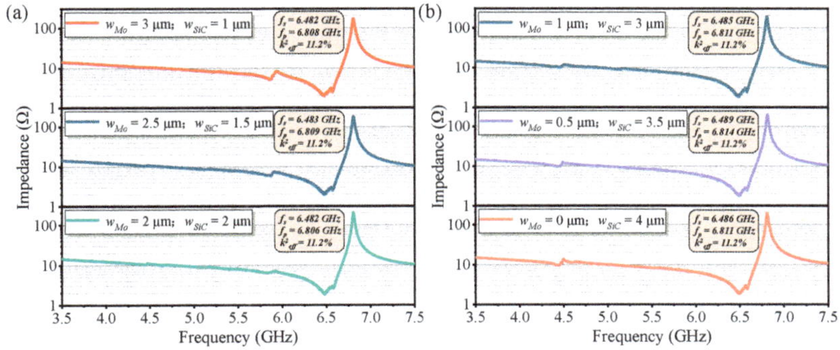

Figure 9. Measured impedance curves of Mo/SiC composite microstructure resonators with different width ratios: (**a**) The width of Mo was 3 μm, 2.5 μm, and 2 μm, respectively. (**b**) The width of Mo was 1 μm, 0.5 μm, and 0 μm, respectively.

Figure 10 shows the measured impedance curves and Qp value of the basic resonator and the optimized FBAR. The basic FBAR exhibits a series resonance frequency (*fs*) of 6.493 GHz and a parallel resonance frequency (*fp*) of 6.891 GHz, corresponding to an effective electromechanical coupling coefficient (k^2_{eff}) of 13.4%. The quality factor at the parallel resonance frequency (Qp) is 205. The optimized FBAR (w_{Mo} = 1.5 μm, w_{SiC} = 2.5 μm) exhibits an *fs* of 6.488 GHz and an *fp* of 6.813 GHz. The Qp value reaches 310, and no significant spurious modes are observed. The Qp value of the optimized FBAR was improved by 51.2% compared to the basic FBAR. Impedance analysis reveals that the effective electromechanical coupling coefficient decreases from 13.4% in the basic FBAR to 11.2% in the optimized FBAR due to the introduction of additional boundary conditions. This is attributed to the presence of Mo/SiC composite microstructures within the active region, which couple a portion of the electrical energy into localized standing waves, resulting in the formation of spurious modes and a reduction in the electromechanical conversion efficiency of the fundamental mode.

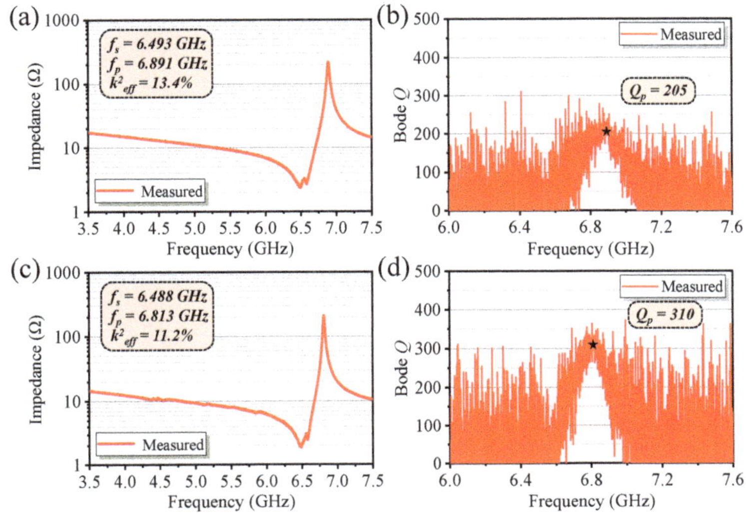

Figure 10. Measured impedance curves and Qp value of the resonator: (**a**,**b**) Basic resonator; (**c**,**d**) Optimized FBAR.

The basic and optimized FBARs were characterized, and the Qp values of five devices from each type are summarized in Table 3. The mean and standard deviation were also calculated accordingly. The results indicate a significant improvement in Qp values for the optimized FBARs after structural optimization. Optimized microstructure dimensions prevent spurious modes, ensuring improved device performance. The quality factor (Qp) of the optimized resonator reaches 310, marking a 51.2% increase from the basic Qp of 205.

Table 3. The Qp values of five basic FBARs and five optimized FBARs, along with their mean and standard deviation.

Type	Device 1	Device 2	Device 3	Device 4	Device 5	Mean Value	Standard Deviation
Basic FBAR	201	201	202	202	205	202.2	1.64317
Optimized FBAR	302	304	309	309	310	306.8	3.56371

In summary, by introducing Mo/SiC composite microstructures within the active region to suppress lateral acoustic energy leakage, the Qp value of the resonator was successfully improved. Meanwhile, to address the spurious modes induced by microstructures in the active area, an optimized width configuration of the Mo/SiC composite structure was proposed, effectively eliminating the spurious responses. While existing studies focus on pushing FBARs toward higher frequencies, the Qp values of high-frequency FBARs are generally limited (typically below 300) due to intrinsic losses in the piezoelectric film and severe lateral acoustic energy leakage, as summarized in Table 4. In comparison, this work demonstrates a high-frequency, low-loss FBAR with a series resonance frequency of 6.488 GHz and a Qp value as high as 310, enabled by the growth of high-quality $Sc_{0.2}Al_{0.8}N$ piezoelectric film on the SiC layer and the implementation of SiC-based composite microstructures.

Table 4. Comparison with other works.

Piezoelectric Material and Thickness	Substrate	Frequency	Quality Factor	Reference
$Sc_{0.15}Al_{0.85}N$	Si	2.15 GHz	410	[24]
$Sc_{0.3}Al_{0.7}N$	SOI	2.93 GHz	210	[25]
$Sc_{0.2}Al_{0.8}N$	Si/SiO_2	3.094 GHz	209	[26]
$Sc_{0.2}Al_{0.8}N$	Si/SiO_2	3.39 GHz	193	[27]
$Sc_{0.2}Al_{0.8}N$	Si/SiO_2	3.521 GHz	255.4	[28]
$Sc_{0.2}Al_{0.8}N$	Si/SiO_2	4.001 GHz	202	[29]
$Sc_{0.2}Al_{0.8}N$	Si/SiO_2	4.235 GHz	287.4	[30]
$Sc_{0.3}Al_{0.7}N$	Si	4.38 GHz	194	[31]
$Sc_{0.2}Al_{0.8}N$	Si	5.611 GHz	193	[32]
AlN	Si/SiO_2	6.84 GHz	135	[33]
$Sc_{0.2}Al_{0.8}N$	SiC	6.488 GHz	310	This work

5. Conclusions

This work successfully demonstrates a high-frequency, low-loss FBAR by suppressing lateral acoustic wave leakage using Mo/SiC microstructures. Through theoretical analysis, numerical simulations, and experimental validation, a high-frequency, low-loss FBAR with a quality factor of 310 and a series resonance frequency of 6.488 GHz was successfully demonstrated. By optimizing the piezoelectric film properties and refining the resonator structure, the dissipation of resonance energy at high frequencies was effectively mitigated, leading to a significant enhancement in the quality factor. The findings provide both theoretical foundations and experimental support for the advancement of FBAR technology in RF applications, facilitating the commercialization of high-frequency acoustic filters.

Author Contributions: Conceptualization, B.L.; methodology, Y.Z. and B.L.; software, H.L.; validation, Y.R. and T.Y.; formal analysis, Z.W.; investigation, Y.C.; writing—original draft preparation, B.L.; writing—review and editing, B.L., Q.X., and C.S.; supervision, Q.X. and C.S. All authors have read and agreed to the published version of the manuscript.

Funding: This research was supported by the National Key R&D Program of China under Grant No. 2023YFB3211301, the Wuhan Science and Technology Major Project under Grant No. 2024051102020828, the Young Scientists Fund of the National Natural Science Foundation of China under Grant No. 62204177, and the Large-Scale Instrument and Equipment Sharing Foundation of Wuhan University.

Data Availability Statement: The data presented in this study are available in this article.

Acknowledgments: The authors would like to thank the Core Facility of Wuhan University (XRD) and the School of Power and Mechanical Engineering of Wuhan University (SEM) for access to analytical equipment.

Conflicts of Interest: The authors declare no conflicts of interest. The funders had no role in the design of this study, in the collection, analysis, or interpretation of data, in the writing of the manuscript, or in the decision to publish the results.

References

1. Shafi, M.; Molisch, A.F.; Smith, P.J.; Haustein, T.; Zhu, P.; De Silva, P.; Tufvesson, F.; Benjebbour, A.; Wunder, G. 5G: A tutorial overview of standards, trials, challenges, deployment, and practice. *IEEE J. Sel. Areas Commun.* **2017**, *35*, 1201–1221. [CrossRef]
2. Ahad, A.; Tahir, M.; Sheikh, M.A.; Ahmed, K.I.; Mughees, A.; Numani, A. Technologies trend towards 5G network for smart health-care using IoT: A review. *Sensors* **2020**, *20*, 4047. [CrossRef] [PubMed]
3. Gustavsson, U.; Frenger, P.; Fager, C.; Eriksson, T.; Zirath, H.; Dielacher, F.; Studer, C.; Parssinen, A.; Correia, R.; Matos, J.N.; et al. Implementation challenges and opportunities in beyond-5G and 6G communication. *IEEE J. Microw.* **2021**, *1*, 86–100. [CrossRef]
4. Giribaldi, G.; Colombo, L.; Simeoni, P.; Rinaldi, M. Compact and wideband nanoacoustic pass-band filters for future 5G and 6G cellular radios. *Nat. Commun.* **2024**, *15*, 304. [CrossRef]
5. Liu, Y.; Sun, K.; Ma, J.; Yu, Z.; Lan, Z. Design and fabrication of temperature-compensated film bulk acoustic resonator filter based on the stress compensation effect. *Coatings* **2022**, *12*, 1126. [CrossRef]
6. Chang, Y.; Hui, Z.; Wang, X.; Qu, H.; Pang, W.; Duan, X. Dual-mode gas sensor composed of a silicon nanoribbon field effect transistor and a bulk acoustic wave resonator: A case study in freons. *Sensors* **2018**, *18*, 343. [CrossRef]
7. Khan, S.; Shahid, M.; Mahmood, A.; Shah, A.; Ahmed, I.; Mehmood, M.; Aziz, U.; Raza, Q.; Alam, M. Texture of the nanocrystalline AlN thin films and the growth conditions in DC magnetron sputtering. *Prog. Nat. Sci. Mater. Int.* **2015**, *25*, 282–290. [CrossRef]
8. Choudhary, R.K.; Mishra, P.; Biswas, A.; Bidaye, A.C. Structural and optical properties of aluminum nitride thin films deposited by pulsed DC magnetron sputtering. *Int. Sch. Res. Not.* **2013**, *2013*, 759462. [CrossRef]
9. Han, J.; Cui, B.; Xing, Y.; Li, T.; Zhao, J.; Cao, X.; Zhang, Y.; Zhang, B. Influence of nitrogen flow ratio on the optical property of AlN deposited by DC magnetron sputtering on Si (100) substrate. *Micro Nano Lett.* **2020**, *15*, 556–560. [CrossRef]
10. Li, M.; Chen, B.; Xie, J.; Song, W.; Zhu, Y. Effects of post-annealing on texture evolution of sputtered ScAlN films. In Proceedings of the 2020 IEEE International Ultrasonics Symposium (IUS), Las Vegas, NV, USA, 7–11 September 2020; IEEE: New York, NY, USA, 2020; pp. 1–3.
11. Liu, B.; Gao, J.; Wu, K.; Liu, C. Preparation and rapid thermal annealing of AlN thin films grown by molecular beam epitaxy. *Solid State Commun.* **2009**, *149*, 715–717. [CrossRef]
12. Gao, C.; Jiang, Y.; Zhang, L.; Liu, B.; Zhang, M. An investigation on efficient acoustic energy reflection of flexible film bulk acoustic resonators. *Nanotechnol. Precis. Eng.* **2018**, *1*, 129–132.
13. Shi, L.; Xuan, W.; Zhang, B.; Dong, S.; Jin, H.; Luo, J. Numerical investigation of phononic crystal based film bulk acoustic wave resonators. *Nanomaterials* **2021**, *11*, 2547. [CrossRef] [PubMed]
14. Ohara, R.; Yanase, N.; Yasumoto, Y.; Kawase, M.; Masuko, S.; Ohno, T.; Sano, K. P2G-4 Suppression of Acoustic Energy Leakage in FBARs with Al Bottom Electrode: FEM Simulation and Experimental Results. In Proceedings of the 2007 IEEE Ultrasonics Symposium Proceedings, New York, NY, USA, 28–31 October 2007; IEEE: New York, NY, USA, 2007; pp. 1657–1660.

15. Hu, Y.; Dong, B.; Lei, L.; Wang, Z.; Ruan, S. High-order harmonic film bulk acoustic resonator based on a polymer reflector. *Sensors* **2022**, *22*, 7439. [CrossRef] [PubMed]
16. Xu, T.; Zou, Y.; Huang, X.; Wu, J.; Wu, S.; Liu, Y.; Xu, X.; Liu, F. The effect of defect charge and parasitic surface conductance on aluminum nitride RF filter circuit loss. *Micromachines* **2023**, *14*, 583. [CrossRef] [PubMed]
17. Hu, W.; Wang, Y.; Feng, D.; Zou, J. Impact of a Trap-Rich Layer on the Performance of D-BAW Resonators. In Proceedings of the 2024 IEEE Ultrasonics, Ferroelectrics, and Frequency Control Joint Symposium (UFFC-JS), Taipei, Taiwan, 22–26 September 2024; IEEE: New York, NY, USA, 2024; pp. 1–4.
18. Fouladi, S.; Wu, J.; Zou, Q.; Barfknecht, A.; Vats, A.; Nasresfahani, M.; Martin, S.; Feld, D.A. FBAR resonators fabricated on insulating substrates with improved RF and nonlinear performance. In Proceedings of the 2019 IEEE International Ultrasonics Symposium (IUS), Glasgow, UK, 6–9 October 2019; IEEE: New York, NY, USA, 2019; pp. 88–92.
19. Qu, Y.; Gu, X.; Zou, Y.; Wen, Z.; Cai, Y.; Soon, B.W.; Liu, Y.; Sun, C. Aluminum Nitride Based Film Bulk Acoustic Resonator With Anchor Column Structure. *J. Microelectromechanical Syst.* **2023**, *32*, 157–163. [CrossRef]
20. Segovia-Fernandez, J.; Yen, E.T.-T. Resonant Confiners for Acoustic Loss Mitigation in Bulk Acoustic Wave Resonators. In Proceedings of the 2023 IEEE 36th International Conference on Micro Electro Mechanical Systems (MEMS), Munich, Germany, 15–19 January 2023; IEEE: New York, NY, USA, 2023; pp. 165–168.
21. Nguyen, N.; Johannessen, A.; Rooth, S.; Hanke, U. A design approach for high-Q FBARs with a dual-step frame. *IEEE Trans. Ultrason. Ferroelectr. Freq. Control* **2018**, *65*, 1717–1725. [CrossRef]
22. Neculoiu, D.; Vasilache, D.; Konstantinidis, G.; Morosanu, C.; Stavrinidis, A.; Kostopoulos, A.; Cismaru, A.; Buiculescu, C.; Petrini, I.; Muller, A. Membrane Supported AlN FBAR Structures Obtained by Micromachining of High Resistivity Silicon. In Proceedings of the 2007 International Semiconductor Conference, Sinaia, Romania, 17 September–15 October 2007; IEEE: New York, NY, USA, 2007; Volume 1, pp. 293–296.
23. Feld, D.A.; Parker, R.; Ruby, R.; Bradley, P.; Dong, S. After 60 years: A new formula for computing quality factor is warranted. In Proceedings of the 2008 IEEE Ultrasonics Symposium, Beijing, China, 2–5 November 2008; IEEE: New York, NY, USA, 2008; pp. 431–436.
24. Moreira, M.; Bjurström, J.; Katardjev, I.; Yantchev, V. Aluminum scandium nitride thin-film bulk acoustic resonators for wide band applications. *Vacuum* **2011**, *86*, 23–26. [CrossRef]
25. Wang, J.; Park, M.; Mertin, S.; Pensala, T.; Ayazi, F.; Ansari, A. A film bulk acoustic resonator based on ferroelectric aluminum scandium nitride films. *J. Microelectromechanical Syst.* **2020**, *29*, 741–747. [CrossRef]
26. Gao, C.; Zheng, Y.; Li, H.; Ren, Y.; Gu, X.; Huang, X.; Wang, Y.; Qu, Y.; Liu, Y.; Cai, Y.; et al. Design and fabrication of a film bulk acoustic wave filter for 3.0 GHz-3.2 GHz S-band. *Sensors* **2024**, *24*, 2939. [CrossRef]
27. Zhou, Y.; Zheng, Y.; Xu, Q.; Qu, Y.; Ren, Y.; Huang, X.; Gao, C.; Liu, Y.; Guo, S.; Cai, Y.; et al. Design and Fabrication of 3.5 GHz Band-Pass Film Bulk Acoustic Resonator Filter. *Micromachines* **2024**, *15*, 563. [CrossRef]
28. Gu, X.; Liu, Y.; Huang, X.; Ding, J.; Qu, Y.; Liu, Z.; Cai, Y.; Liu, W.; Guo, S.; Sun, C. Analysis of Heterostructure Formation of Abnormally Oriented Grains in Sputter-Deposited AlScN Films and Its Impact on Bulk Acoustic Wave Devices. *ACS Appl. Mater. Interfaces* **2024**, *16*, 53106–53115. [CrossRef] [PubMed]
29. Zou, Y.; Gao, C.; Zhou, J.; Liu, Y.; Xu, Q.; Qu, Y.; Liu, W.; Soon, J.B.W.; Cai, Y.; Sun, C. Aluminum scandium nitride thin-film bulk acoustic resonators for 5G wideband applications. *Microsyst. Nanoeng.* **2022**, *8*, 124. [CrossRef] [PubMed]
30. Zou, Y.; Cai, Y.; Gao, C.; Luo, T.; Liu, Y.; Xu, Q.; Wang, Y.; Nian, L.; Liu, W.; Soon, J.B.W.; et al. Design, fabrication, and characterization of aluminum scandium nitride-based thin film bulk acoustic wave filter. *J. Microelectromechanical Syst.* **2023**, *32*, 263–270. [CrossRef]
31. Dou, W.; Zhou, C.; Qin, R.; Yang, Y.; Guo, H.; Mu, Z.; Yu, W. Super-High-Frequency Bulk Acoustic Resonators Based on Aluminum Scandium Nitride for Wideband Applications. *Nanomaterials* **2023**, *13*, 2737. [CrossRef]
32. Wang, Y.; Yang, T.; Gao, C.; Yang, C.; Lin, B.; Liu, Y.; Cai, Y.; Sun, C. $Al_{0.8}Sc_{0.2}N$ film-based BAW filters using transfer process for Wi-Fi6 applications. *J. Appl. Phys.* **2025**, *137*, 085702. [CrossRef]
33. Qi, H.; Zhang, H.; Gao, N.; Fang, Y.; Xie, X.; Bian, L.; Fang, Y.; Chen, G. Simulation and preparation of FBARs based on AlN thin films. *Mater. Sci. Semicond. Process.* **2025**, *185*, 108886. [CrossRef]

micromachines

MDPI

Article

A 3D DC Electric Field Meter Based on Sensor Chips Packaged Using a Highly Sensitive Scheme

Pengfei Yang [1], Xiaolong Wen [2], Xiaonan Li [1], Zhaozhi Chu [3,*], Chunrong Peng [4] and Shuang Wu [5]

1 Beijing Key Laboratory for Sensors, School of Applied Science, Beijing Information Science and Technology University, Beijing 100192, China; pfy@bistu.edu.cn (P.Y.); lixiaonan824@163.com (X.L.)
2 School of Mathematics and Physics, University of Science and Technology Beijing, Beijing 100083, China; xiaolongwen@ustb.edu.cn
3 Institute of Microelectronics of Chinese Academy of Sciences, Beijing 100029, China
4 State Key Laboratory of Transducer Technology, Aerospace Information Research Institute, Chinese Academy of Sciences, Beijing 100094, China; crpeng@mail.ie.ac.cn
5 Beijing Tflying Transducer Technology Co., Ltd., Beijing 100083, China; wushuang@tflying.cn
* Correspondence: chuzhaozhi@ime.ac.cn

Abstract: This study presents a 3D DC electric field meter (EFM) that uses three identical 1D MEMS chips. The shielding electrodes and sensing electrodes of the MEMS chips employ a combination of rigid frames and short strip-type beams to improve vibrational stability. To enhance sensitivity, our MEMS chips feature inner convex packaging covers. Moreover, the integrated design and wireless transmission efficiently eradicate the impact of ground potential on detection results. Detailed simulations have been conducted to analyze the electric field distribution within the chip package and the electric field distribution on the EFM's surface. A prototype was then developed, calibrated, and validated. The test results indicate that the sensitivity of our proposed 3D EFM is at least 4.64 times higher than the highest sensitivity observed in previously reported MEMS 3D EFMs. The maximum relative deviation is a mere 2.2% for any rotation attitude. Remarkably, even in high humidity conditions, the EFM's linearity remains within 1%. Additionally, the resolution of any single axis is less than 10 V/m.

Keywords: MEMS; 3D electric field meter; sensor chips; moistureproof packaging structure; highly sensitive package

1. Introduction

Electric field measurement plays a pivotal role in both scientific research and engineering technologies. For example, in the realm of meteorology, the atmospheric electric fields are monitored for lightning hazard warning and climate change assessment [1,2]. In power systems, measuring electric fields in close proximity to high-voltage equipment is vital. Such measurements ensure the safety of personnel and equipment during live-line maintenance [3] and secure avoidance of power lines by helicopters [4]. They also allow for credible electromagnetic environment assessments [5]. Furthermore, in industrial production lines, electrostatic voltage is a critical factor, obtained via DC electric field measurement. Not only does it provide a direct reading of the electrostatic charge present in products, but it also offers undeniable evidence when industrial static elimination systems fail [6].

Currently, most electric field meters (EFMs) are only capable of measuring one-dimensional (1D) electric fields [7,8]. While they are sufficient for the application needs of electric field detection in many scenarios, there are still scenarios where they fall short in accurately detecting unknown electric fields. For instance, at high altitudes, apart from

https://doi.org/10.3390/mi16040484

the electric field component perpendicular to the ground, a crucial horizontal electric field component parallel to the ground is present, intensifying with increasing altitude. This horizontal component offers valuable information about the direction of thundercloud movement [9,10]. Additionally, buildings surrounding high-voltage DC transmission lines often significantly distort the electric field, making it challenging to determine its direction [11]. If a single-axis EFM is utilized for measuring the distorted electric field of a high-voltage DC line, obtaining a comprehensive and accurate depiction of the field's spatial distribution becomes significantly challenging. As a result, it is essential to pursue the development of a three-dimensional (3D) electric field detection instrument to complement current electric field detection methods.

A variety of techniques have been proposed for measuring the electric field vector, with the field mill being a commonly applied technical method. Predominantly, there are two types of 3D EFMs based on the field mill principle: spherical EFMs [12] and hybrid EFMs [11]. Nevertheless, these field-milled 3D EFMs encounter substantial mechanical wear and power consumption issues. Moreover, despite reports of passive 3D EFMs [13] and integrated optical 3D EFMs [14], they are not fit for DC electric field measurements due to the limitations inherent in their working principles.

Over the past decade, studies have focused on microelectromechanical system (MEMS)-based 3D EFMs due to their notable advantages in wear prevention, reduced power consumption, and compact size. In 2014, Wen et al. proposed a coplanar MEMS 3D EFM, utilizing identical 1D electric field sensor chips (EFSCs) arranged arbitrarily on the same plane for 3D electric field detection [15]. Fast forward to 2016, Li et al. introduced a decoupling calibration method to suppress cross-axis coupling interference, employing a genetic algorithm for the orthogonally assembled 3D EFM [16]. A year later, Ling et al. reported a single-chip 3D electric field microsensor with an in-plane rotary mechanism for the simultaneous detection of the electrostatic field components along the X-, Y-, and Z-axes [17]. By 2018, Ling et al. designed a 1D EFSC presenting a low cross-axis coupling interference and used this 1D EFSC to develop a 3D EFM [18]. Fast forward to last year, 2023, Peng et al. proposed a single-chip 3D electric field microsensor with piezoelectric excitation to minimize driving voltage and feedthrough [19]. Although the issue of cross-axis interference in the aforementioned MEMS 3D field meters has been addressed, the influence of environmental humidity on measurement accuracy remains unconsidered. For packaged EFSCs, measurement accuracy deteriorates when the environmental relative humidity exceeds 60% [1]. In single-chip structures, which lack protective packaging, the impact of ambient humidity on measurement accuracy is even more pronounced. Furthermore, existing MEMS 3D EFMs exhibit low sensitivity; for instance, a maximum uniaxial sensitivity of 3.53 mV/kV/m has been reported [16], which falls short of the requirements for high-sensitivity detection. Additionally, EFSCs typically use elongated strip-type beams, which may cause potential operational stability issues. This study aims to improve the sensitivity and stability of MEMS 3D EFMs, as well as addressing the challenge posed by environmental humidity on measurement accuracy.

This paper proposes a more pragmatic 3D DC EFM utilizing three identical 1D EFSCs, featuring the following enhancements:

1. A moistureproof packaging structure has been incorporated into the design of the new 3D field meter to counter the effects of ambient humidity on the measurement's accuracy.
2. The sensing units of the EFSCs use a combination of hard frames and short strip-type beams to improve vibration stability.

3. Contrary to the flat packaging covers on EFSCs reported previously, the EFSCs in the proposed structure use inward-convex packaging covers, further improving the sensitivity of the 3D field meter.
4. The proposed solution necessitates only the detecting electrodes to be assembled orthogonally, with no restrictions on the arrangement of EFSCs—a welcome flexibility for the EFM's implementation.
5. Unlike 3D EFMs in previous papers which contained probes separated from their signal conditioning circuit, and conveyed signals through wires, the proposed 3D field meter has been designed as an integrated structure. Here the data are transmitted wirelessly, eradicating potential issues of detection results being affected by the ground potential due to changes in the installation location of the signal conditioning circuit.

2. Microsensor and Its Package

The microsensor primarily consists of an electrostatically driven resonator, which detects the incident electric field through lateral vibration modulation [20]. In the previously reported microsensor structures, elongated strip-type beams served as the shielding and sensing electrodes. However, the swing of these long beams presented potential operational stability issues during vibration and could even cause absorption in the device structure. To address the possible vibration instability problems, the shielding electrodes and sensing electrodes of the 1D EFSCs used in this study were redesigned, as shown in Figure 1a. The shielding electrodes of the new sensor chip are structured into a stiffer frame from which short strip-type beams extend. Similarly, short sensing electrodes are also installed on the fixed frame structure. Two arrays of sensing electrodes are symmetrically positioned and coupled to their coplanar shielding electrodes. This new design significantly enhances the stability of the resonator vibration compared to the previously reported microsensor structures.

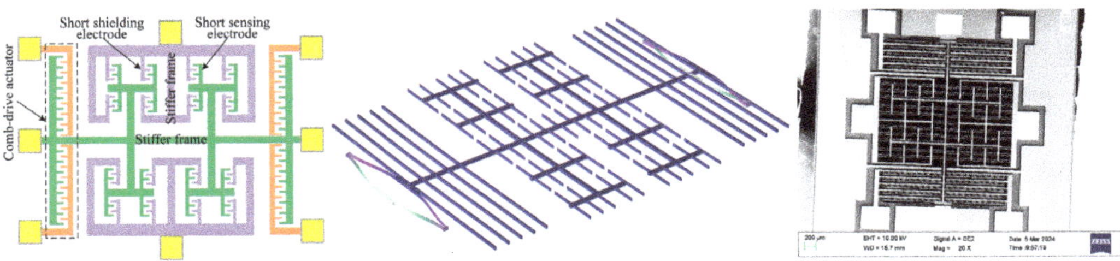

(a) (b) (c)

Figure 1. Improved 1D EFSC. (**a**) Schematic diagram. (**b**) Lateral vibration mode, resonant frequency of 3012 Hz. (**c**) SEM image.

To achieve a large vibration amplitude with low driving voltages, the sensor chip is designed to operate at its resonant frequency. According to the operational principle, the sensor chip requires the lateral vibration mode, as illustrated in Figure 1b. The lateral vibration resonance frequency of the upgraded microsensor is 3012 Hz. The microsensors were fabricated using a silicon-on-insulator process. Figure 1c depicts the scanning electron microscope (SEM) photograph of the fabricated microsensor, and its size is approximately 4.5 mm $\times$ 4.5 mm. The new microsensor is driven by an 18 V common DC bias voltage and antisymmetric 1.6 $V_{p\text{-}p}$ sinusoidal voltages applied on either side of the comb-drive actuator. It can detect electric fields exceeding 600 kV/m.

The sensitivity of the EFSC is closely tied to its packaging structure. Yang et al. introduced the microsensor's packaging solution based on a ceramic base featuring a flat

Kovar cover plate, as depicted in Figure 2a [21]. This sensor chip enables the detection of the incident electric field by measuring the electric field between the cover plate and the ceramic substrate. However, given that the 3D electric field detection employs a floating-ground method, the sensitivity of the reported microsensor after packaging falls short of actual use requirements. Hence, we proposed a highly sensitive packaging structure furnished with an inward-convex cover to enhance the EFSC's detection sensitivity. Figure 2b shows the schematic of the newly designed package.

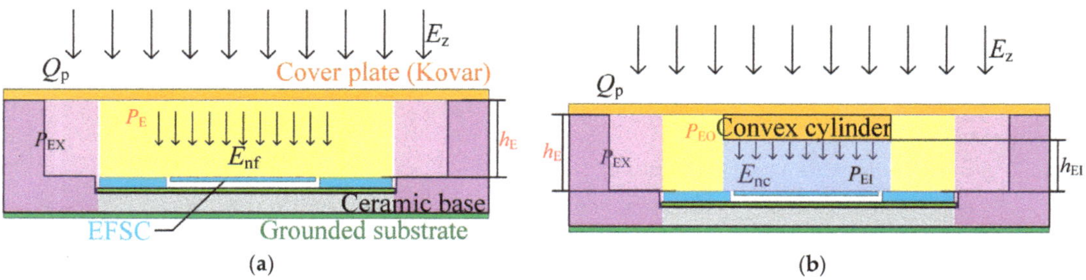

Figure 2. Schematic of the EFSC package. (**a**) Flat cover plate. (**b**) Inward-convex cover plate.

The EFSC package with a flat cover plate can be divided into two regions: P_E (yellow-colored region) and P_{EX} (purple-colored regions), as demonstrated in Figure 2a. The capacitances in the P_E and P_{EX} areas are represented as C_{PE} and C_{PEX}, respectively. Here $C_{PE} + C_{PEX} = C_p$, C_p is the capacitance of the EFSC package. A summary of the capacitance definition for different regions of the EFSC package is listed in Table 1.

Table 1. Capacitance definition for EFSC package.

Capacitance	Packaging Region
C_p	The whole package
C_{PE}	P_E, yellow-colored region in Figure 2a
C_{PEX}	P_{EX}, purple-colored region in Figure 2a
C_{PEO}	P_{EO}, yellow-colored region in Figure 2b
C_{PEI}	P_{EI}, light blue-colored region in Figure 2b

Additionally, C_{PE} can be equivalent to the ideal parallel-plate capacitance without taking into account the fringing effects. Consequently, the voltage V_{pf} on the flat packaging cover plate is expressed as

$$V_{pf} = \frac{-Q_p}{\frac{\varepsilon A_{PE}}{h_E} + C_{PEX}} \quad (1)$$

where Q_p is the amount of charge generated on the packaging cover plate under the incident electric field, A_{PE} is the area of the P_E region when viewed from above, ε is the permittivity of the air inside the package, and h_E is the distance between the packaging cover plate and the microsensor.

The voltage V_{pf} produces an electric field E_{nf} that is perpendicular to the microsensor inside the ceramic package. Because the distribution of the electric field between the flat packaging cover plate and the sensing unit of the microsensor is uniform enough [22], E_{nf} can be solved by

$$E_{nf} = \frac{V_{pf}}{h_E} = \frac{-Q_p}{\varepsilon A_{PE} + h_E C_{PEX}} \quad (2)$$

As illustrated in Figure 2b, the high-sensitivity package's cover plate protrudes a cylinder inwards. This design contrasts with the package structure seen in Figure 2a. The

P_E region in Figure 2b can be also split into two parts: P_{EO} (yellow-colored region) and P_{EI} (light blue-colored region). The capacitances in the P_{EO} and P_{EI} regions are denoted by C_{PEO} and C_{PEI}, respectively. Here, $C_{PEO} + C_{PEI} = C_{PE}$. Neglecting fringing effects, the voltage V_{pc} on the inward-convex packaging cover plate is expressed as

$$V_{pc} = \frac{-Q_p}{\frac{\varepsilon A_{PEO}}{h_E} + \frac{\varepsilon A_{PEI}}{h_{EI}} + C_{PEX}} \quad (3)$$

where A_{PEO} and A_{PEI} represent the areas of the P_{EO} and P_{EI} regions on the top view direction, respectively, and h_{EI} is the distance between the convex cylinder of the packaging cover plate and the microsensor ($0 < h_{EI} < h_E$).

Because of the uniform distribution of the electric field E_{nc} between the inner convex cylinder of the packaging cover plate and the microsensor, as illustrated in Section 4.1, E_{nc} can be derived as

$$E_{nc} = \frac{-Q_p}{\varepsilon\left(A_{PEO}\frac{h_{EI}}{h_E} + A_{PEI}\right) + h_{EI}C_{PEX}} \quad (4)$$

Taking into account that $A_{PEI} = A_{PE} - A_{PEO}$, Equation (4) can be further modified as

$$E_{nc} = \frac{-Q_p}{\varepsilon\left[A_{PE} - A_{PEO}\left(1 - \frac{h_{EI}}{h_E}\right)\right] + h_{EI}C_{PEX}} \quad (5)$$

By comparing Equations (5) and (2), it can be found that $|E_{nc}| > |E_{nf}|$ as a result of $0 < A_{PEO}\ (1 - h_{EI}/h_E) < A_{PE}$, and $h_{EI} < h_E$. This conclusion indicates that under the same condition of the measured electric field E_z, the electric field that exists over the sensor chip inside the protruding cover plate package is greater than that inside the flat cover plate package. Thus, this newly designed packaging structure exhibits higher sensitivity compared to the previously reported flat cover plate package. Moreover, as can be seen from Equation (5), we could further enhance the sensitivity of the new package by expanding A_{PEO} or reducing h_{EI}. Nevertheless, the A_{PEO} is limited to a certain range because A_{PEI} must cover the electric field sensing unit of the sensor chip.

The fabricated high-sensitivity package is presented in Figure 3. The package features an inner convex cylinder with a diameter of 2 mm and a height of 0.78 mm. The distance h_E is 0.92 mm, and h_{EI} is 0.14 mm. The area A_{PE} of the P_E region is 2.025×10^{-5} m^2. A_{PEO} is 1.711×10^{-5} m^2. The overall measured dimensions of the entire package are approximately 7.9 mm × 7.9 mm × 2.2 mm.

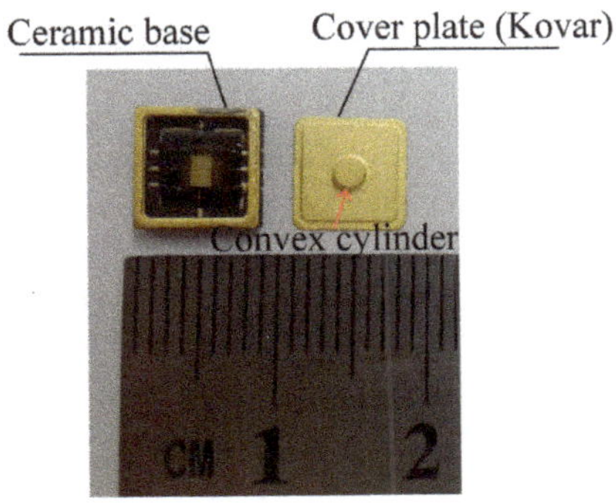

Figure 3. Photograph of the highly sensitive package.

The sensitivity test results of the packaged microsensors are shown in Figure 4. The sensitivity of the EFSC with a flat cover package is 1.95 mV/(kV/m), while the sensitivity of the EFSC with an inner convex cover package is 7.19 mV/(kV/m), which is 3.68 times higher (in the absence of considering the variations among the microsensors). These results further validate the accuracy of the theoretical analysis.

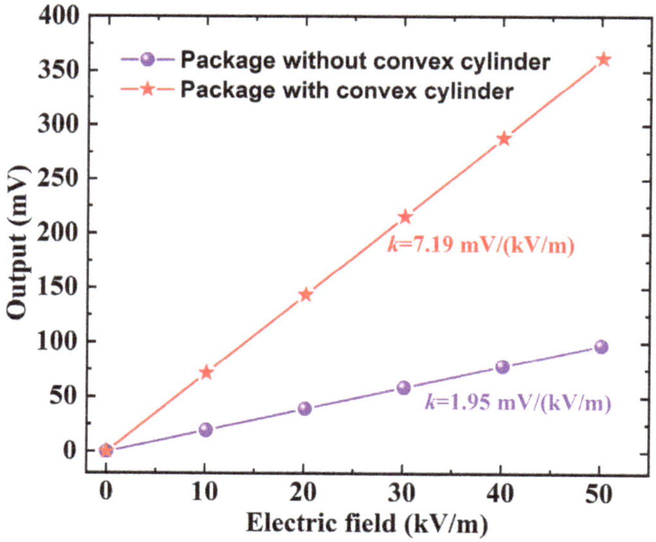

Figure 4. Sensitivity test results of the EFSC with different packaging schemes.

3. Vector Fieldmeter

3.1. Anti-Humidity Scheme and Structural Arrangement

Initial experiments have exposed that in high-humidity environments, the ceramic base of the microsensor package tends to polarize readily under a strong incident field, leading to electric field attenuation inside the package and the zero drift of the EFSC [22]. Thus, it substantiates the need for the highly sensitive packages, such as the one proposed in this study, to be protected against moisture. In the literature [1], our team reported on a moisture-resistant packaging structure for 1D electric field detection. Both indoor environmental experiments and outdoor tests demonstrated that this solution yielded favorable results in moisture resistance. Based on our prior research, a new moistureproof packaging structure for 3D electric field detection is proposed, as depicted in Figure 5a. It consists of a grounded substrate, packaged EFSCs, a tinplate cover, a grounded metal chamber, semi-rigid cables, detecting electrodes, and cured epoxy adhesive. Here, semi-rigid cables are coaxial cables with a solid metal outer conductor, typically made of copper or aluminum, that maintains a fixed shape once bent. For achieving 3D electric field detection, a minimum of three packaged EFSCs are mounted on a grounded substrate. Meanwhile, a tinplate cover is sufficient for the protection of the packaged EFSCs. The detecting electrodes are orthogonally arranged for measuring the three components of the incident electric field. The detecting electrodes for the *X*-axis, *Y*-axis, and *Z*-axis are correspondingly connected electrically to the packaging cover plates of the EFSCs via semi-rigid cables in sequence.

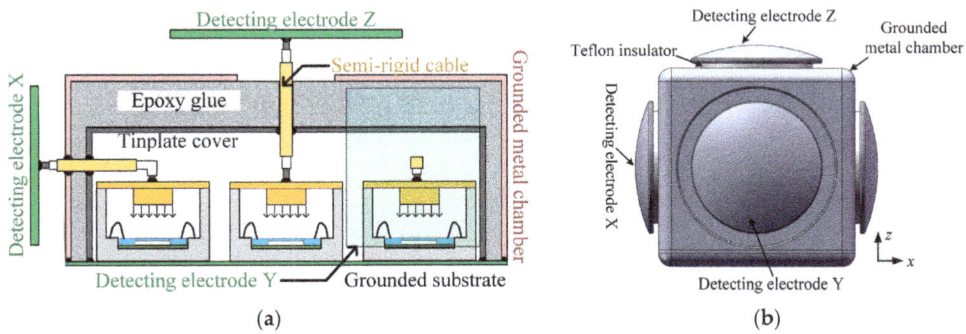

Figure 5. (**a**) Simplified schematic of the proposed 3D DC EFM and (**b**) its arrangement.

Figure 5b illustrates the design layout of the proposed 3D DC EFM. The detecting electrodes, fixed on a cube-shaped grounded metal chamber, are kept separate from the grounded metal chamber by Teflon insulators. Aiming to maintain symmetry in the EFM structure and minimize measurement discrepancies due to structural asymmetry, five detecting electrodes are arranged on the five surfaces of the cube, excluding the bottom. Of these, only three are electrically connected to the package covers of the EFSCs, in a mutually orthogonal arrangement, while the remaining two electrodes are left floating. A noteworthy point from the schematic design and structural configuration discussed above is that the proposed EFM necessitates only three detecting electrodes assembled perpendicularly to each other. However, there exist no constraints concerning the arrangement of the EFSCs. This flexibility greatly benefits the EFM instrument implementation. Since the EFSCs are not required to be oriented orthogonally, the associated interface circuits can be more easily designed, and the internal structure of the 3D EFM can be made more compact and space-efficient.

3.2. Measurement Principle

A local coordinate system is set up, using the center of the 3D EFM as its point of origin, as shown in Figure 6. Here, the X- and Y-axes are in horizontal directions perpendicular to each other, whereas the Z-axis points vertically. These X-, Y-, and Z-axes are perpendicular to the detecting electrodes in various directions. The angle between the incident electric field E and the Z-axis is identified as θ, while φ is the angle between the horizontal projection of E and the X-axis. Considering that the three sensing axes of the proposed EFM measure the electric field in an identical manner, we focus on analyzing one of these sensing axes, referred to here as the i-axis, $i = X, Y, Z$. The output voltage v_i of the i-axis electric field measurement unit can be given by [1].

$$v_i(t) = -\frac{1}{2}\alpha_{\mathrm{i}}\beta R_f G_{\mathrm{INA}} V_a E_i(t) \tag{6}$$

where β is the conversion gain of the microsensor, R_f is the feedback resistance for the microsensor's analog front end (AFE), G_{INA} is the gain of the AFE's instrumentation amplifier, V_a is the amplitude of the microsensor's sinusoidal drive signal, and E_{i} is the electric field distributed vertically on the i-axis detecting electrode.

The ratio α_{i} of the electric field E_{nci} inside the package to E_{i} is defined as

$$\alpha_{\mathrm{i}} = \frac{\varepsilon A_{\mathrm{di}}}{(C_{\mathrm{mi}} + C_{\mathrm{si}} + C_{\mathrm{p}})h_{\mathrm{EI}}} \tag{7}$$

where A_{di} is the effective area of the i-axis detecting electrode, C_{mi} is the capacitance between the i-axis detecting electrode and the metal chamber, and C_{si} is the semi-rigid cable capacitance for the i-axis.

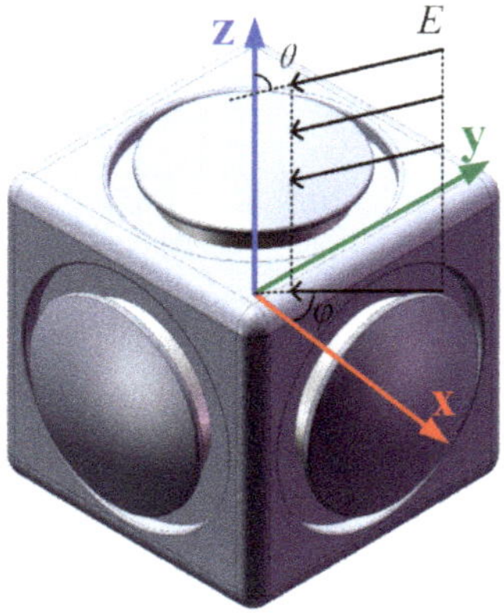

Figure 6. Conceptual drawing of the position of the local coordinate system.

As depicted in Figure 6, when the 3D EFM is situated in a uniform electric field E, the nearby electric field around the EFM is distorted, and the charge induced on different detecting electrodes relies on the direction of E. Unfortunately, due to the distortion effect of the electric field, the X-axis detection electrode measures not just the X component of the evaluated electric field E_x but also experiences cross-axis coupling interferences from the Y and Z components of the evaluated electric field components E_y and E_z. Hence, the output voltage v_x from the X-axis electric field sensing unit can be expressed as

$$v_x = k_{xx}E_x + k_{xy}E_y + k_{xz}E_z \tag{8}$$

Similarly, the outputs v_y and v_z, which originate from the electric field sensing units on the Y-axis and Z-axis, can be written, respectively, as

$$v_y = k_{yx}E_x + k_{yy}E_y + k_{yz}E_z \tag{9}$$

$$v_z = k_{zx}E_x + k_{zy}E_y + k_{zz}E_z \tag{10}$$

where k_{mn} is the sensitivity coefficient of the electric field component E_n at the m-axis detecting electrode, $m = x, y, z$, $n = x, y, z$; when $m = n$, k_{mn} is known as the coaxial sensitivity, but when $m \neq n$, k_{mn} is dubbed the coupling sensitivity. Equations (8)–(10) can be expressed by the following matrix:

$$\begin{bmatrix} v_x \\ v_y \\ v_z \end{bmatrix} = \begin{bmatrix} k_{xx} & k_{xy} & k_{xz} \\ k_{yx} & k_{yy} & k_{yz} \\ k_{zx} & k_{zy} & k_{zz} \end{bmatrix} \begin{bmatrix} E_x \\ E_y \\ E_z \end{bmatrix} \tag{11}$$

Solving Equation (11), the three components of the measured electric field are given by

$$\begin{bmatrix} E_x \\ E_y \\ E_z \end{bmatrix} = C \begin{bmatrix} v_x \\ v_y \\ v_z \end{bmatrix} \tag{12}$$

where the decoupling matrix C is defined as

$$C = \begin{bmatrix} k_{xx} & k_{xy} & k_{xz} \\ k_{yx} & k_{yy} & k_{yz} \\ k_{zx} & k_{zy} & k_{zz} \end{bmatrix}^{-1} \tag{13}$$

Finally, $\hat{E}$, which is the amplitude estimation of the measured electric field E, can be determined by

$$\hat{E} = \sqrt{E_x^2 + E_y^2 + E_z^2} \tag{14}$$

4. Numerical Simulation

4.1. Simulation of Highly Sensitive Package

A numerical simulation model of the package was created using the FEM tool COMSOL Multiphysics (Version 5.4) to validate the uniform distribution of the electric field between the microsensor and the inner convex cylinder of its highly sensitive package. The highly sensitive package is placed in air, which has a relative dielectric constant of 1. The gap h_{EI} between the inner convex cylinder and the microsensor is set at 0.14 mm. The package cover is applied with a voltage of 1 V, while the microsensor is grounded equivalently. Figure 7a presents the simulated results of the electric field cloud map. The curve of the electric field change from point **A**, on one side of the inner convex cylinder, to point **B**, on the opposing side, is shown in Figure 7b. Both points **A** and **B** are located in the middle position between the inner convex cylinder and the microsensor. It can be found from Figure 7b that an unchanging electric field value of 6896.55 V/m prevails over a circular realm of a 0.75 mm radius above the microsensor. However, inside the belt range with a radius extending from 0.75 mm to 1 mm, the electric field value experiences minor fluctuations—a maximum deviation of 214.68 V/m—due to the distortion effect at the boundary of the inner convex cylinder. Despite these fluctuations near the external vicinity of the inner convex cylinder, the observed relative change in the electric field is merely 3%, indicating a fairly uniform electric field distribution between the inner convex cylinder of the package and the microsensor. Additionally, a capacitance value of $C_p = 1.87$ pF for the EFSC's package is obtained from the simulation.

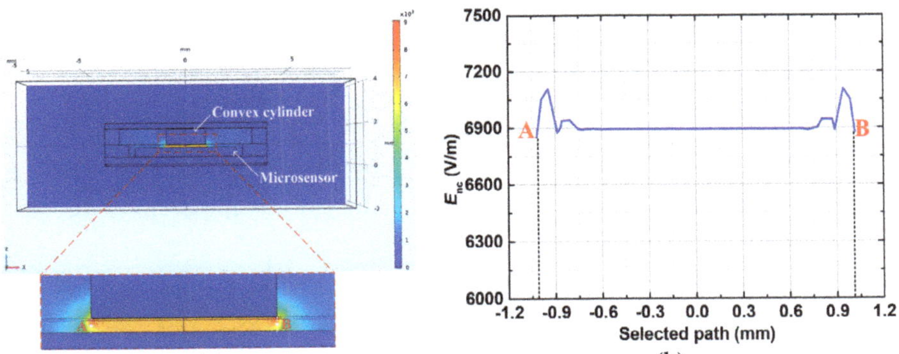

(**a**) (**b**)

Figure 7. Simulation results of the electric field inside the package: (**a**) cloud map; (**b**) the field change curve from points **A** to **B** at the middle position between the inner convex cylinder and the microsensor.

4.2. Simulation of 3D EFM Structure

The 3D EFM structure proposed was analyzed using the same FEM tool from the package simulation to investigate the cross-axis coupling interference and the distribution characteristics of induced charge on the various axes' detecting electrodes when the EFM is rotated. A cubic space model of 2000 mm side length, capable of generating a uniform electric field, was constructed. The cubic space's medium was set to air with a relative permittivity of 1. The EFM model was centrally positioned within this cubic space, with the *X*-, *Y*-, and *Z*-axes of the cubic space placed perpendicular to the *X*-axis, *Y*-axis and *Z*-axis detecting electrodes of the EFM, respectively. Several boundary conditions were imposed on the EFM. They include setting the total charge of each axis's detection electrode to be 0 C, the relative permittivity of the Teflon insulators to be 2.55, and the voltage applied to the grounded metal chamber to be 0 V. The cube space's top and bottom voltages were set at 1 kV and -1 kV, respectively, resulting in a vertical incident field with an intensity of 1 kV/m directed in the *Z*-axis.

Figure 8 shows a simulation of the electric field distribution for the 3D EFM. The results highlight a pronounced concentration of electric field lines on the *Z*-axis detecting electrode. With an incident electric field of 1 kV/m along the *Z*-axis, the maximum field around the *Z*-axis detecting electrode is found to be 2.6 kV/m, and the charge is found to be -3.05×10^{-11} C. Considering the EFM structure's incomplete symmetry, the *X*-axis and *Y*-axis detecting electrodes are seen to induce charges of 2.99×10^{-13} C and 2.84×10^{-13} C respectively. Suppose that the *X*-, *Y*-, and *Z*-axis detecting electrodes of the EFM have an equal ability to induce charges, then the coupling interference ratios of the electric field in the *Z*-axis direction on the *X*- and *Y*-axis detecting electrodes are 0.98% and 0.93%, respectively. This finding further underscores the importance of establishing matrix Equation (11).

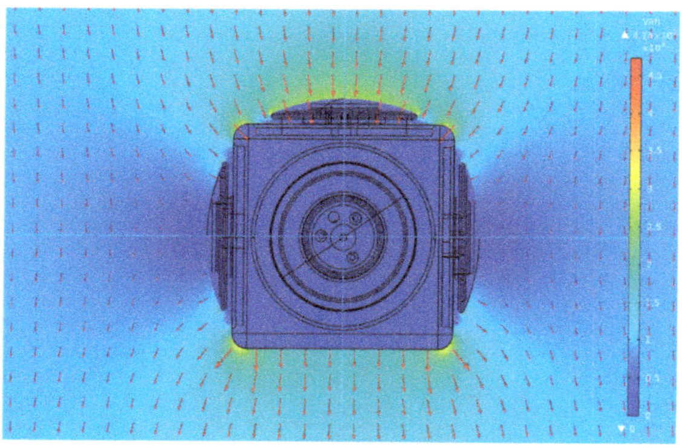

Figure 8. Cloud map and vector map of the 3D EFM. The arrows represent the direction of the electric field.

Figure 9 provides simulated results of the induced charge on the detecting electrodes of different axes when the 3D EFM undergoes rotation. Figure 9a illustrates that with $\varphi = 0^\circ$, the curves concerning the induced charge on the *X*-axis detecting electrode and the *Z*-axis detecting electrode exhibit strong sinusoidal characteristics with a 90° phase difference. This behavior arises because the rotation of the 3D EFM causes the angle between each detecting electrode's normal vector and the electric field direction to vary periodically. The observed 90° phase difference results from the fact that the normal vectors of adjacent detecting electrodes are oriented 90° apart in space, making their electric field projections orthogonal and thus resulting in phase-shifted sinusoidal responses. On the other hand,

Figure 9b–d show that when $\theta \neq 0^{\circ}$, the curves corresponding to the induced charge on the X-axis detection electrode and the Y-axis detection electrode exhibit prominent sinusoidal characteristics, again maintaining a 90° phase difference. However, the Z-axis detection electrode's induced charge stays constant.

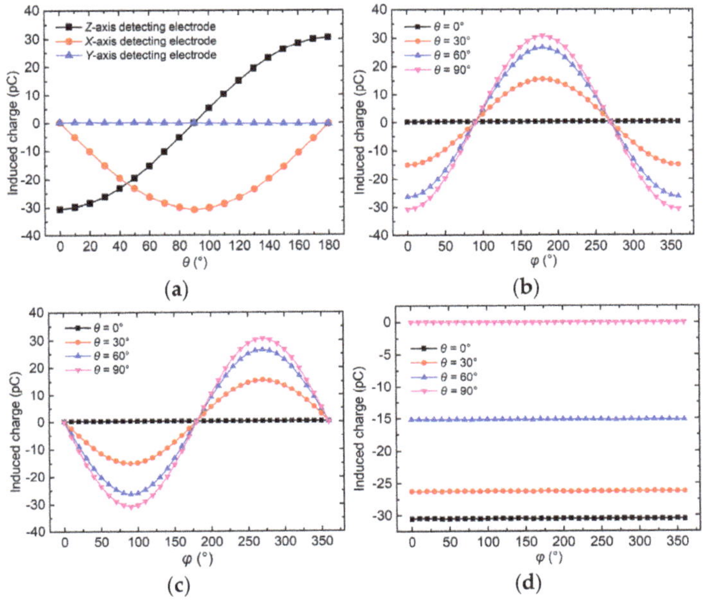

Figure 9. Simulation results of the 3D EFM's response to the incident electric field with different directions. (**a**) Induced charge on each detecting electrode versus θ when $\varphi = 0^{\circ}$; (**b**) induced charge on X-axis detecting electrode versus φ when $\theta = 0^{\circ}, 30^{\circ}, 60^{\circ}$, and 90°; (**c**) induced charge on Y-axis detecting electrode; (**d**) induced charge on Z-axis detecting electrode.

5. Implementation

5.1. Proposed Structure

Figure 10a provides a look at the main features of the 3D EFM prototype. Teflon insulators and mechanical transfer units are mounted on the grounded metal chamber via screws in sequence. Printed circuit boards (PCBs) with the packaged EFSCs protected by tinplate covers are placed inside the metal chamber. One end of the semi-rigid cable extends through the metal chamber, and its core is electrically connected to the mechanical transfer unit using screws. The detecting electrodes then get mounted onto these mechanical transfer units through internal thread connections. The detecting electrodes, mechanical transfer units, and metal chamber are all made of aluminum. The battery is also installed inside the metal chamber, and the switch controls the startup of the EFM. Figure 10b provides a snapshot of the manufactured EFM prototype. The physical dimensions of the EFM prototype are 74 mm × 74 mm × 67 mm.

Since all three measurement units of the manufactured 3D EFM have precisely the same structure, the capacitance C_{mi} between the i-axis detecting electrode and the metal chamber is calculated to be 24.31 pF, and the semi-rigid cable capacitance $C_{si} = 1.96$ pF for all electric field measurement units. The capacitance of the EFSC package is $C_p = 1.87$ pF, and the distance between the inner convex cylinder of the package and the microsensor is $h_{EI} = 0.14$ mm. Referencing the results of the simulation detailed in Section 4.2, the induced charge on the i-axis detecting electrode is -3.05×10^{-11} C under a vertical incident field $E_i = 1$ kV/m. Applying these values into Equation (7), we compute the ratio α_i to be 7.48. However, neglecting humidity influences, if the EFM merely includes highly sensitive packaged EFSCs without external detecting electrodes, the ratio α_i reduces to 2.11. Therefore, the proposed structure accomplishes two vital goals: it mitigates the effect of environmental humidity, and it enhances detection sensitivity.

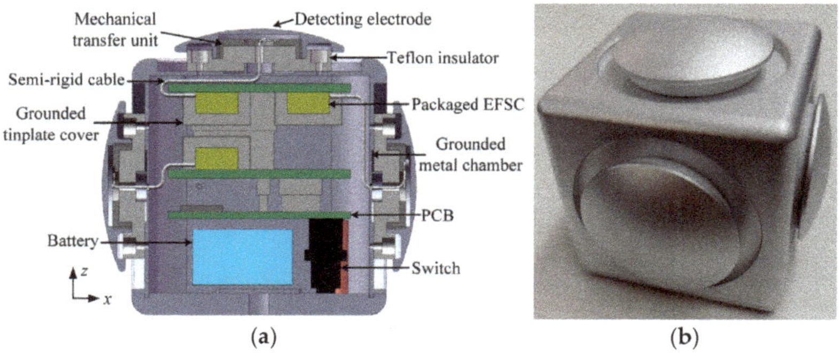

Figure 10. Proposed 3D EFM device. (**a**) Cross section shows instrument arrangements; (**b**) prototype.

5.2. Interface Circuit

Figure 11 shows the circuitry of the 3D EFM. Considering that the three 1D EFSCs are independent of each other and need to be measured simultaneously, the interface circuit is suitably segmented into three identical signal-conditioning channels. This configuration allows for parallel processing and individual readouts from each 1D EFSC. A 16-bit analog-to-digital converter (ADC) digitizes both the output signal of the AFE and the sinusoidal drive voltage. The system uses an embedded 32-bit microcontroller to accomplish synchronous demodulation. To avoid ground potential impact on the measurement, the outputs of the EFM are sent wirelessly from a universal asynchronous receiver–transmitter to an LoRa converter directly to a laptop.

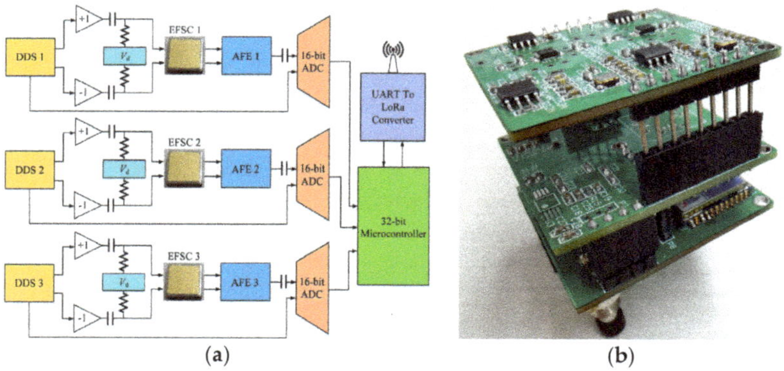

Figure 11. Proposed 3D EFM circuitry. (**a**) Schematic view; (**b**) photograph.

6. Experimental Results

6.1. Experiment Setup

The 3D EFM prototype was calibrated using a parallel-plate apparatus with a side dimension of 90 cm and a spacing of 60 cm, as depicted in Figure 12. However, for finite-sized metal plates, the actual field between parallel conducting plates was not ideally uniform due to the effects of fringing. To control the fringing errors, IEEE Standard

644 [22] recommends the use of iron wires as grading rings, placed to surround the parallel conducting plates at equal intervals. An Iseg HP 500 power supply served to generate the required DC high voltage. We developed a rod-shaped bracket, made of Teflon, to fix the 3D EFM in the parallel-plate apparatus and control its orientation. The axis of this rod-shaped bracket was parallel to the conducting plates and perpendicular to their opposing sides. One end of the rod-shaped bracket, situated in the center of the parallel-plate apparatus, was designated for mounting the 3D EFM. The other end connected to a rotary motor. The unique design of the bracket incorporated two orthogonal and intersecting rotation axes, R1 and R2, and their intersection coincided with the center of the 3D EFM. This arrangement ensured the 3D EFM's center remained unchanged during rotation. Prior to testing, the direction of the Z-axis detecting electrode of the 3D EFM was parallel to the applied electric field, and the direction of its X-axis detecting electrode aligned parallel to the rotation axis R1.

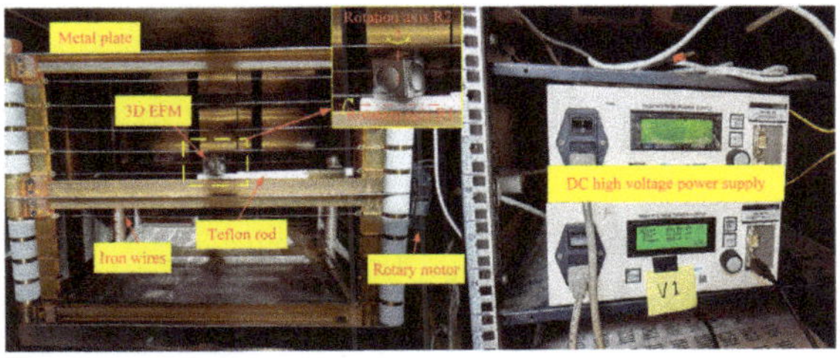

Figure 12. Calibration setup for the proposed 3D EFM.

6.2. Sensitivity Matrix

The 3D EFM underwent uniaxial calibration to determine the sensitivity k_{mn}. The applied voltage to the parallel-plate apparatus was varied from 0 to 20 kV in 2 kV increments, which generated a standard electric field ranging from 0 to 33.33 kV/m, stepped at 3.33 kV/m intervals. The Z-axis measurement unit of the 3D EFM was first calibrated. To enhance calibration accuracy, after testing the Z-axis measurement unit in the upward direction, rotate it and then calibrate it in the downward direction. While calibrating the Z-axis measurement unit, we simultaneously recorded output data from the X- and Y-axis measurement units. The calibration curve of the Z-axis measurement unit is illustrated in Figure 13a. The sensitivity coefficient k_{mz} is determined by averaging the results from the upward and downward calibrations, followed by fitting a straight line using the least squares method, $m = x, y, z$. Next, by rotating the 3D EFM and applying the same calibration method as used on the Z-axis measurement unit, the X- and Y-axis measurement units were calibrated, as represented in Figure 13b,c. The linearity δ_L is within a 1.6% range. This demonstrates that the prototype displays robust linearity in response to the incident electric field. Finally, according to Equation (11), the prototype's sensitivity matrix is given by

$$K = C^{-1} = \begin{bmatrix} 16.77 & 0.19 & 1.53 \\ -0.62 & 15.77 & 1.81 \\ 0.17 & 0.18 & 14.78 \end{bmatrix} \left(\mathrm{mV{\cdot}kV^{-1}{\cdot}m}\right) \tag{15}$$

Besides, the maximum cross-axis sensitivity calculated from Equation (15) is 11.86%.

A uniaxial sensitivity comparison between the proposed scheme and the existing MEMS 3D EFM has been conducted, and the results are listed in Table 2. Owing to the use of highly sensitive packaged EFSCs and the humidity-resistant packaging structure that can improve sensitivity, the sensitivity achieved in this study is at least 4.64 times higher than the highest sensitivity recorded in the previously reported MEMS 3D EFM.

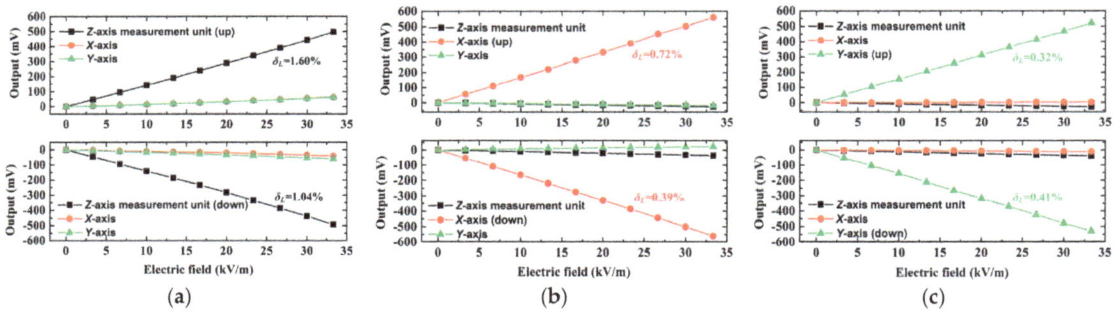

Figure 13. Calibration results. (**a**) The electric field is perpendicular to the *Z*-axis detecting electrode. (**b**) The electric field is perpendicular to the *X*-axis detecting electrode. (**c**) The electric field is perpendicular to the *Y*-axis detecting electrode.

Table 2. Sensitivity comparison of the proposed scheme with existing MEMS 3D EFM.

MEMS-Based 3D EFM	Year	Minimum Sensitivity $\left(\mathrm{mV}\cdot\mathrm{kV}^{-1}\cdot\mathrm{m}\right)$	Maximum Sensitivity $\left(\mathrm{mV}\cdot\mathrm{kV}^{-1}\cdot\mathrm{m}\right)$
Assembled 3D EFM with three 1D EFSCs [16]	2016	3.18	3.53
Single-chip 3D EFM based on in-plane rotary mechanism [17]	2017	0.10	0.14
Assembled 3D EFM with low cross-axis interference [18]	2018	0.33	0.46
Microassembled 3D EFM [23]	2019	0.33	1.48
Single-chip 3D EFM with piezoelectric excitation [19]	2023	0.23	2.19
This work	2024	14.78	16.77

6.3. Vector Electric Field Verification Tests

When exposed to a uniform electric field of 8 kV/m, the 3D EFM rotates around the R1 axis and the resultant electric field outputs from its three measurement units are depicted in Figure 14. It is obvious that the electric field curves of the *X*- and *Z*-axis measurement units display sinusoidal characteristics and are separated by an approximate phase difference of 90°, which is consistent with the simulation results in Figure 9a. One notable contrast is that since Figure 14 represents electric field curves, they are exactly opposite to the induced charge curves shown in Figure 9a. Additional observations from Figure 14 reveal that the *Y*-axis measurement unit's output does not consistently hold at 0. This may be attributed to two likely factors. One, adjusting φ to the ideal 0° during testing could be challenging, and two, the Teflon-made rod-shaped bracket may not be robust enough. A slight bend may be introduced once the 3D EFM is installed, impacting the *Y*-axis output.

Verification tests were conducted to evaluate the measurement accuracy of the developed 3D EFM. The parallel-plate apparatus generated a uniform electric field of 10 kV/m. The 3D EFM was randomly rotated into eight different attitudes. For each attitude, the estimated electric field was recorded, and the relative deviation (RD) between this estimate and the original applied electric field was calculated. The data are tabulated and given in Table 3. The maximum RD recorded was a mere 2.2%. The verification test results show that the estimated electric field is mostly consistent with the applied electric field. However, some measurement discrepancies still exist, which could be attributed to the calibration errors, as well as deviations caused by a slight bending of the rod-shaped bracket after installing the 3D EFM.

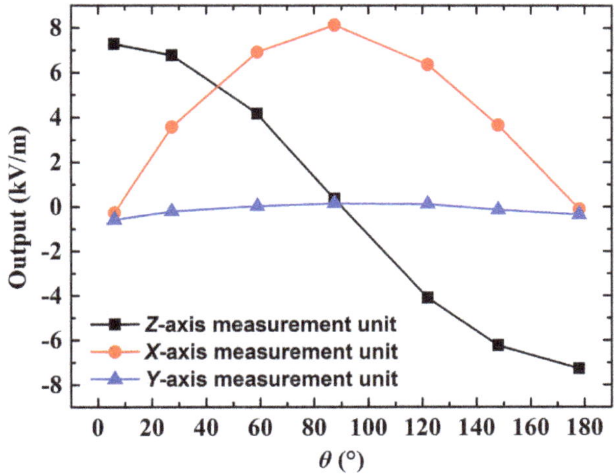

Figure 14. Test results of the 3D EFM rotating around the R1 axis when $E = 8$ kV/m.

Table 3. Verification test results with different attitudes.

θ (°)	φ (°)	Estimated Electric Field (kV/m)	RD (%)
174	10	10.03	0.3
135	86	9.85	−1.5
32	88	10.01	0.1
4	51	10.10	1.0
34	277	9.93	−0.7
71	277	9.80	−2.0
122	275	10.22	2.2
165	301	10.01	0.1

6.4. Moistureproof Verification Tests

Verification experiments were carried out within a temperature and humidity test chamber (THTC) to examine the impact of environmental humidity on the sensitivity and linearity of the 3D EFM. A small parallel-plate calibrator with a side length of 30 cm and an inter-plate space of 20 cm was placed in the THTC. The THTC settings were fixed at 99% relative humidity and a temperature of 25 °C. During the tests, it was difficult to rotate the 3D EFM within the THTC. Due to identical electric field detection and moisture prevention schemes in all three measurement units of the 3D EFM, only the Z-axis measurement unit was selected for the humidity verification test. The first phase of the experiment involved placing the 3D EFM, untreated for moisture resistance, in the THTC. Calibration tests were sequentially conducted after durations of 10, 20, and 30 min. Subsequently, a moisture-resistant 3D EFM was placed in the THTC. In this instance, calibration tests were performed following periods of 30, 60, 120, 180, and 240 min. The calibration curves from both experiments are exhibited in Figure 15, with sensitivity and linearity details tabulated in Table 4. It can be clearly seen from the test results that, without moisture-resistance treatment, 3D EFM's uniaxial sensitivity in a 99% relative humidity environment decays from 15.04 mV·kV^{-1}·m in a low-humidity environment to 1.77 mV·kV^{-1}·m after just 10 min. Sensitivity progressively decreases over time, and linearity deteriorates, rendering the device unable to meet application requirements. Conversely, the moisture-resistant 3D

EFM maintains near-constant uniaxial sensitivity at 15.03 mV·kV^{-1}·m after four hours in the same 99% high-humidity setting and holds linearity within 1%. These experimental results validate that the proposed 3D EFM can achieve accurate measurements, even in high-humidity environments.

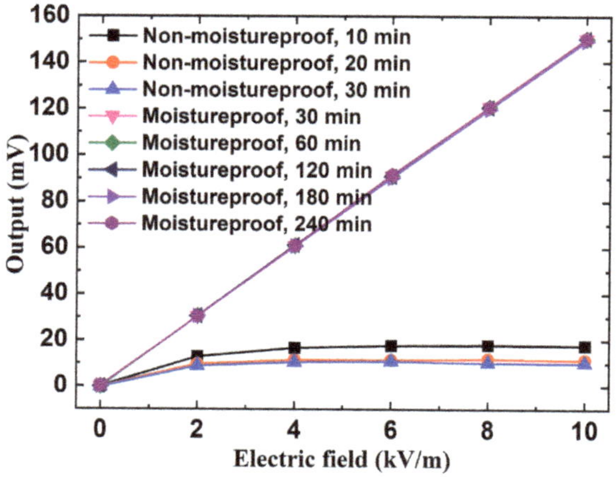

Figure 15. Verification test curves of non-moistureproof and moistureproof 3D EFM at 99% relative humidity for different times.

Table 4. Sensitivities and linearities of non-moistureproof and moistureproof 3D EFM at 99% Relative Humidity for Different Times.

Type	Duration (min)	Sensitivity $k_{zz}\left(\text{mV}\cdot\text{kV}^{-1}\cdot\text{m}\right)$	Linearity δ_L (%)
Non-moistureproof	10	1.77	12.32
	20	1.16	19.55
	30	1.04	29.63
Moistureproof	30	15.04	0.32
	60	15.05	0.18
	120	15.04	0.26
	180	14.95	0.77
	240	15.03	0.75

6.5. Zero Output Tests and Resolution Analysis

To evaluate the noise characteristics and stability of the 3D EFM, we measured the zero output over 8 h at a temperature of 20 °C and a relative humidity of less than 20% after shielding it in a grounded metal cavity. The data update rate of the 3D EFM is 1 s. The test results are plotted in Figure 16, showing that the output of the 3D EFM exhibits no significant drift. Grouping the data into sets of 10 points, the Allan deviations σ_A of the zero output of the 3D EFM were calculated as 0.023 (X), 0.07 (Y), and 0.066 (Z), respectively. The resolution R of any sensing axis i of the 3D EFM can be determined by

$$R_i = \frac{2\sigma_{A-i}}{k_{ii}} \tag{16}$$

where $i = X, Y, Z$, Based on the uniaxial sensitivity test results, the resolutions of the X-, Y-, and Z-axis measurement units of the 3D EFM were calculated as 2.7 V/m, 8.9 V/m, and 8.9 V/m, respectively, all of which are below 10 V/m.

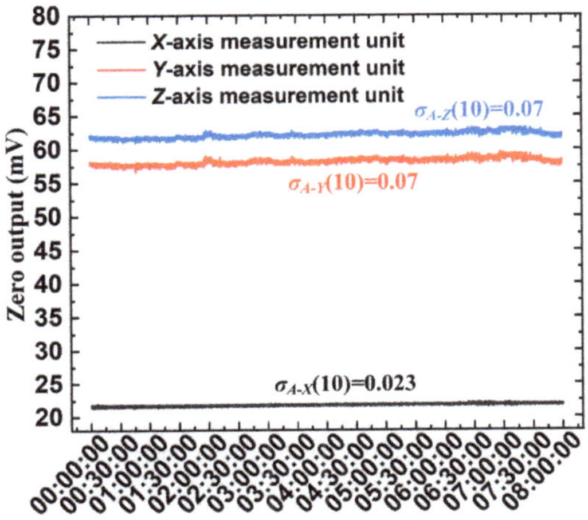

Figure 16. Zero output at a temperature of 20 °C and a relative humidity of less than 20%.

7. Conclusions

In this paper, a 3D EFM has been developed, and the detailed design, simulation, and implementation method are presented. To solve potential vibration instability issues, the shielding electrodes and fixed sensing electrodes of the 1D EFSCs employed in the EFM were redesigned. We proposed a highly sensitive packaging structure, furnished with an inward-convex cover, to augment the EFSC's sensitivity further. The integrated design and wireless transmission effectively suppress the impact of ground potential on detection results. It only necessitates that the three detecting electrodes be arranged perpendicularly to each other and imposes no restrictions on the layout of the EFSCs, which is particularly beneficial for the EFM's implementation. Furthermore, we adopted a humidity-resistant packaging structure in the EFM design, which not only mitigates the effect of environmental humidity but also further improves sensitivity. As a result of the inherent asymmetry of the EFM structure and electric field distortion, a matrix equation was developed to address cross-axis coupling interference. Finite element simulations reveal roughly uniform electric field distribution between the package's inner convex cylinder and the microsensor. Calibration and verification tests have been carried out. The EFM's uniaxial sensitivity is at least 4.64 times higher than the highest recorded sensitivity in previously reported MEMS 3D EFMs. The electric field curves of the *X*- and *Z*-axis measurement units exhibit sinusoidal characteristics separated by an approximate phase difference of 90°, consistent with simulation results. The EFM was rotated arbitrarily to eight different attitudes, with the maximum relative deviation recorded being a mere 2.2%. After four hours in high-humidity environments, the EFM maintains near-constant uniaxial sensitivity at 15.03 $mV \cdot kV^{-1} \cdot m$ and retains linearity within 1%. Based on the sensitivity test results, the uniaxial resolution of the EFM was found to be less than 10 V/m, as determined by analyzing the zero output.

Author Contributions: Conceptualization, P.Y. and X.W.; methodology, P.Y.; software, P.Y.; validation, P.Y., X.W., and Z.C.; formal analysis, P.Y.; investigation, P.Y.; resources, P.Y.; data curation, P.Y.; writing—original draft preparation, P.Y.; writing—review and editing, X.L.; visualization, P.Y.; supervision, C.P.; project administration, S.W.; funding acquisition, P.Y. All authors have read and agreed to the published version of the manuscript.

Funding: This research was funded in part by the National Natural Science Foundation of China, grant numbers 62101054 and 62031025, and in part by Beijing Municipal Science and Technology Project, grant number Z231100006623012.

Data Availability Statement: Data are contained within the article.

Acknowledgments: Part of this research work was carried out at Beijing TFlying Transducer Technology Co., Ltd. The authors would like to express their sincere gratitude to Zhouxi Wang and Zhenzhou Wang from the aforementioned company for their technical support and valuable contributions to the experimental work.

Conflicts of Interest: Author Shuang Wu was employed by the company Beijing TFlying Transducer Technology Co., Ltd. The remaining authors declare that the research was conducted in the absence of any commercial or financial relationships that could be construed as a potential conflict of interest.

References

1. Yang, P.; Wen, X.; Lv, Y.; Chu, Z.; Peng, C.; Liu, Y.; Wu, S. Improved microsensor-based fieldmeter for ground-level atmospheric electric field Measurements. *IEEE Trans. Instrum. Meas.* **2022**, *71*, 2001510. [CrossRef]
2. Fort, A.; Mugnaini, M.; Vignoli, V.; Rocchi, S.; Perini, F.; Monari, J. Design, modeling, and test of a system for atmospheric electric field measurement. *IEEE Trans. Instrum. Meas.* **2011**, *60*, 2778–2785. [CrossRef]
3. Wijeweera, G.; Bahreyni, B.; Shafai, C.; Rajapakse, A.; Swatek, D. Micromachined electric-field sensor to measure AC and DC fields in power systems. *IEEE Trans. Power Delivery.* **2009**, *24*, 988–995. [CrossRef]
4. Williams, K.R.; De Bruyker, D.P.H.; Limb, S.J.; Amendt, E.M.; Overland, D.A. Vacuum steered-electron electric-field sensor. *J. Microelectromech. Syst.* **2014**, *23*, 157–167. [CrossRef]
5. Wang, H.; Zeng, R.; Zhuang, C.; Lyu, G.; Yu, J.; Niu, B.; Li, C. Measuring AC/DC hybrid electric field using an integrated optical electric field sensor. *Electr. Power Syst. Res.* **2020**, *179*, 106087. [CrossRef]
6. Song, B.; Zhou, R.; Yang, X.; Zhang, S.; Yang, N.; Fang, J.; Song, F.; Zhang, G. Surface electrostatic discharge of charged typical space materials induced by strong electromagnetic interference. *J. Phys. D Appl. Phys.* **2021**, *54*, 275002. [CrossRef]
7. Cui, Y.; Yuan, H.; Song, X.; Zhao, L.; Liu, Y.; Lin, L. Model, design, and testing of field mill sensors for measuring electric fields under high-voltage direct-current power lines. *IEEE Trans. Ind. Electron.* **2018**, *65*, 608–615. [CrossRef]
8. Harrison, R.G.; Marlton, G.J. Fair weather electric field meter for atmospheric science platforms. *J. Electrost.* **2020**, *107*, 103489. [CrossRef]
9. Tantisattayakul, T.; Masugata, K.; Kitamura, I.; Kontani, K. Development of the hybrid electric field meter for simultaneous measuring of vertical and horizontal electric fields of the thundercloud. *IEEE Trans. Electromagn. Compat.* **2006**, *48*, 435–438. [CrossRef]
10. Xing, H.; Yang, X.; Zhang, J. Thunderstorm cloud localization algorithm and performance analysis of a three-dimensional atmospheric electric field apparatus. *J. Electr. Eng. Technol.* **2019**, *14*, 2487–2495. [CrossRef]
11. Liu, C.; Yuan, H.; Lv, J.; Zhao, P.; Li, J.; Xu, H. A sensor for 3-D component measurement of synthetic electric field vector in HVDC transmission lines using unidirectional motion. *IEEE Trans. Instrum. Meas.* **2022**, *72*, 1500110. [CrossRef]
12. Ravichandran, M.; Kamra, A.K. Spherical field meter to measure the electric field vector—Measurements in fair weather and inside a dust devil. *Rev. Sci. Instrum.* **1999**, *70*, 2140–2149. [CrossRef]
13. Zhang, Z.; Li, L.; Xie, X.; Xiao, D.; He, W. Optimization design and research character of the passive electric field sensor. *IEEE Sens. J.* **2014**, *14*, 508–513. [CrossRef]
14. Zhang, J.; Chen, F.; Sun, B.; Chen, K.; Li, C. 3D Integrated optical e-field sensor for lightning electromagnetic impulse measurement. *IEEE Photonics Technol. Lett.* **2014**, *26*, 2353–2356. [CrossRef]
15. Wen, X.; Fang, D.; Peng, C.; Yang, P.; Zheng, F.; Xia, S. Three dimensional electric field measurement method based on coplanar decoupling structure. In Proceedings of the 2014 IEEE SENSORS, Valencia, Spain, 2–5 November 2014; pp. 582–585.
16. Li, B.; Peng, C.; Zheng, F.; Ling, B.; Chen, B.; Xia, S. A decoupling calibration method based on genetic algorithm for three dimensional electric field sensor. In Proceedings of the 2016 IEEE SENSORS, Orlando, FL, USA, 30 October–3 November 2016; pp. 1–3.
17. Ling, B.; Wang, Y.; Peng, C.; Li, B.; Chu, Z.; Li, B.; Xia, S. Single-chip 3D electric field microsensor. *Front. Mech. Eng.* **2017**, *12*, 581–590. [CrossRef]
18. Ling, B.; Peng, C.; Ren, R.; Chu, Z.; Zhang, Z.; Lei, H.; Xia, S. Design, fabrication and characterization of a MEMS-based three-dimensional electric field sensor with low cross-axis coupling interference. *Sensors* **2018**, *18*, 870. [CrossRef] [PubMed]
19. Peng, S.; Zhang, Z.; Liu, X.; Gao, Y.; Zhang, W.; Xing, X.; Liu, Y.; Peng, C.; Xia, S. A single-chip 3-D electric field microsensor with piezoelectric excitation. *IEEE Trans. Electron Devices* **2023**, *70*, 4359–4365. [CrossRef]

20. Yang, P.; Wen, X.; Chu, Z.; Ni, X.; Peng, C. AC/DC fields demodulation methods of resonant electric field microsensor. *Micromachines* **2020**, *11*, 511. [CrossRef] [PubMed]
21. Yang, P.; Wen, X.; Chu, Z.; Ni, X.; Peng, C. Non-intrusive DC voltage measurement based on resonant electric field microsensors. *J. Micromech. Microeng.* **2021**, *31*, 064001. [CrossRef]
22. *IEEE Standard 644*; IEEE Standard Procedures for Measurement of Power Frequency Electric and Magnetic Fields from AC Power Lines. IEEE: Piscataway, NY, USA, 2019.
23. Ling, B.; Peng, C.; Ren, R.; Zheng, F.; Chu, Z.; Zhang, Z.; Lei, H.; Xia, S. A microassembled triangular-prism-shape three-dimensional electric field sensor. In Proceedings of the 2019 20th International Conference on Solid-State Sensors, Actuators and Microsystems & Eurosensors XXXIII (TRANSDUCERS & EUROSENSORS XXXIII), Berlin, Germany, 23–27 June 2019; pp. 222–225.

micromachines

MDPI

Article

Accelerated Electro-Optic Switching in Liquid Crystal Devices via Ion Trapping by Dispersed Helical Carbon Nanotubes

Rajratan Basu * and Christian C. Kehr

Soft Matter and Nanomaterials Laboratory, Department of Physics, The United States Naval Academy, Annapolis, MD 21402, USA
* Correspondence: basu@usna.edu

Abstract: Free ion impurities in liquid crystals significantly impact the dynamic electro-optic performance of liquid crystal displays, leading to slow switching times, short-term flickering, and long-term image sticking. These ionic contaminants originate from various sources, including LC cell fabrication, electrode degradation, and organic alignment layers. This study demonstrates that doping LCs with a small concentration of helical carbon nanotubes effectively reduces free ion concentrations by approximately 70%. The resulting reduction in ionic impurities lowers the rotational viscosity of the LC, facilitating faster electro-optic switching. Additionally, the purified LC exhibits enhanced dielectric anisotropy, further improving its performance in display applications. These findings suggest that helical carbon nanotubes doping offers a promising approach for mitigating ion-related issues in liquid crystals without the need for additional chemical treatments, paving the way for an efficient liquid crystal display technology.

Keywords: liquid crystals; ionic impurities; helical carbon nanotubes; electro-optic effects

1. Introduction

Excess ionic impurities in liquid crystals (LCs) present significant challenges [1,2] in liquid crystal display (LCD) technology, affecting electro-optic performance through slow response times, short-term flickering, and long-term image retention (commonly known as image sticking or dead pixels) [3–9]. These free ions originate from multiple sources, including LC chemical synthesis, polyimide (PI) alignment layers, and the degradation of indium tin oxide (ITO) conductive electrodes [10–12]. Additionally, ionic contamination can arise from the chemical decomposition and self-dissociation of LC materials [13]. Furthermore, LC mixtures containing cyano compounds, such as E7, exhibit a tendency to adsorb more ions due to the cyano polar group. Consequently, for applications in thin-film transistor liquid crystal displays (TFT-LCDs), the predominant LC compounds are fluorinated to mitigate ion absorption [14,15]. Understanding the effects of free ion impurities on the electrical, mechanical, and electro-optic properties of LCs is crucial for both fundamental research and practical applications, as the presence of ions can influence the anchoring energy, elastic constants, and switching responses [16–23].

Several methods exist to mitigate ionic contamination in LCs [24]. Traditional purification techniques, such as vacuum sublimation, chromatography, extraction, electrodialysis, multiple recrystallizations, vacuum distillation, zone refining, and ion exchange, are effective but often costly, time consuming, and labor intensive [25–28]. Furthermore, even highly purified LCs can acquire new ionic impurities during device fabrication due to interactions with electrodes, alignment layers, and adhesives [10–12]. Thus, developing alternative approaches to ion removal remains a key area of ongoing research.

Recent efforts have explored nanomaterials as an alternative to conventional chemical purification. Studies have shown that dispersing nanomaterials such as ferroelectric nanoparticles [29,30], titanium nanoparticles [31], carbon nanotubes [32,33], graphene [34–38], fullerenes [36,39,40], and gold nano-urchins [41] in LCs can effectively reduce free ion concentrations via ion trapping. Moreover, two-dimensional materials such as graphene [42] and hexagonal boron nitride [43,44], as alignment agents, have shown ion-capturing abilities that improve the operation of electro-optic LC devices.

Helical carbon nanotubes (*h*CNTs) [45] represent a unique class of chiral materials with a distinctive spring-like morphology, high modulus, and superior electronic properties compared to conventional graphite structures [46,47]. This study demonstrates that doping LCs with a small concentration of *h*CNTs results in a substantial (~70%) reduction of free ions. The trapping mechanism is attributed to the anisotropic, spring-like outer surface of the *h*CNTs, which effectively captures ionic impurities. Experimental results reveal that this ion reduction leads to notable changes in the intrinsic properties of the LC, including decreased rotational viscosity, improved electro-optic switching dynamics, and enhanced dielectric anisotropy.

From a microscale technology perspective, this approach offers a tunable pathway for improving LC-based displays and photonic devices by controlling ion transport dynamics at the nanoscale. The experimental validation of this mechanism within a well-characterized LC cell architecture advances the understanding and engineering of microstructured LC devices with enhanced electro-optic properties. By presenting an effective strategy for ion management in LC systems, this work paves the way for next-generation high-speed liquid crystal devices, making it relevant to micro and nanoscale technologies.

2. Materials

In this study, non-functionalized helical multiwalled carbon nanotubes (*h*CNTs) in powder form, obtained from US Research Nanomaterials, Inc. (Houston, TX, USA), were utilized to investigate their effect on ionic impurity reduction and electro-optic behavior in a nematic LC system. The *h*CNTs had an outer diameter of 130 nm and an average length of 6 μm. Figure 1a is an SEM image of *h*CNTs.

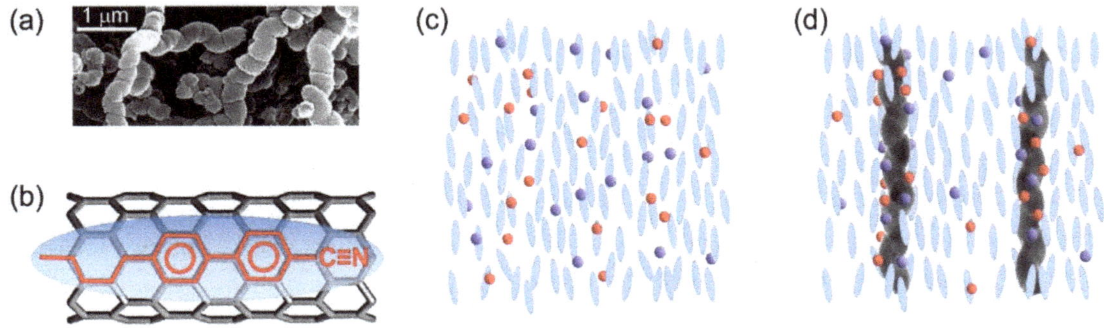

Figure 1. (**a**) An SEM image of *h*CNTs. (**b**) A schematic representation of the LC-CNT interaction: anchoring of a liquid crystal molecule on a carbon nanotube surface due to π–π electron stacking. The blue ellipsoid represents a generic liquid crystal molecule, while the black cylindrical honeycomb structure depicts the carbon nanotube surface. (**c**) Random distribution of free ions in a nematic phase. (**d**) *h*CNTs' ion trapping process in a nematic phase.

The *h*CNTs were initially dispersed in ethanol to ensure a uniform dispersion of *h*CNTs in the LC medium. The ethanol+*h*CNTs suspension was first shaken using a vortex mixer for 1 h and subsequently sonicated for 6 h. The nematic LC E7 (EMD Millipore Corporation,

Billerica, MA, T_{NI} = 60 °C) was then added to the ethanol+hCNTs mixture and sonicated for an additional hour to achieve complete dissolution. The ethanol was slowly evaporated at an elevated temperature, leaving behind a purified E7 + hCNTs mixture with a final hCNT concentration of approximately 3.1×10^{-3} wt.%. To eliminate residual solvent and air bubbles, the mixture was degassed under vacuum for 4 h, followed by further sonication for 6 h.

For consistency, the pure LC E7 was subjected to the same treatment process—dissolution in ethanol, controlled evaporation, and degassing—before experimental analysis. The LC E7 and E7 + hCNTs mixtures were then introduced into commercially manufactured antiparallel-rubbed capacitive type planar LC cells (Instec, Inc., Boulder, CO, USA), featuring an ITO-coated area of 1×1 cm^2 and a cell gap of $d = 20 \pm 0.20$ µm. An optical interferometric measurement was used to accurately measure the cell thickness using the equation, $d = k\,\lambda_1 \lambda_2 / 2(\lambda_2 - \lambda_1)$, where k represents the number of interference cycles between wavelengths λ_1 and λ_2. The cells were filled via capillary action at an elevated temperature (T = 65 °C) in the isotropic phase and subsequently cooled to room temperature for characterization.

It is well established that LC molecules anchor to the surface of *regular* CNTs via π–π stacking interactions, maximizing the hexagon–hexagon interaction between the LC molecules' benzene rings and the honeycomb lattice of the CNTs [48]. This interaction is schematically illustrated in Figure 1b, where the ellipsoid depicts a generic LC molecule, and the black honeycomb cylindrical structure denotes the CNT surface. The molecular structure of a representative LC molecule is depicted within the ellipsoid, positioned on the CNT surface to highlight the π–π electron stacking interaction. The alignment of the benzene rings of the LC molecules with the hexagonal lattice of the CNTs characterizes this interaction. The anchoring energy associated with this π–π stacking interaction is estimated to be $|U_{anchor}| \approx 2.0$ eV per molecule [48]. When CNTs are dispersed as colloidal inclusions in an LC medium, this π–π stacking interaction drives the CNTs to align along the nematic director while simultaneously inducing the LC molecules to orient along the CNT long axis on the CNT surface [49–52]. This stable anchoring mechanism leads to several intriguing phenomena, including an enhancement of the nematic orientational order parameter in LCs [51], the formation of pseudo-nematic LC domains around the CNTs even in the isotropic phase [52], the transfer of CNT surface chirality to otherwise achiral LCs [53–55], an increase in the polar anchoring energy at the LC–CNT interface [56], and vertically aligned CNTs-induced homeotropic LC alignment [57,58].

All those prior studies have focused on regular (non-helical) CNTs. The findings presented in this work demonstrate the potential of hCNTs as a promising approach for mitigating ion-related issues in LCs without necessitating extensive chemical purification. Figure 1c illustrates ions in a nematic phase. Figure 1d depicts ion trapping by dispersed hCNTs in the LC.

3. Experiments, Results, and Discussion

3.1. Ion Concentration

The free ion concentration, n_i, in LC E7 and E7 + hCNTs, was determined by analyzing the transient ion current, I_{ion}, generated when the applied voltage polarity across the cell is inverted [30,59]. When a square-wave voltage alternates between $+V$ and $-V$, the LC molecules do not rotate since their reorientation depends on the electric field magnitude, not its polarity [49–52,60]. However, voltage polarity inversion induces free ions to migrate toward opposite electrodes, resulting in a transient ion current within the LC cell.

To generate I_{ion}, a square-wave voltage with a peak-to-peak amplitude of 30 V (ranging from +15 V to −15 V) at 1 Hz was applied using an automatic liquid crystal tester (Instec, Inc., Boulder, Colorado, USA). The resulting ion current as a function of time was recorded at T = 25 °C, as shown in the inset in Figure 2. The transient ion current reaches its peak value when the positive and negative ions meet approximately at the center of the cell. The peak time is given by the following: $t_{ion-peak} = \frac{d^2}{2\mu E}$, where μ is the ion mobility [59]. As the ions continue their migration, I_{ion} eventually decays to zero when they reach the opposite electrodes, as observed in the inset in Figure 2. The total free ion transport was calculated by integrating the area under the I_{ion} vs. time curve, and the free ion concentration was extracted using the following: $n_i = \left(\int_0^t I_{ion}dt\right) / A.d$, where A is the active electrode area, and d is the cell gap. At elevated temperatures, both the free ion concentration and ion mobility increase. Figure 2 shows n_{i} for both samples as a function of temperature, revealing a substantial suppression of ion concentration in E7 + *h*CNTs compared to pure E7. Notably, for the E7 + *h*CNTs sample, n_{i} is reduced by approximately 60% to 70% in the temperature range, 25 °C to 60 °C.

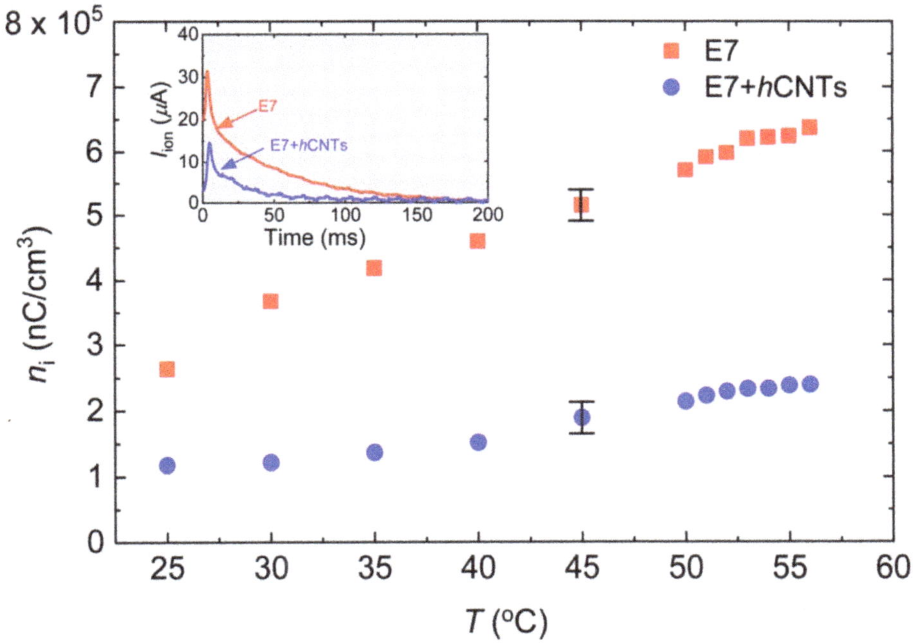

Figure 2. Free ion concentration, n_{i}, as a function of temperature for E7 and E7 + *h*CNTs samples listed in the legend. Typical error bars are shown. Inset: ion current, I_{ion} as a function of time for E7, and E7 + *h*CNTs at 25 °C after the voltage is inverted across the cells. The peak represents the ion bump when positive and negative ions meet in the middle of the cell.

Literature reports [61] indicate that non-helical CNTs at a concentration of 0.05 wt.% reduce the ion concentration in liquid crystal E7 by approximately 50% at 30 °C. However, at temperatures exceeding 45 °C, this reduction decreases to around 20% or less [61]. In the present study, *h*CNTs at an ultralow concentration of ~10^{-3} wt.% achieved a reduction in ion concentration exceeding 65% over a broad temperature range. Since no known attractive force exists between *h*CNTs and free ions, this enhanced ion trapping is likely facilitated by the asymmetric spring-like morphology of *h*CNT-walls, which effectively capture both positive and negative ions within the LC.

The ion-trapping efficiency of *h*CNTs can be better understood by comparing it with other nanomaterials reported in the literature. Previous studies have demonstrated that different nanoparticles exhibit varying degrees of ion-trapping capabilities in liquid crystals. For instance, graphene nanoplatelets at a concentration of 0.5 wt.% in LC 8OCB resulted in a 30% reduction in ion concentration [24], while a similar concentration of graphene platelets in cholesteric LCs led to an approximately 32% reduction [24]. A much higher efficiency was observed in ferroelectric LCs doped with 0.5 wt.% fullerenes (C_{60}), where an 80% decrease in ion concentration was reported [40]. Ferroelectric $BaTiO_3$ nanoparticles, when incorporated at 0.275 wt.% in LC 5CB, were found to trap about 50% of the free ions [30]. In another study [24], titanium dioxide (TiO_2) nanoparticles at a concentration of 0.1 wt% were observed to reduce ion impurities in LC E7 by 53%.

Thus, the ion trapping by *h*CNTs in the current study is comparable to, or in some cases even exceeds, that of other nanomaterials reported in prior studies. However, what sets *h*CNTs apart is their ability to achieve a significant reduction in ion concentration at substantially lower doping concentrations than other nanoparticles.

The *h*CNTs utilized in this study had an average length of 6 μm. The cell gap employed is 20 μm, which is larger than the length of the *h*CNTs. However, to achieve a rapid response time, the LC cell gap is typically 5 μm in most LCD applications. Consequently, shorter *h*CNTs are required for thinner cells. It has been demonstrated that the electrochemical shortening process can effectively truncate CNT lengths to as small as 500 nm [62]. This process is also applicable to *h*CNTs since it is not dependent on any specific type of CNT. Therefore, electrochemically shortened *h*CNTs can be used for thin LC cells.

3.2. Rotational Viscosity

The rotational viscosity, γ_1, of an aligned LC quantifies the internal friction among LC molecules during their rotational motion. To investigate γ_1 in our LC samples, we conducted experiments using the same planar-aligned capacitive cell configuration, where the transient current induced by a DC field was measured [63–65]. When a DC field, significantly exceeding the threshold field, is applied across the LC cell, the induced current $I(t)$ exhibits a characteristic time response as the nematic director undergoes dynamic rotation. The time-dependent current response follows the relation

$$I(t) = \frac{A(\Delta\varepsilon\varepsilon_o)^2E^3}{\gamma_1}sin^2[2\theta(t)] \quad (1)$$

where A is the area of the cell, E is the applied electric field, $\Delta\varepsilon$ is the dielectric anisotropy, ε_o is the free space permittivity, and θ represents the director angle relative to the electrodes at a given time. The peak current occurs at θ = 45°, yielding

$$I_p = \frac{A(\Delta\varepsilon\varepsilon_o)^2E^3}{\gamma_1} \quad (2)$$

at the peak time,

$$t_p = \left[\frac{\gamma_1(-ln(tan\,\theta_o))}{\Delta\varepsilon\varepsilon_o}\right]\frac{1}{E^2} \quad (3)$$

where θ_o is the pre-tilt angle. A DC voltage pulse of 35 V with a 1 Hz interval was applied across the cell to generate $I(t)$, which was subsequently detected as a function of time via a digital storage oscilloscope through a load resistor in series. The inset in Figure 3 presents an example of $I(t)$ for two test cells, E7 and E7 + *h*CNTs, at T = 25 °C. The peak current I_p was extracted from the $I(t)$ versus time graph, and γ_1 was determined using the known values of E, $\Delta\varepsilon$, and A. The method for measuring $\Delta\varepsilon$ is discussed later.

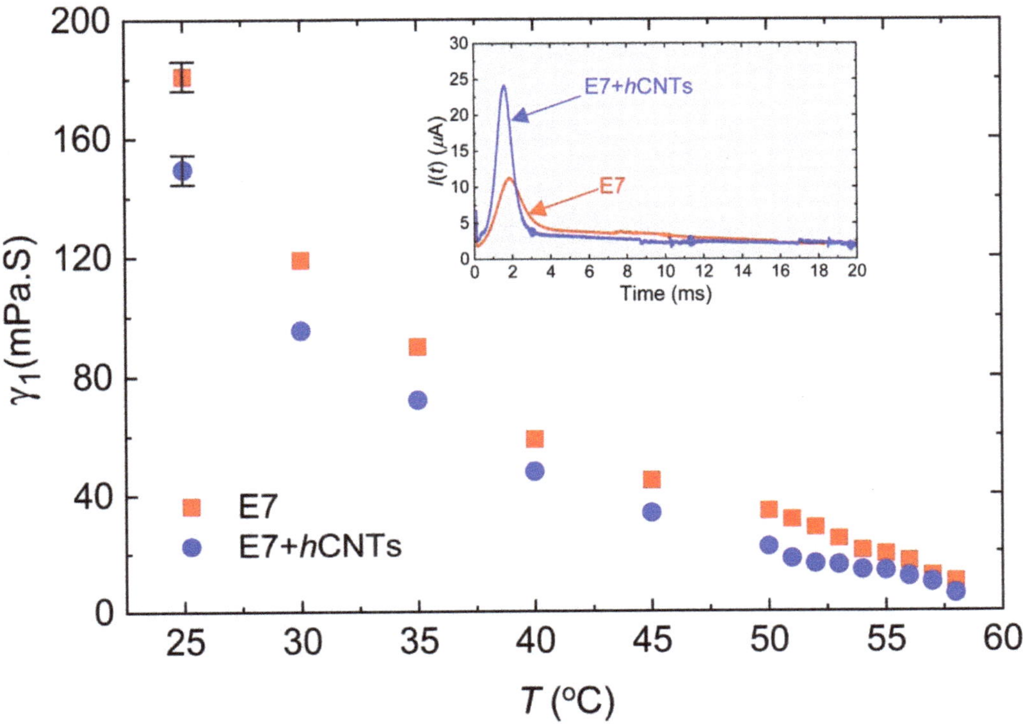

Figure 3. Rotational viscosity, γ_1 as a function of temperature for E7 and E7 + *h*CNTs samples, listed in the legend. Typical error bars are shown. Inset: transient current, $I(t)$ as a function of time for E7 and E7 + *h*CNTs at T = 25 °C.

Figure 3 depicts γ_1 as a function of temperature for the two cells. A clear pre-transitional behavior is observed for both samples, with E7 + *h*CNTs exhibiting a significant reduction in γ_1 compared to pure E7. The reduction in γ_1 for the E7 + *h*CNTs sample is approximately 17% at T = 25 °C.

Previous studies have suggested that the suppression of ionic impurities can lead to a decrease in the rotational viscosity of an LC. For instance, quantum dot-doped LCs have been shown to exhibit reduced γ_1 due to the trapping of ionic impurities, which lowers the overall ionic density and resistance of the nematic medium [66]. Similarly, our group previously reported that graphene flakes in a ferroelectric LC reduce rotational viscosity by effectively trapping ionic impurities [34]. Another study demonstrated that titanium nanoparticles suppress free ions, leading to stronger van der Waals dispersion interactions between LC molecules and the alignment layers, thereby reducing the pre-tilt angle of the LC molecules [31]. According to Equation (3), $\gamma_1 \propto 1/(-ln(tan\,\theta_o))$, for a constant applied field, implying that a decrease in the pre-tilt angle, θ_o, results in a corresponding decrease in γ_1. In our study, we attribute the observed reduction in γ_1 for the E7 + *h*CNTs sample to the suppression of ionic impurities. The decrease in internal resistance and friction, coupled with enhanced van der Waals interactions between LC molecules and the alignment layers, likely leads to a reduction in the pre-tilt angle, thereby decreasing γ_1. While a direct quantitative theoretical model linking ion concentration to γ_1 is not yet available, our findings are in agreement with previous reports, suggesting a coherent relationship between reduced ionic impurities and lower rotational viscosity.

3.3. Dielectric Anisotropy

The nematic phase exhibits dielectric anisotropy, defined as $\Delta\varepsilon = \varepsilon_{\parallel} - \varepsilon_{\perp}$, $\varepsilon_{\parallel}$ and $\varepsilon_{\perp}$, and are the dielectric permittivities parallel and perpendicular to the nematic director, respectively. To measure the dielectric constant ε as a function of the electric field, we employed an automatic liquid crystal tester (Instec, Inc., Boulder, CO, USA) operating at 1000 Hz for both pure E7 and E7 + *h*CNT samples. Figure 4a shows ε vs. V_{rms} at 25 °C. The values of $\varepsilon_{\parallel}$ and $\varepsilon_{\perp}$ were then used to determine Δε, as described in detail elsewhere [67]. Figure 4b illustrates the temperature dependence of Δε for the two test cells. Notably, Δε is higher in E7 + *h*CNTs sample.

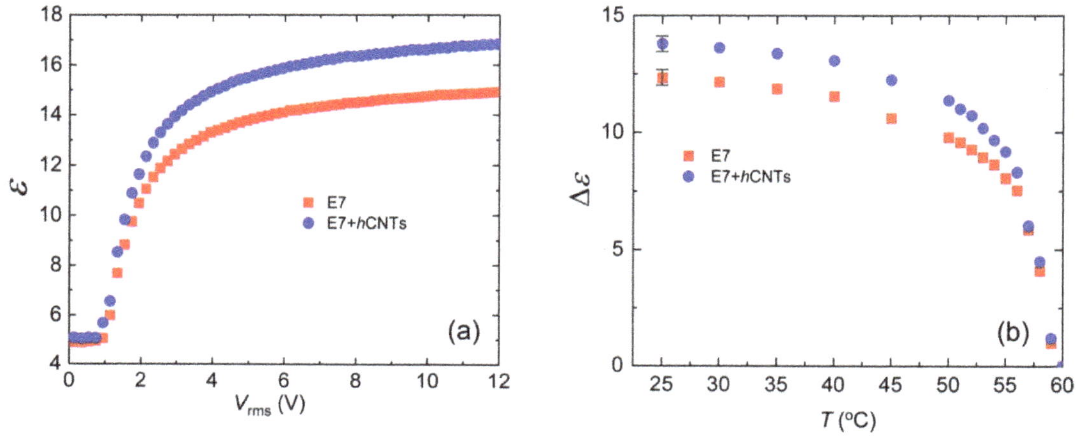

Figure 4. (**a**) Dielectric constant, ε as a function of V_{rms} for E7 and E7 + *h*CNTs samples, listed in the legend at T = 25 °C. (**b**) Dielectric anisotropy, Δε as a function of temperature for E7 and E7 + *h*CNTs samples, listed in the legend. Typical error bars are shown.

According to Maier and Meier's theory [68], the dielectric anisotropy Δε in the nematic phase is given by

$$\Delta\varepsilon = \frac{NhFS}{\varepsilon_0}\left[\Delta\alpha + \frac{\mu^2 F}{2k_B T}\left(3\cos^2\beta - 1\right)\right] \tag{4}$$

where N is the number density of LC molecules, μ is the resultant dipole moment, $\Delta\alpha$ is the polarizability anisotropy, S is the orientational order parameter, β is the angle between the molecular long axis and the dipole moment, h is the cavity field factor, and F is the feedback factor [68].

The presence of excess ionic impurities in the LC can influence the effective dipole moment of the LC molecules. Since LC molecules possess permanent dipole moments, negatively charged ions may accumulate near the positive ends of the molecules, while positively charged ions may cluster near the negative ends. This ionic accumulation effectively reduces the net dipole moment μ, leading to a decrease in Δε according to Equation (4). Thus, we propose that the suppression of ionic impurities in the E7 + *h*CNTs sample enhances the effective polarity of the LC molecules, resulting in an increased Δε.

Previous studies support this interpretation. For example, it has been reported that doping a nematic LC with 0.2 wt.% TiO_2 nanoparticles increases Δε, an effect attributed to the ion-trapping capability of TiO_2 [69]. Another study [66] demonstrated that doping nematic LCs with quantum dots (0.05 wt.%) captures free ions, enhancing the LC's birefringence. This suggests that suppressing ionic impurities improves the orientational order parameter S, which, in turn, leads to an increase in Δε. Our observations are consistent with these findings, further supporting the role of ion trapping in enhancing the dielectric anisotropy of liquid crystal systems.

3.4. Electro-Optic Effect

The rotational viscosity, γ_1, plays a critical role in determining the dynamic electro-optic response of aligned nematic LCs. Since γ_1 is altered in the presence of *h*CNTs, we investigated the dynamic electro-optic response of the E7 + *h*CNTs composite and compared it to that of pure E7. First, Figure 5a shows the field-off bright state and field-on dark state with their corresponding micrographs for the E7 and E7 + *h*CNTs composites. No significant changes were observed in the micrographs as there was no visible aggregation for the E7 + *h*CNTs composite. Then, an experimental approach, schematically shown in Figure 5b was employed using an optical setup where a 5-mW He-Ne laser beam (λ = 633 nm) was directed through a polarizer, an LC cell (with the director oriented at 45° relative to the polarizer), a crossed analyzer, and a photodetector. The detected intensity was recorded using a digital storage oscilloscope to analyze the temporal variation in transmittance upon applying a modulated square-wave driving voltage at 25 °C. Upon voltage application, the transmitted intensity decreases, with the optical switching on time (τ_{on}) defined as the interval required for intensity to drop from 90% to 10% of its maximum value. Conversely, upon voltage removal, the transmitted intensity increases, with the optical switching off time (τ_{off}) defined as the time required for intensity to rise from 10% to 90% of its maximum value. These switching times are governed by the following relations [70]:

$$\tau_{on} \propto \frac{\gamma_1}{\Delta\varepsilon\varepsilon_o V^2 - K_{11}\pi^2}\left(d^2 + \frac{4dK_{11}}{W_\theta}\right); \tau_{off} \propto \frac{\gamma_1}{\Delta\varepsilon\varepsilon_o V_b^2 - K_{11}\pi^2}\left(d^2 + \frac{4dK_{11}}{W_\theta}\right) \quad (5)$$

where d represents the cell gap, $\Delta\varepsilon$ is the dielectric anisotropy, V ($>>V_{th}$) is the driving applied voltage, V_b is the bias voltage, ε_0 is the permittivity of free space, K_{11} is the splay elastic constant, and W_θ is the polar anchoring strength coefficient.

Figure 5c illustrates the normalized transmitted intensity (left-hand *y*-axis) for E7 and E7 + *h*CNTs as a function of time upon applying a modulated square-wave driving voltage (right-hand *y*-axis) with a $|V_b| = 5$ V and a $|V| = 25$ V ($>>V_{th} = 0.90$ V) and $f = 20$ Hz.

Table 1 reveals a small decrease (~4%) in τ_{on} for the E7 + *h*CNT sample, whereas τ_{off} exhibits a significant acceleration (~25%) for the E7 + *h*CNT sample, as corroborated by Figure 5c. This enhancement in switching response is primarily attributed to the substantial reduction in γ_1 in the E7 + *h*CNT system. In this study, $V_{th} \approx 0.90$ V and the driving voltage, $|V| = 25$ V, ensuring $V >> V_{th}$. Under such high-voltage conditions, the electro-optic switching operates in the transient nematic relaxation mode, where τ_{off} is inherently fast (~milliseconds), even for large cell gaps [71,72]. Furthermore, since $V >> V_{th}$, the driving voltage predominantly dictates τ_{on}, as indicated by Equation (5), resulting in a minimal change in τ_{on} for the E7 + *h*CNT sample compared to pure E7. However, upon deactivation of the driving voltage, τ_{off} is a diffusion-type relaxation, mainly governed by the elastic interactions between the LC and the planar alignment layers and influenced by γ_1. Consequently, the observed reduction in γ_1 leads to a significantly faster τ_{off} in the E7 + *h*CNT sample. Additionally, previous studies [56] indicate that incorporating regular CNTs into an LC can enhance W_θ. Similarly, we propose that *h*CNTs can also enhance W_θ, leading to a faster τ_{off} according to Equation (5).

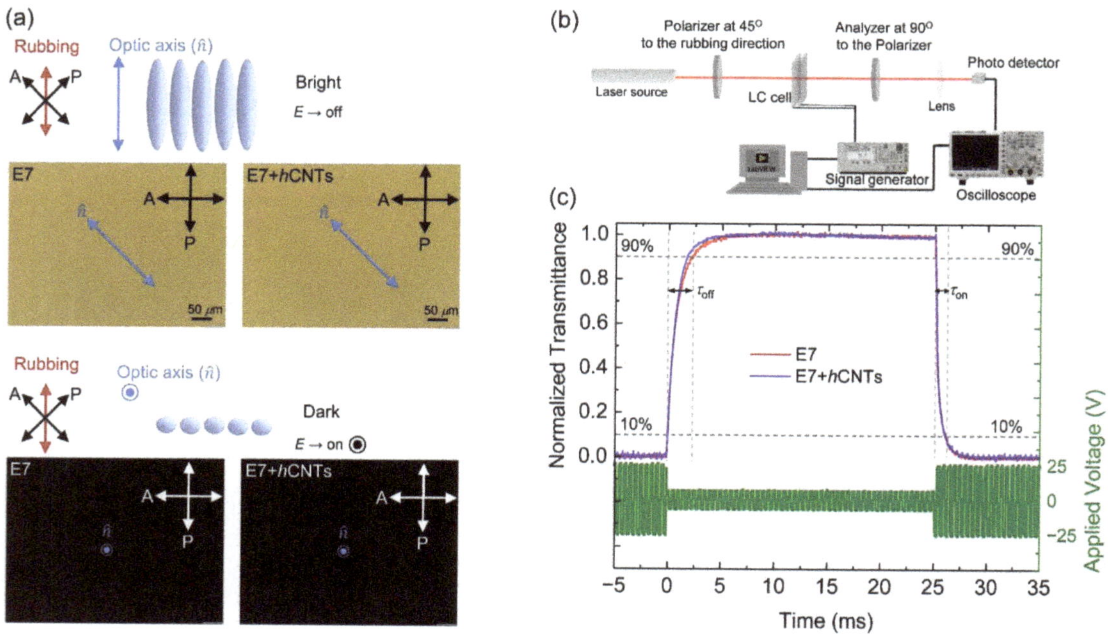

Figure 5. (**a**) A schematic representation of the field-off bright state and field-on dark state for a nematic LC. The micrographs for E7 and E7 + *h*CNTs cells show the field-off bright state and the field-on dark state, respectively. (**b**) A schematic representation of the electro-optic experimental setup. (**c**) Dynamic electro-optic response in E7 and E7 + *h*CNTs filled test cells. The driving modulated square wave voltage profile at f = 20 Hz is indicated on the right-hand y-axis. The left-hand y-axis shows the normalized transmitted intensity over time as V is turned off (at t = 0) and then turned on (at t = 25 ms), for the two test cells, as identified in the legend at T = 25 °C.

Table 1. The two characteristic times for the pure E7 and E7 + *h*CNTs from Figure 5c.

Samples	τ_{on} (μs)	τ_{off} (ms)
E7	970	2.40
E7 + *h*CNTs	930	1.80

Figure 5a presents micrographs of E7 and E7 + *h*CNTs cells under a cross-polarized microscope, capturing their voltage-off and voltage-on states. The E7 + *h*CNTs cell's micrograph at the voltage-off state shows a uniform texture like the E7 cell, with no visible *h*CNT aggregates. This confirms a homogeneous dispersion of *h*CNTs in the LC at the visible length scale.

3.5. The Effect of a Higher hCNTs Concentration

As discussed earlier, all presented experiments here thus far have been conducted at a *h*CNT concentration of 3.1×10^{-3} wt.% in E7. A key question that arises is how the ion-trapping phenomenon is affected by an increase in *h*CNT concentration. To investigate this, we prepared an additional sample with an *h*CNT concentration of 6.4×10^{-3} wt.% in E7, referred to as E7 + *h*CNTs-2. Experimental results indicate that the ion-trapping efficiency at this higher concentration remains largely unchanged, with only a ±4% variation observed over the studied temperature range.

However, a distinct and intriguing phenomenon was observed at this elevated concentration. It is well established that when conventional CNTs are dispersed as colloidal inclusions in an LC medium, their long axes align along the nematic director [49–52]. We propose that this alignment mechanism also holds for *h*CNTs. Upon the application of an external electric field, the nematic director reorients along the field direction, and due to their embedding within the nematic matrix, the *h*CNTs follow this field-induced director rotation.

As the *h*CNTs begin to rotate under the applied field, they exhibit a natural tendency to form wire-like aggregates due to entanglement. At sufficiently high concentrations, such as in E7 + *h*CNTs-2, these *h*CNT-wires become significantly extended. Eventually, under a sufficiently high electric field, the elongated *h*CNT-wires bridge the two electrodes of the LC cell (separated by 20 μm), leading to an insulator-to-conductor transition. Figure 6a shows a micrograph of the E7 + *h*CNTs-2 cell under a cross-polarized microscope, where several small black dots represent *h*CNT aggregates. Figure 6b,c schematically depicts the presence of *h*CNTs in the LC when the field is off, as well as the formation of *h*CNT wires that bridge the two electrodes of the cell under a high field. Figure 6d presents ε as a function of V_{rms} for E7 + *h*CNTs-2. After reaching a maximum value, ε decreases to zero, indicating that the LC cell capacitor has been shorted.

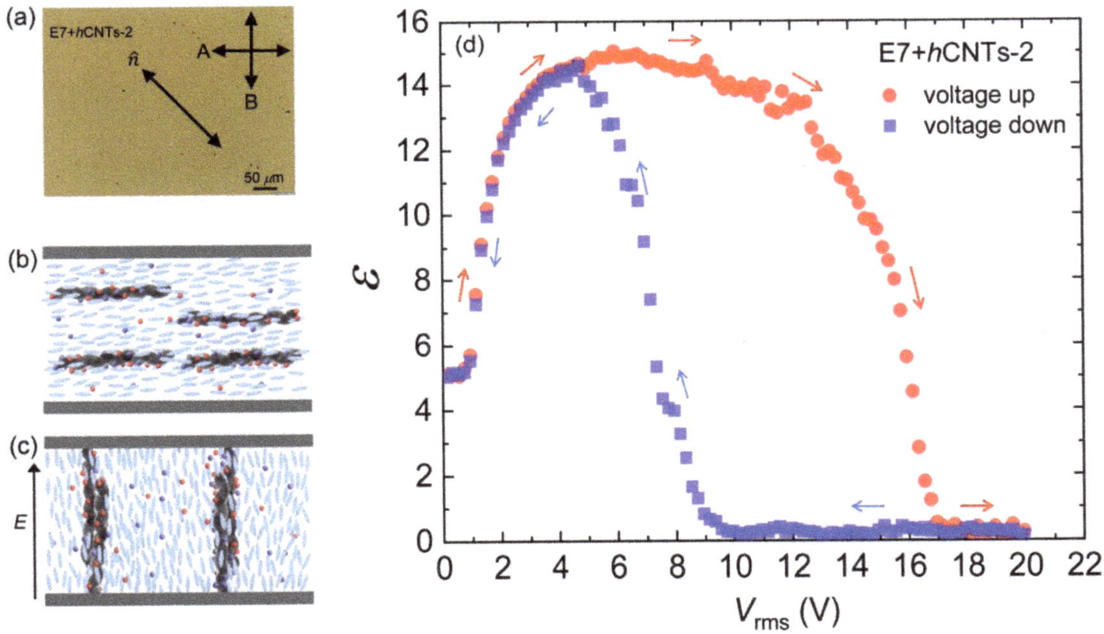

Figure 6. (**a**) A micrograph for the E7 + *h*CNTs-2 cell under a cross-polarized microscope. Small black dots are *h*CNT aggregates. Schematic representation of the presence of *h*CNTs in the LC when (**b**) the field is off and (**c**) the construction of *h*CNT wires bridging the two electrodes of the cell at a high field. (**d**) Dielectric constant, ε as a function of V_{rms} [voltage cycle up (red arrows) and down (blue arrows)] for E7 + *h*CNTs-2 at T = 25 °C.

This insulator-to-conductor transition of the LC cell is a reversible process. When the applied field is reduced to zero, the LC cell regains its original capacitance, as illustrated in the voltage up/down cycle shown in Figure 6d. This hysteresis behavior indicates that bridging the two electrodes by the *h*CNT wires occurs at different voltages during the voltage increase and decrease cycles. These findings suggest that 3.1×10^{-3} wt.% represents an optimal *h*CNT concentration for stable LC operation. Any concentration approaching or exceeding 6.4×10^{-3} wt.% leads to excessive *h*CNT aggregation, ultimately shorting the cell and preventing normal device functionality. It is important to observe that Figure 4a does not exhibit any hysteresis behavior for either pure E7 or E7 + *h*CNTs.

4. Conclusions

In this study, we have experimentally demonstrated that the incorporation of a small concentration of *h*CNTs into an LC medium effectively reduces the free ion concentration via an ion-trapping mechanism. The resulting decrease in mobile ion density lowers the internal friction within the nematic phase, thereby facilitating faster reorientation of the nematic director under an applied electric field. However, when the *h*CNT concentration exceeds a critical threshold (optimal concentration), the intrinsic entanglement tendency of *h*CNTs leads to the formation of wire-like aggregates, which ultimately short the LC cell.

These findings are significant for mitigating excess ionic impurities in LC materials and highlight the existence of an optimal *h*CNT concentration that balances ion trapping and electro-optic stability. The observed enhancement in optical switching speed is primarily attributed to the reduction in free ion density. This work provides valuable insights into the development of high-performance electro-optic devices based on nanostructured LC composites, offering a potential pathway towards faster and more efficient display and photonic applications.

Author Contributions: Conceptualization, R.B.; methodology, R.B. and C.C.K.; software, R.B. and C.C.K.; validation, R.B. and C.C.K.; formal analysis, R.B. and C.C.K. investigation, R.B. and C.C.K.; resources, R.B. and C.C.K.; data curation, R.B. and C.C.K.; writing—original draft preparation, R.B.; writing—review and editing, R.B. and C.C.K.; visualization, R.B. and C.C.K.; supervision, R.B.; project administration, R.B.; funding acquisition, R.B. All authors have read and agreed to the published version of the manuscript.

Funding: This research was funded by the Office of Naval Research (Program: Nano-engineered Materials, Award No. N0001424WX01991).

Data Availability Statement: The data that support the findings of this study are available from the corresponding author, R.B., upon reasonable request.

Conflicts of Interest: The authors declare no conflict of interest.

References

1. Heilmeier, G.H.; Heyman, P.M. Note on Transient Current Measurements in Liquid Crystals and Related Systems. *Phys. Rev. Lett.* **1967**, *18*, 583–585. [CrossRef]
2. Briere, G.; Gaspard, F.; Herino, R. Ionic residual conduction in the isotropic phase of a nematic liquid crystal. *Chem. Phys. Lett.* **1971**, *9*, 285–288. [CrossRef]
3. Takahashi, S. The investigation of a dc induced transient optical 30-Hz element in twisted nematic liquid-crystal displays. *J. Appl. Phys.* **1991**, *70*, 5346–5350. [CrossRef]
4. De Vleeschouwer, H.; Verweire, B.; D'Have, K.; Zhang, H. Electrical and Optical Measurements of the Image Sticking Effect in Nematic LCD'S. *Mol. Cryst. Liq. Cryst.* **1999**, *331*, 567–574. [CrossRef]
5. De Vleeschouwer, H.; Bougrioua, F.; Pauwels, H. Importance of Ion Transport in Industrial LCD Applications. *Mol. Cryst. Liq. Cryst.* **2001**, *360*, 29–39. [CrossRef]
6. Xu, D.; Peng, F.; Chen, H.; Yuan, J.; Wu, S.-T.; Li, M.-C.; Lee, S.-L.; Tsai, W.-C. Image sticking in liquid crystal displays with lateral electric fields. *J. Appl. Phys.* **2014**, *116*, 193102. [CrossRef]
7. De Vleeschouwer, H.; Verschueren, A.; Bougrioua, F.; van Asselt, R.; Alexander, E.; Vermael, S.; Neyts, K.; Pauwels, H. Long-term Ion Transport in Nematic Liquid Crystal Displays. *Jpn. J. Appl. Phys.* **2001**, *40*, 3272–3276. [CrossRef]
8. Yang, K.H. Charge retention of twisted nematic liquid-crystal displays. *J. Appl. Phys.* **1990**, *67*, 36–39. [CrossRef]
9. Sasaki, N.A. New Measurement Method for Ion Density in TFT-LCD Panels. *Mol. Cryst. Liq. Cryst.* **2001**, *367*, 671–679. [CrossRef]
10. Murakami, S.; Naito, H. Charge Injection and Generation in Nematic Liquid Crystal Cells. *Jpn. J. Appl. Phys.* **1997**, *36*, 773–776. [CrossRef]

11. Naemura, S.; Sawada, A. Ion Generation in Liquid Crystals under Electric Field. *Mol. Cryst. Liq. Cryst.* **2000**, *346*, 155–168. [CrossRef]
12. Van Aerle, N.A.J.M. Influence of Polyimide Orientation Layer Material on the Liquid Crystal Resistivity in LCDs. *Mol. Cryst. Liq. Cryst.* **1994**, *257*, 193–208. [CrossRef]
13. Sierakowski, M. Ionic interface-effects in electro-optical LC-cells. *Mol. Cryst. Liquid Cryst.* **2002**, *375*, 659–677. [CrossRef]
14. Hird, M. Fluorinated liquid crystals—Properties and applications. *Chem. Soc. Rev.* **2007**, *36*, 2070–2095. [CrossRef]
15. Yang, D.-K.; Wu, S.-T. *Fundamentals of Liquid Crystal Devices*, 2nd ed.; John Wiley & Sons, Ltd.: Chichester, UK, 2014.
16. Neyts, K.; Vermael, S.; Desimpel, C.; Stojmenovik, G.; van Asselt, R.; Verschueren, A.R.M.; de Boer, D.K.G.; Snijkers, R.; Machiels, P.; van Brandenburg, A. Lateral ion transport in nematic liquid-crystal devices. *J. Appl. Phys.* **2003**, *94*, 3891–3896. [CrossRef]
17. Yamashita, M.; Amemiya, Y. Drift Mobility of Positive Ions in Nematic MBBA at Low Electric Field. *Jpn. J. Appl. Phys.* **1978**, *17*, 1513–1517. [CrossRef]
18. Novotny, V. Measurement of mobilities of particles in liquids by optical and electrical transients. *J. Appl. Phys.* **1979**, *50*, 2787–2794. [CrossRef]
19. Sugimura, A.; Matsui, N.; Takahashi, Y.; Sonomura, H.; Naito, H.; Okuda, M. Transient currents in nematic liquid crystals. *Phys. Rev. B* **1991**, *43*, 8272–8276. [CrossRef]
20. Naito, H.; Okuda, M.; Sugimura, A. Transient discharging processes in nematic liquid crystals. *Phys. Rev. A* **1991**, *44*, R3434–R3497. [CrossRef]
21. Naito, H.; Yoshida, K.; Okuda, M. Transient charging current in nematic liquid crystals. *J. Appl. Phys.* **1993**, *73*, 1119–1125. [CrossRef]
22. Colpaert, C.; Maximus, B.; Meyere, A.D. Adequate measuring techniques for ions in liquid crystal layers. *Liq. Cryst.* **1996**, *21*, 133–142. [CrossRef]
23. Sawada, A.; Manabe, A.; Nameura, S. A Comparative Study on the Attributes of Ions in Nematic and Isotropic Phases. *Jpn. J. Appl. Phys.* **2001**, *40*, 220–224. [CrossRef]
24. Garbovskiy, Y.; Glushchenko, I. Nano-Objects and Ions in Liquid Crystals: Ion Trapping Effect and Related Phenomena. *Crystals* **2015**, *5*, 501–533. [CrossRef]
25. Kumar, S. *Liquid Crystals: Experimental Studies of Physical Properties and Phase Transitions*, 1st ed.; Cambridge University Press: Cambridge, UK, 2000.
26. Keller, P.; Liebert, L. Liquid crystal synthesis for physicists. In *Liquid Crystals*; Supplement, 14; Liebert, L., Ed.; Academic Press: New York, NY, USA, 1978; pp. 20–75.
27. Haberfeld, J.L.; Hsu, E.C.; Johnson, J.F. Liquid crystal purification by zone refining. *Mol. Cryst. Liquid Cryst.* **1973**, *24*, 1–5. [CrossRef]
28. Gaspard, F.; Herino, R.; Mondon, F. Low Field Conduction of Nematic Liquid Crystals Studied by Means of Electrodialysis. *Mol. Cryst. Liquid Cryst.* **1973**, *24*, 145–161. [CrossRef]
29. Garbovskiy, Y.; Glushchenko, I. Ion trapping by means of ferroelectric nanoparticles, and the quantification of this process in liquid crystals. *Appl. Phys. Lett.* **2015**, *107*, 041106. [CrossRef]
30. Basu, R.; Garvey, A. Effects of ferroelectric nanoparticles on ion transport in a liquid crystal. *Appl. Phys. Lett.* **2014**, *105*, 151905. [CrossRef]
31. Ha, Y.-S.; Kim, H.-J.; Park, H.-G.; Seo, D.-S. Enhancement of electro-optic properties in liquid crystal devices via titanium nanoparticle doping. *Opt. Express* **2012**, *20*, 6448–6455. [CrossRef]
32. Lee, C.-W.; Shih, W.-P. Quantification of ion trapping effect of carbon nanomaterials in liquid crystals. *Mater. Lett.* **2010**, *64*, 466–668. [CrossRef]
33. Chen, H.Y.; Lee, W.; Clark, N.A. Faster electro-optical response characteristics of a carbon-nanotube-nematic suspension. *Appl. Phys. Lett.* **2007**, *90*, 033510. [CrossRef]
34. Basu, R. Effects of graphene on electro-optic switching and spontaneous polarization of a ferroelectric liquid crystal. *Appl. Phys. Lett.* **2014**, *105*, 112905. [CrossRef]
35. Basu, R.; Garvey, A.; Kinnamon, D. Effects of graphene on electro-optic response and ion-transport in a nematic liquid crystal. *J. Appl. Phys.* **2015**, *117*, 074301. [CrossRef]
36. Wu, P.-W.; Lee, W. Phase and dielectric behaviors of a polymorphic liquid crystal doped with graphene nanoplatelets. *Appl. Phys. Lett.* **2013**, *102*, 162904. [CrossRef]
37. Wu, P.-C.; Lisetski, L.N.; Lee, W. Suppressed ionic effect and low-frequency texture transitions in a cholesteric liquid crystal doped with graphene nanoplatelets. *Opt. Express* **2015**, *23*, 11195–11204. [CrossRef]
38. Singh, D.P.; Gupta, S.K.; Vimal, T.; Manohar, R. Dielectric, electro-optical, and photoluminescence characteristics of ferroelectric liquid crystals on a graphene-coated indium tin oxide substrate. *Phys. Rev. E* **2014**, *90*, 022501. [CrossRef]
39. Lee, W.; Wang, C.-Y.; Shih, Y.-C. Effects of carbon nanosolids on the electro-optical properties of a twisted nematic liquid-crystal host. *Appl. Phys. Lett.* **2004**, *85*, 513. [CrossRef]

40. Shukla, R.K.; Raina, K.K.; Haase, W. Fast switching response and dielectric behaviour of fullerene/ferroelectric liquid crystal nanocolloids. *Liq. Cryst.* **2014**, *41*, 1726–1732. [CrossRef]
41. Basu, R.; Gess, D. Ion trapping, reduced rotational viscosity, and accelerated electro-optic response characteristics in gold nano-urchin-nematic suspensions. *Phys. Rev. E* **2023**, *107*, 024705.
42. Basu, R.; Lee, A. Ion trapping by the graphene electrode in a graphene-ITO hybrid liquid crystal cell. *Appl. Phys. Lett.* **2017**, *111*, 161905. [CrossRef]
43. Basu, R.; Atwood, L. Reduced ionic effect and accelerated electro-optic response in a 2D hexagonal boron nitride planar-alignment agent based liquid crystal device. *Opt. Mater. Express.* **2019**, *9*, 1441–1449. [CrossRef]
44. Basu, R. Reduced ionic effects and enhanced spontaneous polarization in a ferroelectric liquid crystal device employing a two-dimensional hexagonal boron nitride planar-alignment agent. *Phys. Rev. E* **2025**, *111*, 025402. [CrossRef] [PubMed]
45. Li, Y.; Xu, Z.; Jia, A.; Yang, X.; Feng, W.; Wang, P.; Li, K.; Lei, W.; He, H.; Tian, Y.; et al. Controllable modification of helical carbon nanotubes for high-performance microwave absorption. *Nanotechnol. Rev.* **2021**, *10*, 671–679. [CrossRef]
46. Vijayan, R.; Ghazinezami, A.; Taklimi, S.R.; Khan, M.Y.; Askari, D. The geometrical advantages of helical carbon nanotubes for high-performance multifunctional polymeric nanocomposites. *Compos. Part B Eng.* **2019**, *156*, 28–42. [CrossRef]
47. Liu, M.; Cowley, J.M. Structures of the helical carbon nanotubes. *Carbon* **1994**, *32*, 393–403. [CrossRef]
48. Park, K.A.; Lee, S.M.; Lee, S.H.; Lee, Y.H. Anchoring a Liquid Crystal Molecule on a Single-Walled Carbon Nanotube. *J. Phys. Chem. C* **2007**, *111*, 1620–1624. [CrossRef]
49. Lynch, M.D.; Patrick, D.L. Organizing Carbon Nanotubes with Liquid Crystals. *Nano Lett.* **2002**, *2*, 1197–1201. [CrossRef]
50. Dierking, I.; Scalia, G.; Morales, P. Liquid crystal–carbon nanotube dispersions. *J. Appl. Phys.* **2005**, *97*, 044309. [CrossRef]
51. Basu, R.; Iannacchione, G.S. Orientational coupling enhancement in a carbon nanotube dispersed liquid crystal. *Phys. Rev. E.* **2010**, *81*, 051705. [CrossRef]
52. Basu, R.; Garvey, A. Insulator-to-conductor transition in liquid crystal-carbon nanotube nanocomposites. *J. Appl. Phys.* **2016**, *120*, 164309. [CrossRef]
53. Basu, R.; Boccuzzi, K.; Ferjani, S.; Rosenblatt, C. Carbon nanotube induced chirality in an achiral liquid crystal. *Appl. Phys. Lett.* **2010**, *97*, 121908. [CrossRef]
54. Kalakonda, P.; Basu, R.; Nemitz, I.R.; Rosenblatt, C.; Iannacchione, G.S. Studies of nanocomposites of carbon nanotubes and a negative dielectric anisotropy liquid crystal. *J. Chem. Phys.* **2014**, *140*, 104908. [CrossRef] [PubMed]
55. Basu, R.; Chen, C.-L.; Rosenblatt, C. Carbon nanotube-induced macroscopic helical twist in an achiral nematic liquid crystal. *J. Appl. Phys.* **2011**, *109*, 083518. [CrossRef]
56. Lu, Y.; Chien, L.C. Carbon nanotube doped liquid crystal OCB cells: Physical and electro-optical properties. *Opt. Express* **2008**, *16*, 12777–12785. [CrossRef] [PubMed]
57. Basu, R.; Atwood, L. Homeotropic liquid crystal device employing vertically aligned carbon nanotube arrays as the alignment agent. *Phys. Rev. E* **2020**, *102*, 022701. [CrossRef]
58. Basu, R.; Gess, D. Electro-optic hybrid aligned nematic device utilizing carbon nanotube arrays and two-dimensional hexagonal boron nitride nanosheet as alignment substrates. *Phys. Rev. E* **2021**, *104*, 054702. [CrossRef]
59. Zou, Z.; Clark, N.A.; Handschy, M.A. Ionic transport effects in SSFLC cells. *Ferroelectrics* **1991**, *121*, 147–158. [CrossRef]
60. De Gennes, P.G.; Prost, J. *The Physics of Liquid Crystals*, 1st ed.; Oxford University Press: New York, NY, USA, 1994.
61. Liu, H.-H.; Lee, W. Ionic properties of liquid crystals dispersed with carbon nanotubes and montmorillonite nanoplatelets. *Appl. Phys. Lett.* **2010**, *97*, 173501. [CrossRef]
62. Zhang, Y.; Li, Y.; Wen, H.; Cao, L.; Xiao, W.; Zhao, J.; Nan, Z.; Zhang, S.; Qiu, S.; Li, Q. Length-controlled sorting and length-dependent properties of short semiconducting single-walled carbon nanotubes. *Carbon* **2023**, *215*, 118468. [CrossRef]
63. Imai, M.; Naito, H.; Okuda, M.; Okuda, M.; Sugimura, A. Determination of rotational viscosity of nematic liquid crystals from transient current numerical analysis and experiment. *Jpn. J. Appl. Phys.* **1994**, *33*, 3482–3487. [CrossRef]
64. Imai, M.; Naito, H.; Okuda, M.; Sugimura, A. Determination of rotational viscosity and pretilt angle in nematic liquid crystals from transient current influence of ionic conduction. *Mol. Cryst. Liq. Cryst. Sci. Technol. Sect. A* **1995**, *259*, 37–46. [CrossRef]
65. Imai, M.; Naito, H.; Okuda, M.; Sugimura, A. A method for determination of rotational viscosity and pretilt angle from transient current in twisted nematic liquid crystal cells. *Jpn. J. Appl. Phys.* **1995**, *34*, 3170–3176. [CrossRef]
66. Rastogi, A.; Pandey, F.; Manohar, R.; Singh, S. Effect of Doping of Cd1_xZnxS/ZnS Core/Shell Quantum Dots in Negative Dielectric Anisotropy Nematic Liquid Crystal p-Methoxybenzylidene p-Decylaniline. *Crystals* **2021**, *11*, 605. [CrossRef]
67. Basu, R.; Atwood, L.J.; Sterling, G.W. Dielectric and electro-optic effects in a nematic liquid crystal doped with *h*-BN flakes. *Crystals* **2020**, *10*, 123. [CrossRef]
68. Demus, D.; Goodby, J.; Gary, G.W.; Spiess, H.-W.; Vill, V. *Physical Properties of Liquid Crystals*, 1st ed.; Wiley VCH: Weinheim, Germany, 1999.
69. Yadav, G.; Katiyar, R.; Pathak, G.; Manohar, R. Effect of ion trapping behavior of TiO_2 nanoparticles on different parameters of weakly polar nematic liquid crystal. *J. Theor. Appl. Phys.* **2018**, *12*, 191–198. [CrossRef]

70. Nie, X.; Lu, R.; Xianyu, H.; Wu, T.X.; Wu, S.T. Anchoring energy and cell gap effects on liquid crystal response time. *J. Appl. Phys.* **2007**, *101*, 103110. [CrossRef]
71. Wu, S.T.; Wu, C.S. High-speed liquid-crystal modulators using transient nematic effect. *J. Appl. Phys.* **1989**, *65*, 527–532. [CrossRef]
72. Wu, S.T.; Wu, C.S. Small angle relaxation of highly deformed nematic liquid crystals. *Appl. Phys. Lett.* **1988**, *53*, 1794–1796. [CrossRef]

micromachines

MDPI

Review

Examination of Impact of NBTIs on Commercial Power P-Channel VDMOS Transistors in Practical Applications

Danijel Danković *, Emilija Živanović, Nevena Veselinović, Dunja Đorđević, Marija Petrović, Lana Tasić, Miloš Marjanović, Sandra Veljković, Nikola Mitrović, Vojkan Davidović and Goran Ristić

Faculty of Electronic Engineering, University of Niš, Aleksandra Medvedeva 4, 18000 Niš, Serbia; emilija.zivanovic@elfak.ni.ac.rs (E.Ž.); nevena.veselinovic@elfak.rs (N.V.); dunja.djordjevic@elfak.rs (D.Đ.); marija.petrovic@elfak.rs (M.P.); lana.tasic@elfak.rs (L.T.); milos.marjanovic@elfak.ni.ac.rs (M.M.); sandra.veljkovic@elfak.ni.ac.rs (S.V.); nikola.i.mitrovic@elfak.ni.ac.rs (N.M.); vojkan.davidovic@elfak.ni.ac.rs (V.D.); goran.ristic@elfak.ni.ac.rs (G.R.)

* Correspondence: danijel.dankovic@elfak.ni.ac.rs

Abstract

In this paper, the impact of negative bias temperature instabilities (NBTIs) on commercial power p-channel Vertical Double-Diffused MOS (VDMOS) transistors from the standpoint of practical applications was analyzed. The effects of NBTI are one of the main reliability concerns for this type of device, so it is necessary to investigate how these effects influence various applications. A series of experiments were carried out including negative bias temperature stressing, infra-red thermographic recording and circuit characterization, with the goal of evaluating the effects of negative bias temperature stressing on the self-heating of samples in load-driving circuits operating with higher currents and circuit performance of a CMOS inverter circuit containing the examined samples. The findings suggest that negative bias temperature stressing-induced threshold voltage shift directly affects increased self-heating in load-driving circuits and that it also directly affects transfer and dynamics characteristics in CMOS inverters.

Keywords: VDMOS; NBTI; self-heating effects; CMOS inverter; threshold voltage shift

1. Introduction

Through research and development, power VDMOS (Vertical Double-Diffused Metal Oxide Semiconductor) transistors have been proposed as devices suitable for multiple practical applications in different industries. Their wide range of utilization includes applications in the automotive industry and aerospace industry, as well as in switching power supplies and motor drivers [1–3]. This vast scope of use cases enables application in different types of circuits and various environments (presence of different types of irradiation or extreme temperatures). Although there are many reports in the literature investigating the reliability of VDMOS devices, constant expansion of application areas in different industry branches demands continuous attention to reliability considerations, such as degradation mechanisms, operating limitations and lifetime estimations [2,4–7].

The main reliability concerns regarding power VDMOS devices are BTIs (bias temperature instabilities), SHEs (self-heating effects), HCEs (Hot Carrier Effects) and TDDB (Time-Dependent Dielectric Breakdown) [4,8,9]. These effects represent the main parametric failure mechanisms in practical applications.

1.1. Effect of Negative Bias Temperature Instabilities

One of the most serious effects that alters the parameters of VDMOS transistors is represented by bias temperature instabilities (BTIs). Depending on whether the bias is negative or positive, two different groups of effects (NBTIs and PBTIs) are reported in the literature; the effects of NBTI on p-channel devices are the most significant ones [4,5]. In many studies devoted to understanding NBTIs, it has been found that this phenomenon is a result of the generation of oxide-trapped charge (N_{ot}) and interface traps (N_{it}), which occur during the operation of devices. NBTIs have been found to occur mostly in p-channel devices operated under negative gate oxide fields in the range 2–6 MV/cm and at elevated temperatures (100–250 °C) [5,10,11].

In p-channel devices, NBTI leads to an increase in the absolute value of the threshold voltage, which is one of the crucial parameters of the MOS transistor, and an increase in drain-source resistance when the device is in an on-state ($R_{DS(ON)}$); it also results in a decrease in the device transconductance (g_m) and the absolute drain current (I_{Dsat}). Even though the first publication describing the effects of NBTI from Jeppson and Svenson dates back to 1977, scaling and miniaturization of devices led to an increasing number of studies on this topic in the previous decade [10–12]. The main degradation mechanisms still remain a subject of debate [11], while many different analytical and compact models covering various transistor structures have been developed [6,13–15]. Many reports in the literature, including this one, are focused on investigating the physical mechanisms and responsible processes that occur during negative bias temperature stressing (NBTS) in VDMOS devices [4,5,7,16,17]. Effects are typically observed under accelerated stress conditions, meaning that they are under elevated gate voltage and temperature values. In previous decades, our research included a great number of different combinations of gate voltages and temperatures, each producing corresponding threshold voltage shifts. Experiments have been carried out for combinations of five different gate voltages (−30 V, −35 V, −40 V, −45 V and −50 V) and three different temperatures (125 °C, 150 °C and 175 °C). Part of these previous results, which show ΔV_T for the conditions producing the largest shift (V_G = −50 V, T = 175 °C) and the smallest shift (V_G = −30 V, T = 125 °C), as well as the most frequently used stress conditions (V_G = −45 V, T = 175 °C), are presented in Figure 1. It should be noted that an investigation of pulsed NBT stressing for different frequency and duty cycle values, under the described gate voltages and temperature conditions, has been reported in our previous publications [18,19], with experiment durations from 24 h to 2000 h [18–21].

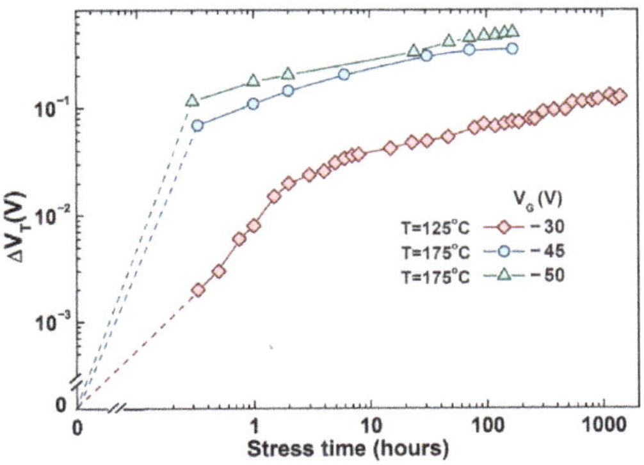

Figure 1. Threshold voltage shift in p-channel power VDMOSFET subjected to NBT stress under different stressing conditions.

https://doi.org/10.3390/mi17010052

Figure 1 shows the threshold voltage shift in devices subjected to NBT stress under selected conditions. It should be emphasized that the group stressed at 125 °C was exposed to NBT stress for 1416 h, whereas the groups stressed at 175 °C were exposed to NBT stress for 168 h. A pronounced difference in threshold voltage shift is observed between the stress conditions at 125 °C and 175 °C, with significantly larger shifts recorded at 175 °C. It is also clearly visible that the applied gate voltage has a substantial influence on the threshold voltage variation. Specifically, devices exposed to a gate voltage of −50 V exhibit larger threshold voltage shifts compared to those stressed at −45 V under the same temperature conditions.

Furthermore, in the initial phase of NBT stress, differences in threshold voltage shift rates can be observed between devices stressed at the same temperature but under different stress voltages. The data show that the rate of threshold voltage shift is higher for devices exposed to the higher gate voltage (−50 V) compared to those stressed at lower stress voltages (−45 V).

Recent reports also suggest that the effects of NBTI can have a direct impact on the practical applications of VDMOS devices [2,3,5,16]. Their superior switching characteristics, which allow operation in the megahertz range, tend to suffer from NBTI-induced pulse narrowing. In switching applications, the p-channel VDMOS operates as a switch, where turning on the transistor (with negative voltage in p-channel devices) leads to channel inducing and allows for conductance of current. Turning off the transistor thereby disallowing the conductance of current. The turning on/off of the transistor is controlled with the signal that is being fed to the gate of the transistor.

Continuous operation of the devices leads to eventual degradation; a change in the threshold voltage is one of the most significant changes. During longer operation intervals, the controlling signal, which is being generated from some external circuit, usually keeps its initial characteristics, while the characteristics of the transistor alter. This leads to some limitations in operations and some unwanted effects, which are considered in the literature as NBTI pulse narrowing [3]. An illustration depicting this effect is shown in Figure 2.

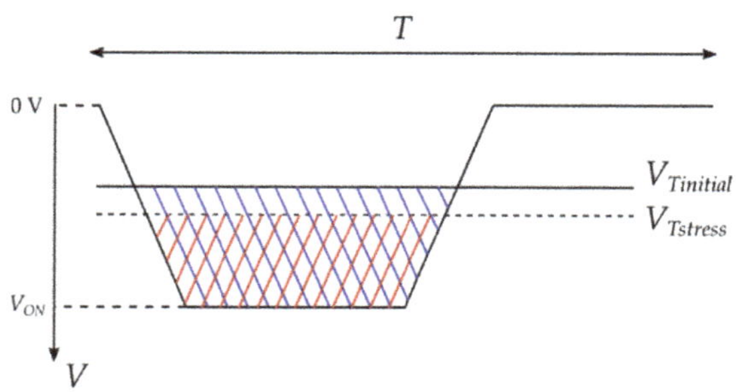

Figure 2. Illustration of NBTI-induced pulse narrowing and reduced operation area during on-time of one period.

When the threshold voltage is changed (increased in absolute value), the transistor will turn on later than expected, meaning later than it will turn on if the threshold voltage sustains its initial value. Therefore, the transistor will sustain the generated channel between the drain and the source terminals for a shorter interval of time than expected. This also shortens the interval of the conduction of current, which can make a direct impact on some of the applications. These effects are more emphasized with longer t_{rise} and t_{fall} intervals, whereas in practical application of the device, longer transition times are typical. Conclusions from the literature suggest that further examination is needed to assess the impact of NBTI in direct practical applications [2,3,5].

https://doi.org/10.3390/mi17010052

1.2. Self-Heating Effects

Besides NBTI, the significant impact of the parameters of VDMOS devices is caused by self-heating effects. Power p-channel devices are able to conduct relatively high currents (for example, commercial power p-channel VDMOSFET IRF9520 (Vishay, Malvern, PA, USA) can conduct a maximum drain current of 6.8 A, as stated in the datasheet [22]). This characteristic is widely used in motor driver applications. While there are many different motor driving circuits reported in the literature, one of the most basic circuits is presented in Figure 3 [23].

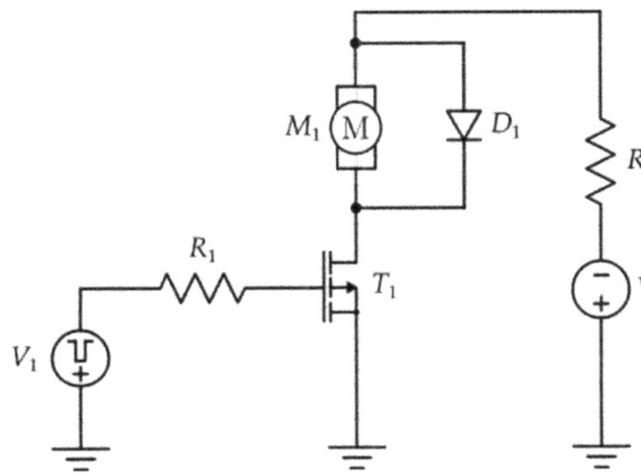

Figure 3. An example of the Basic DC motor driving circuit with p-channel power VDMOSFET operating as a high-end switch [23].

When conducting higher currents, the power dissipation of the device increases, leading to an increase in self-heating. There are several investigations reporting that excessive self-heating wears out power VDMOS devices and degrades their parameters [5,23–26]. A combination of self-heating and higher ambient temperature can lead to the parametric failure of the device.

One of the parameters that deviates due to either aging or accelerated NBT stress is the change in the turn-on resistance ($R_{DS(ON)}$). One of the consequences of $R_{DS(ON)}$ increase is increased channel resistance (R_{ch}), which affects power dissipation during the change between the turned-on and turned-off state. Previous investigations suggested that, even though the area of operation is reduced after threshold voltage shift, as is illustrated in Figure 2, self-heating of the devices is not necessarily reduced accordingly. This is caused by increased power dissipation during state changes.

Figure 4 schematically illustrates the internal architecture of a power VDMOS transistor, consisting of hundreds of hexagonal unit cells, with two representative half-cells highlighted to emphasize the current flow distribution and the locations of the dominant resistances. The figure provides an expanded view of the structural regions that contribute to the overall on-state resistance of the device.

In addition to the resistances originating from the active regions, such as the channel, p+ source region, JFET and drift region, the total on-state resistance is also significantly influenced by elements located outside the chip but still within the TO-220 package. These include the bond wire resistance (R_{BW}), bonding solder resistance (R_{Die}) and contact resistance R_C [27]. Combined, these form the drain contact resistance (R_{CD}). These packaging-related resistances become particularly relevant at high currents and elevated operating temperatures.

https://doi.org/10.3390/mi17010052

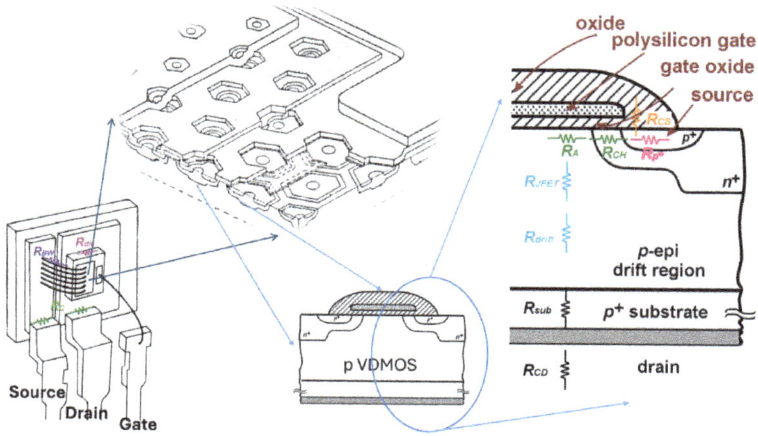

Figure 4. Illustration of cross-section of the commercial p-channel power VDMOSFET with marked internal resistances.

Therefore, many previous studies have concluded that more investigation is needed to better assess the self-heating effects in p-channel VDMOS devices and determine the effects of prior treatment on self-heating [5,24–26].

1.3. Impact on the Circuit Performance

Besides self-heating effects, additional attention should be given to the impact of NBT stressing on the performance of the circuits containing p-channel VDMOS devices. Although many studies have been performed with the goal of evaluating responsible physical mechanisms occurring during the operation and stressing of VDMOS devices, focus on only device-level degradation does not fully describe the impact of NBTI. When in circuit, the p-channel VDMOS device interacts with other devices, with power supplies and connections. This type of circuit-level reliability investigation is needed in order to faithfully reflect the impact of NBTI on circuit performance [28,29].

There are several reports in the literature targeting the impact of NBTI on circuit performance [28–30]. In most of the available studies, the authors investigate effects on rather simple circuits, such as a CMOS inverter [28–32]. As shown in Figure 2, NBT stress-induced degradation of a p-channel VDMOSFET comprising a CMOS inverter and NBTI-induced pulse narrowing directly affects the transient times of a CMOS inverter, as it would impact the transient times of even more complex circuits. Thorough analysis is needed in order to assess the impact of NBTI on circuit performance [28].

In order to tackle ongoing open questions regarding examination and impact of the NBTI on self-heating effects and circuit performance, a series of extensive experiment sessions have been carried out. The rest of the paper will present a detailed explanation, analysis of the experimental setup and the obtained results.

2. Experiment

The experimental investigation started with a large number of samples, analyzing different conditions to obtain results. Samples are commercial p-channel VDMOS devices IRF9520 with polysilicon gate technology and hexagonal cell geometry [22]. These devices, encapsulated in TO-220 packages, have a gate oxide thickness of approximately 100 nm. Maximum continuous drain current is 6.8 A, while the value of the threshold voltage is in the range from −2 V to −4 V. Although the threshold voltage lies within this range, previously conducted experiments have shown that identical stress conditions consistently produce the same threshold-voltage shifts, regardless of the initial value. The samples from this experiment had an initial value of V_T = 3 V ± 0.005 V.

A block schematic of the entire experiment is given in Figure 5.

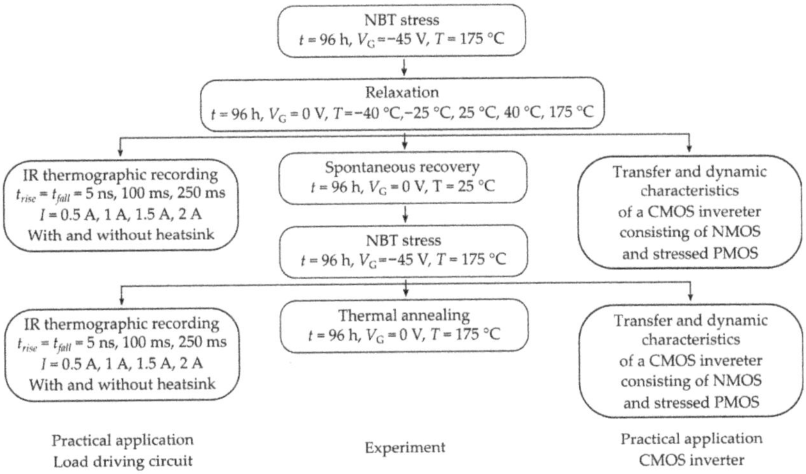

Figure 5. Block schematics of all three lines of the experiment.

Each of the experimental phases will be explained separately, in more detail, in the following sections.

2.1. Stressing

2.1.1. Negative Bias Temperature Stressing

The central line of the experiment is the negative bias temperature stressing of the samples. This part of the experiment has five different phases. The first phase is NBT stressing.

In this phase, samples are subjected to the accelerated NBT stressing conditions. Continuous voltage of −45 V was applied to the gate terminal, while the source and drain terminals were grounded. Stressing was performed in the heating chamber at a temperature of 175 °C for a duration of t = 96 h.

Accelerated stressing demands a voltage magnitude of V_G = −45 V, which is not typical within the generic signal source units. Because of that, a high voltage stress circuit was designed, which is detailly described in our previous paper [33]. The laboratory setup for NBT stressing is shown in Figure 6.

The main measuring equipment is the Keysight B2901A Source Measure Unit (SMU), which is controlled with a laptop, performing measurement of I-V transfer characteristics in the saturation region. All of the measurements are carried out at room temperature, outside the heating chamber. The measured characteristics are obtained with a short-circuited gate and drain terminals. Controlled voltage is sourced from the SMU, and the drain current was measured.

The measurement procedure was repeated for all samples at each phase of the experiment according to predefined time intervals. The I-V transfer characteristics in the saturation region, presented in Figure 7, are used to determine the threshold voltage values.

The experimental methodology adopted in this work is based on long-term NBT stressing at room temperature, measuring the transfer characteristics, which is commonly used in reliability studies of power VDMOS transistors operating under high gate voltages and elevated temperatures. Following similar approach as proposed in the JEDEC procedure, which is primarily defined for planar p-channel MOSFETs and focuses on shorter stress

https://doi.org/10.3390/mi17010052

times and specific measure–stress–measure or in vivo measurement sequences. Also, this standard has recently been superseded by the updated JESD241 standard, which focuses on advanced CMOS technologies and scaled devices. Consequently, some research groups did similar methodological adaptations for their research [34]. Our approach is modified for commercial TO-220 power devices and is intended to emulate realistic long-duration operating conditions in practical applications. Following a similar approach as proposed in the JEDEC procedure, it provides additional insights into the combined influence of extended stress duration, wide temperature range and subsequent relaxation/annealing phases on the threshold voltage shift and the behavior of complete circuits such as load drivers and CMOS inverters.

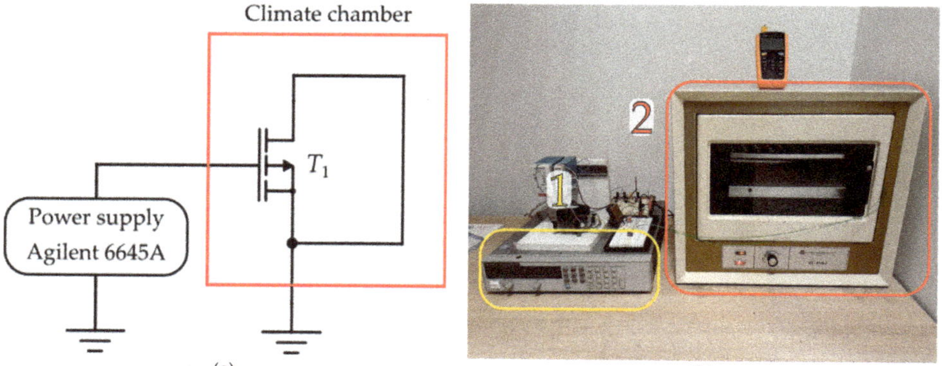

Figure 6. Laboratory setup used for NBT stressing of devices: (**a**) schematic; (**b**) photo during the experiment in the laboratory: (1) Agilent 6645A (Santa Clara, CA, USA) power supply, (2) climate chamber used for temperature control.

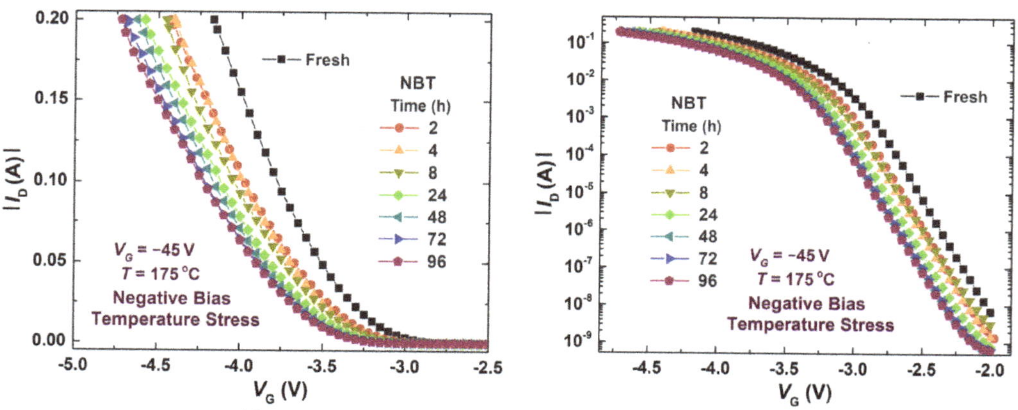

Figure 7. An example of measured transfer characteristics for one of the samples during the first experiment phase (**a**) above threshold transfer characteristics; (**b**) subthreshold transfer characteristics.

2.1.2. Relaxation

The second phase is relaxation. During this phase, no bias is applied to the samples, but the samples stay inside the heating chamber. Here, the samples are divided into several groups. Part of the samples are subjected to the temperature of 175 °C, 40 °C, 25 °C, −25 °C and −40 °C. Additional attention should be given to the behavior of devices operating in below-zero temperatures. While the application range of the VDMOS devices also covers

https://doi.org/10.3390/mi17010052

many cases where the devices and their comprising circuits operate in temperatures below 0 °C, there are not many studies investigating reliability under these conditions. Because of that, a part of this experiment was dedicated to analyzing the impact of below-zero temperatures on the behavior and reliability of VDMOS devices.

Constant temperatures of −25 °C and −40 °C were maintained using the climate chamber BINDER MKF 56 (Shanghai Jianheng Instrument Ltd., Shanghai, China), while the other temperatures were maintained using a standard heating chamber. The duration of relaxation is also t = 96 h, and transfer characteristics in the saturation region are measured in an identical manner to the first phase of the experiment.

After the second phase of the experiment, some samples are used for IR thermographic recordings, some for CMOS inverter circuit characteristics, and some for the NBTS experiment.

2.1.3. Continuation of the Experiment

The third phase of the experiment involves spontaneous recovery with no bias, and no direct temperature change is performed. Samples are kept at a room temperature of 25 °C.

The fourth phase of the experiment is the second cycle of NBT stressing. Both the stressing procedure and the measuring procedure are identical to those in the first phase of the experiment (V_G, T, t), as explained in Section 2.1.1. After the fourth phase, some samples are used for IR thermographic recordings and CMOS inverter circuit characteristics measurement, while other samples are subjected to the fifth phase of the experiment.

The fifth phase of the experiment is thermal annealing at the elevated temperature of 175 °C without bias applied. Measuring is performed at room temperature using an identical routine to that in the previous phases of the experiment. A tabular presentation of the used samples classification is given in Table 1.

Table 1. Groups of devices under test and treatment of each group of devices.

	D1	D2	D3	D4	D5	D6	D7	D8	D9	D10	D11	D12	D13	D14	D15
NBTS, t = 96 h V_G = −45 V, T = 175 °C	X	X	X	X	X	X	X	X	X	X	X	X	X	X	X
Relaxation, t = 96 h T = 175 °C	X	X	X												
Relaxation, t = 96 h T = 40 °C				X	X										
Relaxation, t = 96 h T = 25 °C						X	X	X							
Relaxation, t = 96 h T = −25 °C									X	X	X				
Relaxation, t = 96 h T = −40 °C												X	X	X	X
IR and CMOS	X			X		X			X			X			
Spont. Recovery, t = 96 h, T = 25 °C		X	X		X		X	X		X	X		X	X	X
NBTS, t = 96 h V_G = −45 V, T = 175 °C		X	X		X		X	X		X	X		X	X	X
IR and CMOS		X					X			X					X
Annealing, t = 96 h T = 175 °C			X		X			X			X		X		

https://doi.org/10.3390/mi17010052

2.2. Infra-Red Thermographic Recordings

The second line of the experiment involves the investigation of the self-heating effects evaluated using infra-red thermographic recordings. To examine the thermal behavior of the tested p-channel VDMOS transistors under different operating conditions, infra-red thermographic measurements were performed using a FLIR E8 thermal camera. The experimental arrangement, shown in Figure 8, was designed to emulate realistic operating scenarios in which power VDMOS transistors are commonly used. In such applications, devices are frequently exposed to rapid transitions between on/off-states and thermally demanding environments.

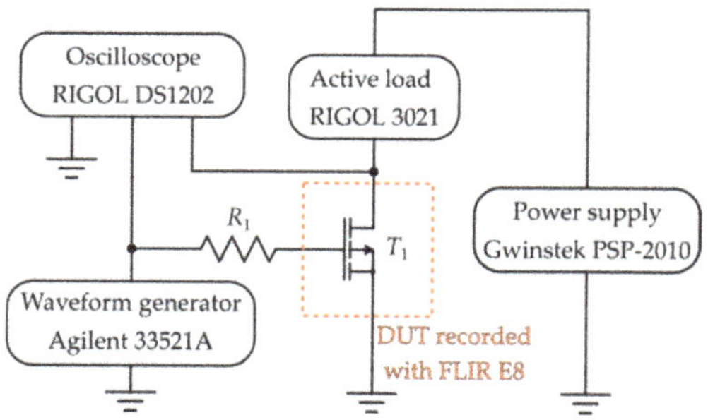

(a)

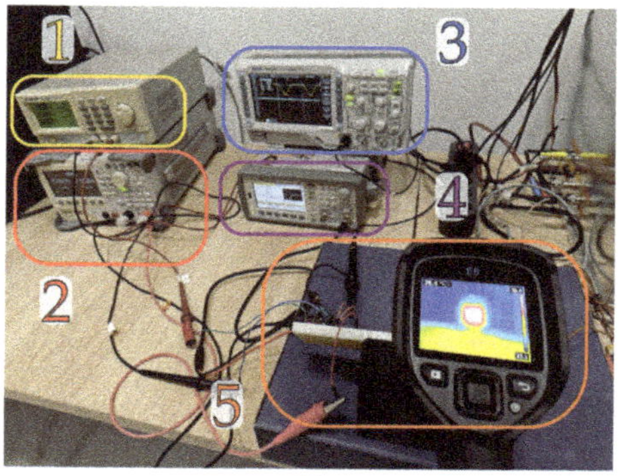

(b)

Figure 8. Measurement setup used to measure transistor self-heating: (**a**) schematic; (**b**) photo during measurement in the laboratory: (1) Gwinstek PSP-2010 (Taipei, Taiwan) power supply, (2) RIGOL DL3021A (Beijing, China) active load, (3) RIGOL DS1202 (Beijing, China) oscilloscope, (4) Agilent 33521A (Santa Clara, CA, USA) arbitrary waveform generator, (5) FLIR E8 thermal camera (Wilsonville, OR, USA) recording device used for test.

In this experiment, the drain terminal was connected to a DC supply voltage, while the transistor gate was driven by pulsed signals that replicate typical switching conditions. Three different control waveforms, illustrated in Figure 9, were applied, each defined by distinct rise and fall times (5 ns, 100 ms and 250 ms) at a frequency of 1 Hz, amplitude of −10 V and a 50% duty cycle [25]. The test signals were generated using an Agilent 33921A (Santa Clara, CA, USA) arbitrary waveform generator. A Rigol DS1202 oscilloscope was used to enable accurate waveform delivery to the device gate.

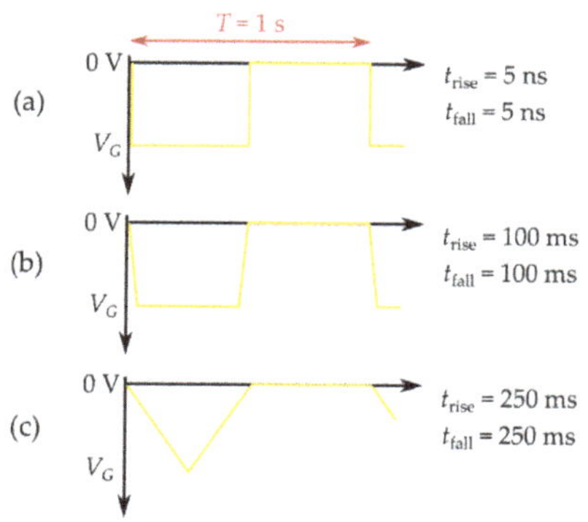

Figure 9. Controlling signals used in the experiment (f = 1 Hz, V_G = 10 V): (**a**) $t_{rise} = t_{fall} = 5$ ns; (**b**) $t_{rise} = t_{fall} = 100$ ms; (**c**) $t_{rise} = t_{fall} = 250$ ms.

https://doi.org/10.3390/mi17010052

To evaluate the impact of different load conditions, a Rigol DL3021 active load was connected to the drain circuit and configured to draw four different constant current levels: 0.5 A, 1 A, 1.5 A and 2 A. This enabled investigation of the thermal response as a function of power dissipation.

The camera was connected to a computer and controlled via dedicated software, FLIR Thermal Studio (version 2024), enabling continuous acquisition of temperature profiles throughout the predefined measurement interval. The interface of the software during the measurement is presented in Figure 10. The image at the center illustrates the levels of the device temperature. Different colors represent different temperature levels, providing a visual indication of how the transistor heats up during operation. This setup provided precise monitoring of the device surface temperature during both heating and cooling phases.

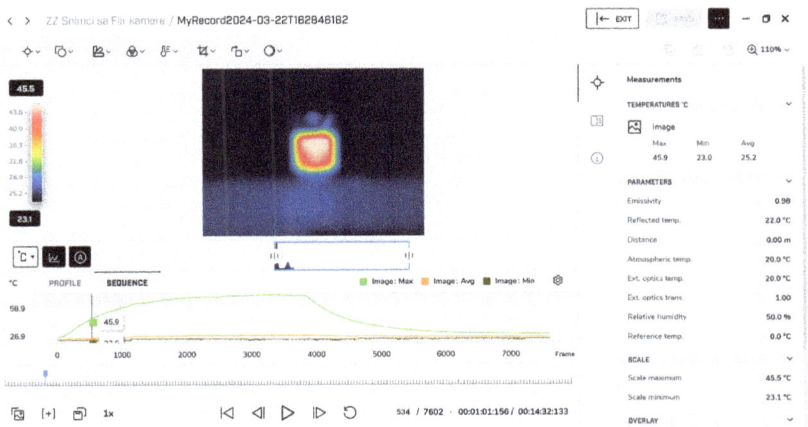

Figure 10. Screenshot of software FLIR Thermal Studio, which was used for extraction of temperature, and setup used to measure transistor self-heating.

During each test, the transistor was switched on for approximately 6 min (heating phase) and turned off for the following 6 min (cooling phase). This 12 min cycle was selected based on preliminary observations showing that the device temperature reaches a maximum within this interval [5,25]. Temperature evolution was recorded for both fresh samples and devices previously subjected to NBT stress and the subsequent relaxation process (after the second phase of the experiment and after the fourth phase of the experiment), allowing comparison of thermal behavior before and after degradation.

This measurement methodology provides detailed insight into the influence of dynamic switching conditions, gate-signal characteristics, and load variations on the self-heating behavior of VDMOS transistors, which is an essential aspect when analyzing their reliability and long-term stability.

This methodology has previously proven effective for evaluating self-heating effects in power MOSFETs [5,25], and it provides valuable insight into the thermal response of p-channel VDMOS transistors, particularly considering their typically higher on-state resistance and different carrier mobility degradation mechanisms.

2.3. Characterization of CMOS Inverter Containing NBT Stressed Device

The third line of the experiment is focused on eliminating the impact of NBTI on circuit performance, more precisely on the performance of the CMOS inverter containing an NBT-stressed p-channel VDMOSFET. A CMOS inverter is selected as one of the most

 https://doi.org/10.3390/mi17010052

fundamental logic gates and a part of numerous circuits. It consists of PMOS and NMOS transistors that are connected in a complementary configuration, as shown in Figure 11a.

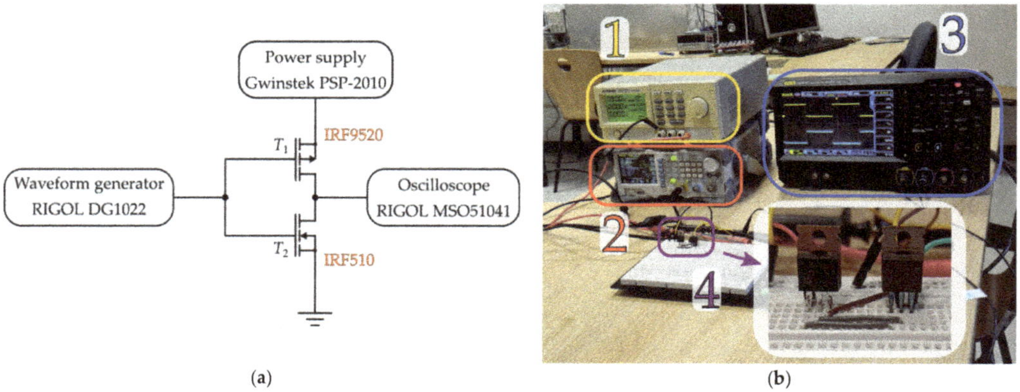

Figure 11. Measurement setup used for CMOS inverter characterization: (**a**) schematic; (**b**) photo during measurement in the laboratory: (1) Gwinstek PSP-2010 (Taipei, Taiwan) power supply, (2) RIGOL DG1022 (Beijing, China) waveform generator, (3) RIGOL DS1202 (Beijing, China) oscilloscope, (4) CMOS inverter circuit designed using n-channel power VDMOSFET IRF510 (Vishay, Malvern, PA, USA) and p-channel power VDMOSFET IRF9520 (Vishay, Malvern, PA, USA) .

If the input is at a high voltage level, the NMOS transistors turn on, pulling the output to the low voltage level. On the other hand, if the input is on a low voltage level, the PMOS turns on and pulls the output to the high voltage level. The behavior of the CMOS inverter is typically characterized through transfer and dynamic characteristics [28,29,32]. The transfer characteristic of a CMOS inverter presents the output voltage changes in response to the input voltage, while the dynamic characteristics of a CMOS inverter describe its response to changes over time, including rise time, fall time, and propagation delay.

In this experiment, the behavior of the CMOS inverter that comprises a previously stressed p-channel power VDMOS transistor IRF9520 and a non-stressed n-channel power VDMOS transistor IRF510 was analyzed (illustrated in Figure 12). Commercial n-channel power VDMOSFET IRF510 is selected for the CMOS inverter because of its almost compatible characteristics with IRF9520 [35]. The maximum drain current is 5.6 A, and the threshold voltage is in the identical range, between 2 and 4 V. It is also encapsulated in TO-220 packages with an identical pin order.

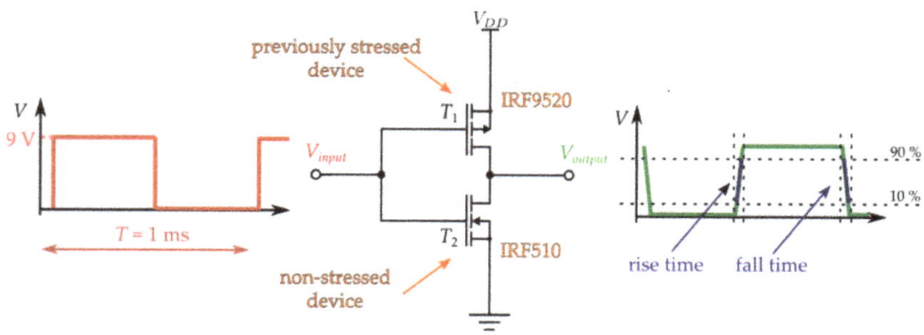

Figure 12. Signal and circuit setup for CMOS inverter characterization.

Besides these devices, as illustrated in Figure 11, a programmable DC power supply, the GW Instek PSP-2010 (Taipei, Taiwan), was used to provide bias voltages for the transis-

 https://doi.org/10.3390/mi17010052

tors. The inverter input square-wave signal was generated by a RIGOL DG1022 waveform generator, while a RIGOL MSO5104 (Beijing, China) oscilloscope was used to capture input and output signal waveforms. The supply voltage for the CMOS inverter was V_{DD} = 9 V, while characteristics of the input signal were V_{HIGH} = 9 V, V_{LOW} = 0 V, f = 1 kHz, DTC = 50% and t_{rise} = t_{fall} = 5 ns. Measuring transfer and dynamic characteristics involves acquiring signal waveforms from the oscilloscope in the form of a .csv file. The sample of the data acquired with the oscilloscope is given in Figure 13.

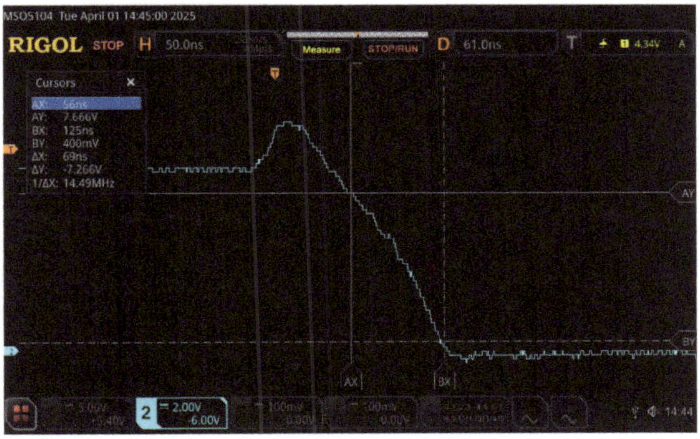

Figure 13. The transition between the high voltage level and the low voltage level captured using an oscilloscope and used to determine the transfer and dynamic characteristics of a CMOS inverter.

The data from the captured file is then processed using a specially designed Python script (Python version 3.13) with the goal of obtaining transfer characteristics as well as the t_{rise} and t_{fall} of the CMOS inverter output signal.

3. Results and Discussion

3.1. Stress-Induced Threshold Voltage Shift

For every transfer characteristic measured during the experiment, the threshold voltage value is determined. The threshold voltage shifts throughout the experiment phases for all of the groups of devices are presented in Figure 14.

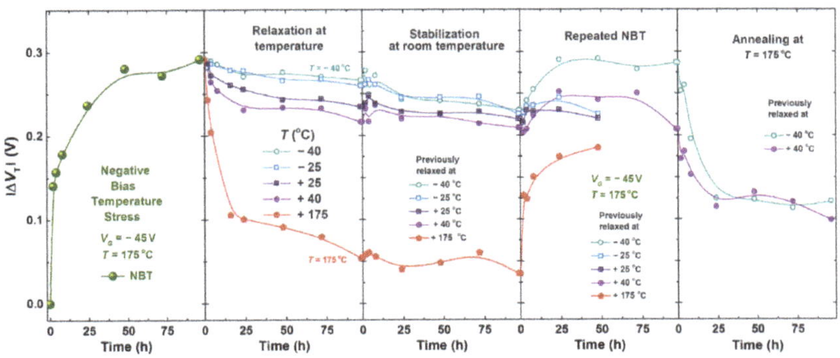

Figure 14. Threshold voltage shifts (average value per group) for all of the samples subjected to different experiment phases presented in Section 2.1.

Negative bias temperature (NBT) stress has a well-established impact on charge trapping mechanisms within the gate oxide and at the oxide–semiconductor interface.

https://doi.org/10.3390/mi17010052

These processes directly influence the electrical behavior of MOSFETs, where the most notable change is the increase in the value of the threshold voltage, as can be seen in the first part of Figure 14. In p-channel devices, prolonged NBT stress promotes the trapping of positive charge in the oxide and the capture of channel holes at interface traps. Both mechanisms effectively act against the applied gate voltage, resulting in an increased magnitude of the threshold voltage [5,21,36].

During the NBT stress, V_T varies, which naturally suggests the involvement of different underlying degradation mechanisms. The increase in positively charged oxide charge under high negative oxide fields and elevated temperatures can be attributed to hole trapping and oxygen vacancy-related defects. These processes may proceed through reactions such as

$$O_3 \equiv Si^{\bullet\bullet}Si \equiv O_3 + h^+ \rightarrow O_3 \equiv Si^{+\bullet}Si \equiv O_3 \quad (1)$$

Also, this could happen at the weakened Si–H bonds near the $SiOi_2$/Si interface.

Similarly, the buildup of interface traps N_{it} observed during NBT stressing resulted from the combined action of the strong electric field and elevated temperature, which enables the dissociation of Si–H bonds at the SiO_2/Si interface:

$$Si_3 \equiv Si - H \leftrightarrow Si_3 \equiv Si^{\bullet} + H^{\bullet} \quad (2)$$

The released hydrogen species are highly reactive and may further contribute to interface degradation by breaking additional interface Si–H bonds, enabling the formation of new interface traps.

Following the NBT stress phase, the devices undergo a relaxation period during which partial recovery of electrical parameters is expected, as can be seen in the second part of Figure 14. Higher temperatures generally accelerate detrapping processes and thus enhance recovery. However, the experimental findings show that the threshold voltage does not fully return to its initial value, indicating the presence of permanent degradation.

The third part of Figure 14 presents threshold voltage shifts during relaxation at room temperature without biasing, when the spontaneous recovery occurs. Relaxation at room temperature exhibits only a minor influence on the previously stressed device, confirming that thermal activation is essential for effective recovery. Still, the change during this phase is the most pronounced with the group of samples previously relaxed at the lowest temperatures (−40 °C and −25 °C). Room temperature (25 °C) is a significant increase in the temperature for these groups of samples, leading to thermal activation-induced recovery.

The second NBT stress cycle produces only a modest additional threshold voltage shift, as can be seen in the fourth part of Figure 14. These results reinforce the conclusion that the degradation mechanisms approach saturation after the initial stress, since the total increment of the NBT stress-induced threshold voltage shift during the second cycle is less pronounced.

The last experiment phase, thermal annealing at 175 °C, leads again to the partial recovery of the NBT stress-induced threshold voltage shift, presented in the fifth part of Figure 14. Still, similar to the relaxation phase, recovery seems to saturate after a specific time. Since all samples are annealed at the same annealing temperature, the magnitude and shape of the curve of the threshold voltage shift during this phase are similar for all groups of devices, regardless of previous treatment. This behavior is consistent with the strong temperature dependence of detrapping processes.

3.2. Self-Heating in Practical Applications

The results of the infra-red thermographic recording can be observed through three comparisons. The first one is the comparison of the devices' self-heating when the same

https://doi.org/10.3390/mi17010052

controlling signal waveform is applied for different drain current values. The second one is the comparison when the same drain current value is applied using the different controlling signal waveforms, while the third one is focused on how the previous treatment of the samples (different experiment groups) affects self-heating. Figure 15 shows the temperature response of the devices to a trapezoidal gate-controlling signal (t_{rise} = t_{fall} = 100 ms) at four different drain current levels (0.5 A, 1 A, 1.5 A and 2 A). The temperature response of the device after the first two experiment phases (D12 from Table 1) is presented in Figure 15a, while the temperature response of the devices after the first four experiment phases (D15 from Table 1) is presented in Figure 15b.

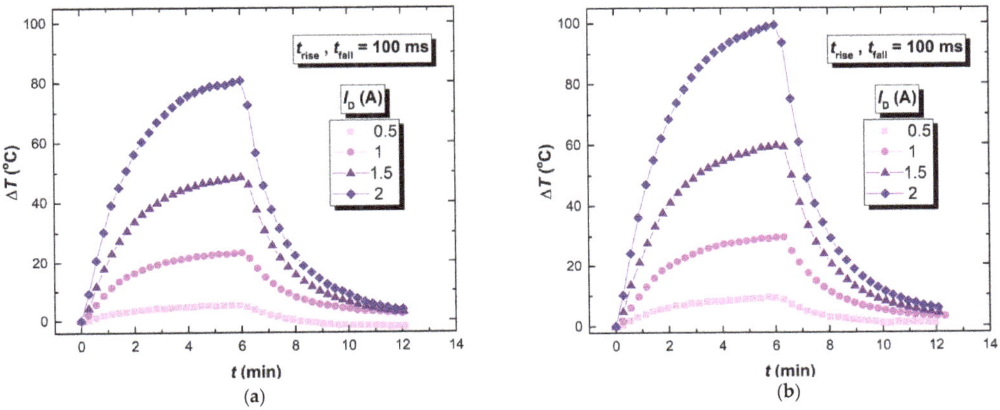

Figure 15. Thermal response of devices to a trapezoidal gate-controlling signal (t_{rise} = t_{fall} = 100 ms) at four different drain current levels 0.5 A, 1 A, 1.5 A and 2 A: (**a**) D12 (NBT stressed and relaxed at −40 °C); (**b**) D15 (NBT stressed, relaxed at −40 °C, relaxed at room temperature and again NBT stressed).

In Figure 15, the self-heating temperature, given as a difference between the measured temperature and the room temperature, is presented as a function of time (t). A consistent trend was observed across all measurement conditions. The higher value of the drain current resulted in more pronounced self-heating. The highest device temperature is accomplished with the 2 A drain current, while the lowest device temperature is accomplished with the 0.5 A drain current.

The most pronounced increase in the device temperature, for all of the analyzed groups of samples, is shown within the first two and a half minutes of the experiment, where the temperature increase rate reached its maximum.

Compared to the previously non-stressed device, NBT stressed devices exhibited greater self-heating. In Figure 15b, the self-heating is even more pronounced, since the device group D15 went through the identical first two phases of the experiment like group D12 (NBT stress and relaxation at 175 °C), but also two additional phases, room temperature relaxation and an additional cycle of NBT stressing. Very similar behavior is also noticed within the other pairs of device groups subjected to the thermographic recording when the first two experiment phases are identical, after which one group proceeds to additional treatment (D6 and D7, D9 and D10). These results point out the direct impact of NBT stressing and relaxation on this type of device application. After the devices were switched off, the temperature of the devices gradually decreased and spontaneous cooling occurred. For all devices and drain current values, a similar shape of the cooling curve is observed. More details can be given with the comparison of the heating and cooling rate values, as given in Table 2.

https://doi.org/10.3390/mi17010052

Table 2. Heating rate for samples NBT stressed ($V_G = -45$ V, $T = 175$ °C) and relaxed at −40 °C (D12).

Drain Current [A]	Initial Rates (First 60 s) [°C/min]		Average Rates [°C/min]	
	Heating	Cooling	Heating	Cooling
0.5	2.42	3.39	0.87	0.52
1 A	11.2	10.92	3.88	3.50
1.5 A	23.00	22.77	8.06	7.62
2 A	39.12	43.04	13.43	12.83

Values of the initial heating and cooling rates are significantly higher than the average ones. The relation between the initial and the average heating rate is similar for all of the drain currents, with the initial being approximately three times greater than the average heating rate. The results for the device group D12 presented in Table 2 are quantitatively and qualitatively in line with the results obtained for other device groups subjected to thermographic recording.

Self-heating of the devices when driven with different controlling signal waveforms needs to be analyzed in more detail. Total self-heating recorded in devices is a combination of two effects. The first effect is the heating caused by power dissipation when the transistor is turned on and in its current-conducting state. As the interval during which the transistor is turned on becomes longer, this type of self-heating becomes more pronounced. The second effect is the increase in the self-heating caused by the power dissipation, where the transistor switches states between on and off.

While the first effect is not directly related to the consequences of the NBT stressing, the second effect is directly related to the change in the R_{ch}, which is increased because of the NBT stressing. Still, the magnitude of impact of this effect is dependent on the switching frequency, duty cycle and the rise and fall times of the controlling signal [37]. For lower frequencies, with shorter transient times, conductive self-heating will be dominant over the transient self-heating, meaning that, on a larger scale, transient self-heating could be negligible. But, during high-frequency switching, which is present in many VDMOS transistor applications, transient self-heating effects become relevant. Similar studies show that this type of SHE can be more emphasized than the active SHE for specific working conditions [25]. Therefore, in this case, it is not possible to give clear conclusions, unlike in the case of Figure 15, since the change in the self-heating is caused by multiple factors.

In many real-world practical applications of p-channel power VDMOSFET, these devices are used connected to a heatsink. Application of the heatsink is typical for power devices, including power VDMOSFETs, since power devices are designed to conduct higher values of currents and tend to suffer from heating. In order to examine this type of practical application, additional rounds of IR thermographic recordings have been carried out under groups of devices D15 with two different heatsinks.

The first heatsink (HS1) is an extruded aluminum heatsink, SK 104 50.8 STS, manufactured by Fischer Elektronik (Lüdenscheid, Germany) [38], while the other smaller heatsink (HS2) is also an aluminum heatsink, Alutronic PR17/35II/SE (Halver, Germany) [39]. Both of the heatsinks are compatible with the TO-220 package. Figure 16a presents samples with mounted HS1 and HS2. IR thermographic recording was performed four times using the routine presented in Section 2.2. For both of the heatsinks, recording was performed twice: first, recording the device temperature, and second, recording the heatsink temperature. Figure 16b presents the temperature response of the device group D15 with mounted HS1 and with mounted HS2, operating with a drain current of 2 A and a $t_{rise} = t_{fall} = 100$ ms gate-controlling signal.

https://doi.org/10.3390/mi17010052

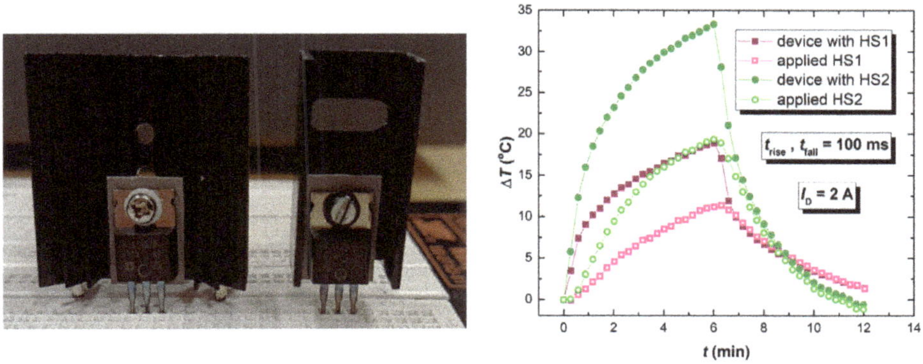

(a) (b)

Figure 16. (**a**) IRF9520 transistor with mounted HS1 [38] and with mounted HS2 [39]; (**b**) Temperature response of the device group D15 with mounted HS1, of HS1, with mounted HS2 and HS2, operating with the drain current of 2 A and with $t_{\mathrm{rise}} = t_{\mathrm{fall}}$ = 100 ms gate-controlling signal.

As can be seen from Figure 16b, the application of a heatsink drastically reduced the temperature of the device. Even with the smaller heatsink (HS2), device temperature does not rise beyond 35 °C, whereas without the heatsink, device temperature exceeds 80 °C. With the usage of an appropriate heatsink, the negative effects of prior NBT treatment can be overcome or reduced.

3.3. Impact of Stressing on the CMOS Inverter Circuit Performance

Figure 17 presents transfer characteristics of a CMOS inverter designed using a previously stressed p-channel power VDMOSFET (D1, D6, D9) and a non-stressed n-channel power VDMOSFET. It is worth mentioning that each oscilloscope captured a very large number of points (10^6). During results processing, the number of presented data points is reduced in order to improve clarity and visibility.

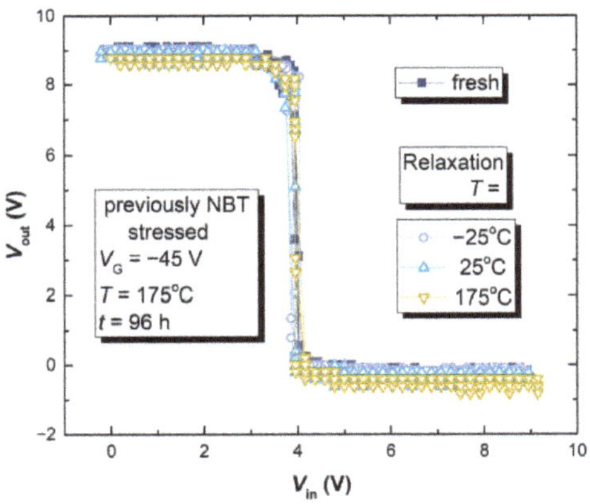

Figure 17. Transfer characteristics of a CMOS inverter designed using a previously stressed p-channel power VDMOSFET (D1, D6, D9) and a non-stressed n-channel power VDMOSFET.

As can be seen from Figure 17, the impact of NBT stressing followed by relaxation does not affect CMOS inverter transfer characteristics drastically. Similar behavior is noted for the groups of devices, D2, D7 and, D10 that passed two additional experiment phases. From these results, it was not possible to clearly examine the impact of NBT stressing on the performance of the CMOS inverter circuit. Therefore, an additional experiment has

https://doi.org/10.3390/mi17010052

been carried out where p-channel VDMOS devices were subjected to the more intense NBT stress ($V_G = -50$ V, $T = 175$ °C, the critical conditions from Figure 1) in a manner explained in Section 2.1.1, but for a longer period of time (240 h instead of 96 h). The measured transfer characteristics of the CMOS inverter built with p-channel VDMOSFETs that passed this treatment and non-stressed n-channel VDMOSFET are presented in Figure 18.

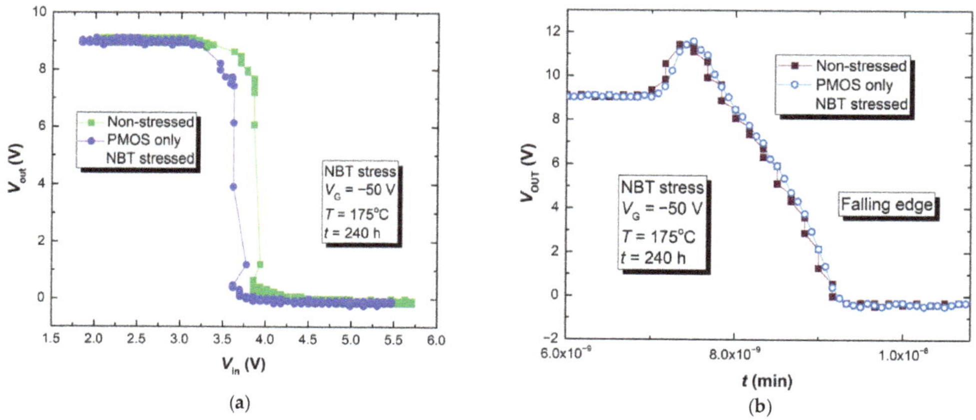

Figure 18. The results of the CMOS inverter built with a ten-day NBT-stressed p-channel VDMOSFET IRF9520 and a non-stressed n-channel VDMOSFET IRF510: (**a**) transfer characteristics; (**b**) falling edge.

The transfer characteristics of a CMOS inverter with a p-channel device that was NBT-stressed for ten days show a clear shift to the left. During the ten days of NBT stressing, the threshold voltage value of the p-channel VDMOSFET shifted enough so that it could affect the transfer characteristic of a CMOS inverter. Similar results can be seen from Figure 18b, where the dynamic characteristics, more precisely the fall time, are compared. The fall time increased, and similar results are obtained for the rise time. It can be concluded that, under harsher operating conditions, NBT stress-induced threshold voltage shift in the p-channel power VDMOSFET presents additional deterioration of the circuit operation.

However, for the analysis of the circuit performance, it is important to analyze one additional factor. From the standpoint of practical application, it would be rare to use a circuit where only a single transistor is stressed, while the other components in the circuit are completely fresh. Ambient conditions usually affect the entire circuit, whereas some of the devices are more susceptible to the environmental conditions than others. Therefore, an additional round of experiments regarding the examination of the NBTI on the circuit performance has been carried out. Special attention is given to the analysis of the circuit's performance over a broad temperature range.

First, thermal characterization of both the p-channel power VDMOSFET IRF9520 and n-channel power VDMOSFET IRF510 was performed using the BINDER MKF 56 climate chamber. Transfer characteristics in the saturation region at different temperatures in the range from −40 to 180 °C, at 20 °C intervals, were measured. The threshold voltage value was determined for each of the measured characteristics. The temperature shift in the threshold voltage for both of the analyzed devices is presented in Figure 19.

The average value of the threshold voltage shift on the observed temperature range, calculated from the results presented in Figure 19, for the p-channel VDMOSFET IRF9520 is 4.5 mV/°C, while for the n-channel VDMOSFET IRF510, it is 4.8 mV/°C. These devices show a similar temperature-induced threshold voltage shift, suggesting that the circuit containing the analyzed devices should exhibit stability at elevated temperatures.

https://doi.org/10.3390/mi17010052

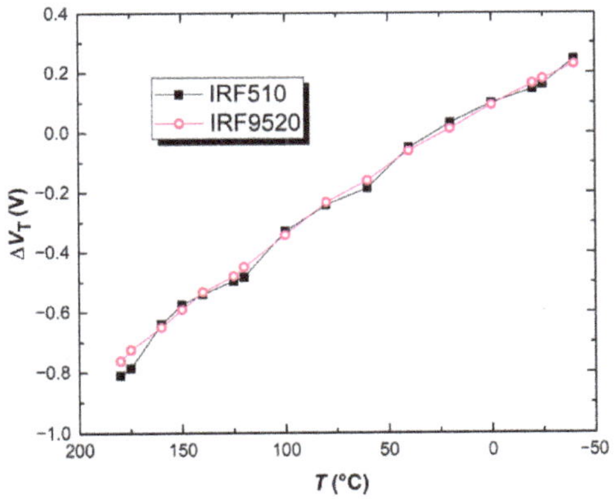

Figure 19. Temperature dependence of V_T for p-channel VDMOSFET IRF9520 and n-channel VDMOSFET IRF510, which comprise the analyzed CMOS inverter.

Therefore, a circuit with a CMOS inverter consisting of analyzed devices was put into the BINDER MKF 56 climate chamber in order to perform thermal characterization on a circuit level. While in the climate chamber, the circuit of the CMOS inverter remains connected to the measuring setup in an identical manner, as shown in Figure 11. Thermal characterization was performed in an identical temperature range for the single samples (from −40 °C to 180 °C), while the dynamic and transient characteristics were measured at 20 °C intervals. The determined transfer characteristics of the CMOS inverter circuit under temperatures of −40 °C, 0 °C, 100 °C and 180 °C during the additional experiment are presented in Figure 20.

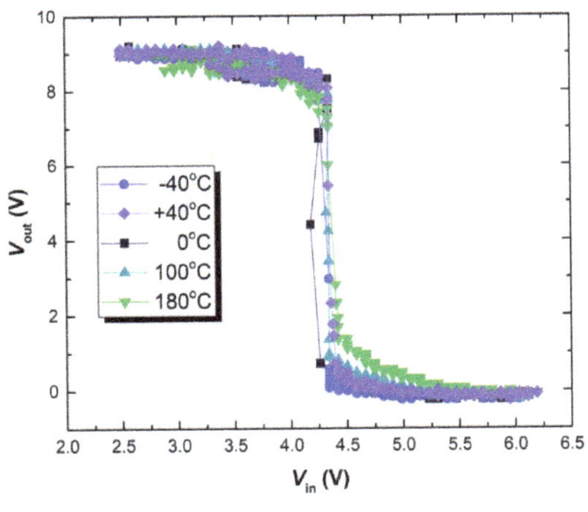

Figure 20. Determined transfer characteristics of the CMOS inverter circuit under different temperatures during the additional experiment.

In the temperature range above 40 °C, a clear shift in the transfer characteristics to the right is noted for all of the analyzed samples. As can be seen for the characteristics measured at −40 °C and at 0 °C, there is no distinct change between the determined results in the lower temperature ranges. In the range from −40 °C to +40 °C, as can be seen from Figure 19, the temperature-induced threshold voltage shift is not drastic. In accordance with the results from Figure 17, the threshold voltage shift needs to be more pronounced in order to affect the transfer characteristics of the CMOS inverter circuit. When the threshold voltage shifts in the transistors comprising the CMOS inverter circuit

https://doi.org/10.3390/mi17010052

become more pronounced (at temperatures above +40 °C), a constant significant shift in the transfer characteristics can be noted. Similar considerations can be given for the dynamic characteristics of the CMOS inverter circuit, as can be seen from Figure 21.

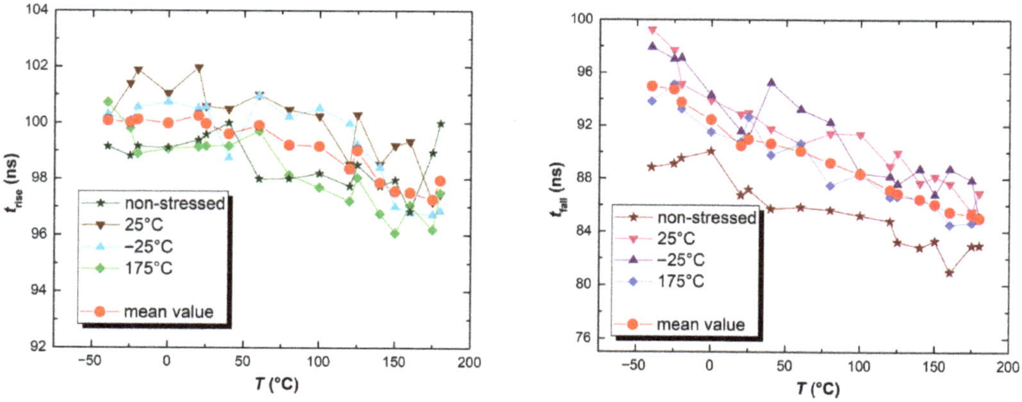

(a) (b)

Figure 21. Comparison of rise (**a**) and fall (**b**) times for output CMOS inverter circuits designed with previously stressed p-channel devices and non-stressed n-channel devices, at different temperatures.

Generally, in the CMOS inverter circuit, the fall time is shorter than the rise time. This is caused by the greater mobility of the electrons, which are charge carriers in the n-channel device, when compared with holes, which are charge carriers in the p-channel devices. The results presented in Figure 21 show agreement with the previous reports in that manner [28,31,32]. The threshold-voltage shift in the p-channel device results in a significant mismatch between the p- and n-channel threshold voltages within the CMOS inverter. During the fall time (transition from high voltage level to low voltage level), the p-channel is turning off, and the n-channel device is turning on. Since the absolute value of the threshold voltage of the p-channel device is increased, the p-channel device will start turning off before the n-channel device starts turning on. Therefore, more time is needed to perform the transition, demonstrating that the impact of stressing is far more pronounced during the fall time, presented in Figure 21b, than the rise time, presented in Figure 21a. This is a direct impact of the NBT stress-induced threshold voltage shift on the performance of the CMOS inverter circuit. It is interesting that although the four groups of devices shown in Figure 21 have undergone distinct prior treatments, some tendencies could be noted. When examining the mean value of the rise time, it can be seen that, up to approximately 40 °C, all components maintain almost a constant value. However, above 40 °C, a noticeable decrease in rise time is observed as the temperature increases. Analyzing Figure 21b, it can be observed that over the entire investigated temperature range, the fall time exhibits a decreasing trend.

Because of the described effects, the difference between the rise and fall times at elevated temperatures is more pronounced. For example, for a CMOS inverter circuit containing D1 that was NBT-stressed and relaxed at 175 °C, the rise time and fall time at 0 °C are 99.3 ns and 91.6 ns, respectively (a difference of 7.7 ns). An identical circuit, at 100 °C, shows a rise time of 97.9 ns, while the fall time is 88.2 ns (a difference of 9.7 ns), which is an increase of more than 20%. It can be concluded that the impact of stressing on the performance of the CMOS inverter circuits increases with the increase in the operating temperature.

https://doi.org/10.3390/mi17010052

4. Conclusions

In this paper, the impact of NBTI on commercial power p-channel VDMOS transistors IRF9520 in practical applications of a load driver circuit and a CMOS inverter circuit is examined. NBT stress induces a threshold voltage shift that is more pronounced with higher gate voltages and higher temperatures. Pulsed NBT stress-induced threshold voltage shift is shown to be commonly lower than in the case of static NBT stressing because of the partial recovery of degradation that occurs during pulsed NBT stressing.

The induced threshold voltage shift during pulsed stressing is also shown to be dependent on the characteristics of the pulsed gate signal, such as amplitude, frequency and the duty cycle, pointing out that these gate signal characteristics affect practical applications of the devices. Various forms of stress are applied to the samples with the goal of emulating real-world conditions of operation. The stressing and measuring process in each part of the experiment is explained in detail.

The results suggest that NBT stressing directly affects the examined samples operating in the load driving circuit through increased self-heating. For all of the tested samples, self-heating during operation is increased in value, corresponding to the NBT stressing and relaxation-induced threshold voltage shift. Increased self-heating could be partially mitigated with the use of a heatsink. Future research should be directed to examining how self-heating affects the signal rise and fall times of the gate-controlling signal.

The results also suggest that NBT stressing directly affects the characteristics of the CMOS inverter comprising the examined samples. An impact is seen with the shift in transfer characteristics and increase in transient times, and it is far more pronounced at elevated temperatures during operation. Future work should focus on investigating the dependence of the input signal frequency on the characteristics of the CMOS inverter containing an NBT-stressed p-channel device and the impact on more complex circuits.

Author Contributions: Conceptualization, D.D., E.Ž. and S.V.; methodology, D.D.; software, N.M., N.V., D.Đ., M.P. and L.T.; validation, N.M. and V.D.; formal analysis, M.M. and S.V.; investigation, E.Ž., S.V. and N.M.; resources, G.R. and V.D.; data curation, N.V., D.Đ., M.P. and L.T.; writing—original draft preparation, N.M. and S.V.; writing—review and editing, M.M., E.Ž. and D.D.; visualization, S.V.; supervision, D.D.; project administration, D.D.; funding acquisition, D.D. All authors have read and agreed to the published version of the manuscript.

Funding: This work has been supported in part by the Horizon Europe Twinning project AIDA4Edge (Grant Agreement No. 101160293), by the UK Research and Innovation organization, and by the Ministry of Science, Technological Development and Innovation of the Republic of Serbia under Grant No. 451-03-137/2025-03/200102. Part of this work was supported by the German Research Foundation-DFG under Project No. 551672220 (RESIST).

Data Availability Statement: The original contributions presented in the study are included in the article; further inquiries can be directed to the corresponding author.

Conflicts of Interest: The authors declare no conflicts of interest.

References

1. Chee, K.W.A.; Ye, T. In *Complementary Metal Oxide Semiconductor*; Yeap, K.H., Nisar, H., Eds.; Towards New Generation Power MOSFETs for Automotive Electric Control Units. InTech: Rijeka, Croatia, 2018.
2. Singh, K.; Kalra, S.; Mahur, J. Evaluating NBTI and HCI effects on device reliability for high-performance applications in advanced CMOS technologies. *FU Electron. Energetics* **2024**, *37*, 581–597. [CrossRef]
3. Bhattacharjee, A.; Das, A.; Sahu, D.K.; Pradhan, S.N.; Das, K. A Meta-Heuristic Search-Based Input Vector Control Approach to Co-Optimize NBTI Effect, PBTI Effect, and Leakage Power Simultaneously. *Microelectron. Reliab.* **2023**, *144*, 114979. [CrossRef]
4. Zhang, J.F.; Gao, R.; Duan, M.; Ji, Z.; Zhang, W.; Marsland, J. Bias Temperature Instability of MOSFETs: Physical Processes, Models, and Prediction. *Electronics* **2022**, *11*, 1420. [CrossRef]

https://doi.org/10.3390/mi17010052

5. Živanović, E.; Veljković, S.; Mitrović, N.; Jovanović, I.; Djorić-Veljković, S.; Paskaleva, A.; Spassov, D.; Danković, D. A Reliability Investigation of VDMOS Transistors: Performance and Degradation Caused by Bias Temperature Stress. *Micromachines* **2024**, *15*, 503. [CrossRef] [PubMed]
6. Sun, Z.; Chen, S.; Zhang, L.; Huang, R.; Wang, R. The Understanding and Compact Modeling of Reliability in Modern Metal–Oxide–Semiconductor Field-Effect Transistors: From Single-Mode to Mixed-Mode Mechanisms. *Micromachines* **2024**, *15*, 127. [CrossRef]
7. Bhattacharjee, A.; Pradhan, S.N. NBTI-Aware Power Gating Design with Dynamically Varying Stress Probability Control on Sleep Transistor. *J. Circ. Syst. Comput.* **2021**, *30*, 2120004.
8. Grasser, T. *Bias Temperature Instability for Devices and Circuits*; Springer: New York, NY, USA, 2014.
9. Li, X.; Shao, Y.; Wang, Y.; Liu, F.; Kuang, F.; Zhuang, Y.; Li, C. Interaction of Negative Bias Instability and Self-Heating Effect on Threshold Voltage and SRAM (Static Random-Access Memory) Stability of Nanosheet Field-Effect Transistors. *Micromachines* **2024**, *15*, 420. [CrossRef]
10. Schroder, D.K.; Babcock, J.A. Negative bias temperature instability: Road to cross in deep submicron silicon semiconductor manufacturing. *J. Appl. Phys.* **2003**, *94*, 1–18. [CrossRef]
11. Stathis, J.H.; Mahapatra, S.; Grasser, T. Controversial issues in negative bias temperature instability. *Microelectron. Reliab.* **2018**, *81*, 244–251. [CrossRef]
12. Jeppson, K.O.; Svensson, C.M. Negative bias stress of MOS devices at high electric fields and degradation of MNOS devices. *J. Appl. Phys.* **1977**, *48*, 2004–2014. [CrossRef]
13. Ye, X.R. VDMOSFET HEF Degradation Modelling Considering Turn-Around Phenomenon. *Microelectron. Reliab.* **2018**, *80*, 37–41. [CrossRef]
14. Otomański, P.; Pawłowski, E.; Szlachta, A. In Situ Evaluation of the Self-Heating Effect in Resistance Temperature Sensors. *Sensors* **2025**, *25*, 3374. [CrossRef] [PubMed]
15. Broche, A.; Cerdeira, A.; Iñiguez, B.; Estrada, M. Determining the Characteristics of the Localized Density of States Distribution Present in MoS_2 2D FETs. *FU Electron. Energetics* **2024**, *37*, 609–617. [CrossRef]
16. Veljković, S.; Mitrović, N.; Davidović, V.; Živanović, E.; Ristić, G.; Danković, D. Successive irradiation and bias temperature stress induced effects on commercial p-channel power VDMOS transistors. *FU Electron. Energetics* **2024**, *37*, 561–579. [CrossRef]
17. Petrović, M.; Veselinović, N.; Tasić, L.; Đorđević, D.; Veljković, S.; Mitrović, N.; Marjanović, M.; Živanović, E.; Danković, D. Investigation of NBTI and Relaxation Effects in P-Channel Power VDMOS transistors. In Proceedings of the IEEE 34th International Conference on Microelectron (MIEL), Niš, Serbia, 13–16 October 2025; pp. 73–77.
18. Steinmann, P.; Lichtenwalner, D.J.; Stein, S.; Park, J.-H.; Das, S.; Ryu, S.-H. Measurement of the Dit Changes Under BTI-Stress in 4H-SiC FETs Using the Subthreshold Slope Method. In Proceedings of the IEEE International Reliability Physics Symposium (IRPS), Grapevine, TX, USA, 14–18 April 2024; p. P58.SiC.
19. Danković, D.; Manić, I.; Prijić, A.; Davidović, V.; Prijić, Z.; Golubović, S.; Djorić-Veljković, S.; Paskaleva, A.; Spassov, D.; Stojadinović, N. A review of pulsed NBTI in P-channel power VDMOSFETs. *Microelectron. Reliab.* **2018**, *82*, 28–36. [CrossRef]
20. Volosov, V.; Bevilacqua, S.; Anoldo, L.; Tosto, G.; Fontana, E.; Russo, A.-L.; Fiegna, C.; Sangiorgi, E.; Tallarico, A.N. Positive Bias Temperature Instability in SiC-Based Power MOSFETs. *Micromachines* **2024**, *15*, 872. [CrossRef]
21. Avraham, T.; Dhyani, M.; Bernstein, J. Reliability Challenges, Models, and Physics of Silicon Carbide and Gallium Nitride Power Devices. *Energies* **2025**, *18*, 1046. [CrossRef]
22. International Rectifier. *IRF9520 Datasheet*; International Rectifier: El Segundo, CA, USA, 1998.
23. Vishay. *P-Channel MOSFETs, the Best Choice for High-Side Switching*; Technical Report AN804; Vishay: Malvern, PA, USA, 1997.
24. Moens, P.; Cano, J.F.; De Keukeleire, C.; Desoete, B.; Aresu, S.; De Ceuninck, W.; De Vleeschouwer, H.; Tack, M. Self-Heating Driven VthShifts in Integrated VDMOS Transistors. In Proceedings of the 2006 IEEE International Symposium on Power Semiconductor Devices & IC's, Naples, Italy, 4–8 June 2006; pp. 1–4.
25. Aviñó-Salvadó, O.; Buttay, C.; Bonet, F.; Raynaud, C.; Bevilacqua, P.; Rebollo, J.; Morel, H.; Perpiñà, X. Physics-Based Strategies *for* Fast TDDB Testing and Lifetime Estimation in SiC Power MOSFETs. *IEEE Trans. Ind. Electron.* **2023**, *71*, 5285–5295. [CrossRef]
26. Đorđević, D.; Tasić, L.; Veselinović, N.; Petrović, M.; Veljković, S.; Mitrović, N.; Marjanović, M.; Živanović, E.; Ristić, G.; Danković, D. Investigation of the Self-Heating Effect in P-Channel VDMOS Transistor after NBT Stress and Relaxation. In Proceedings of the IEEE 34th International Conference on Microelectron (MIEL), Niš, Serbia, 13–16 October 2025; pp. 79–82.
27. Gonzalez-Hernando, F.; San-Sebastian, J.; Garcia-Bediaga, A.; Arias, M.; Iannuzzo, F.; Blaabjerg, F. Wear-Out Condition Monitoring of IGBT and MOSFET Power Modules in Inverter Operation. *IEEE Trans. Ind. Appl.* **2019**, *55*, 6184–6192. [CrossRef]
28. Chenouf, A.; Djezzar, B.; Benadelmoumene, A.; Tahi, H. Deep experimental investigation of NBTI impact on CMOS inverter reliability. In Proceedings of the 24th International Conference on Microelectronics (ICM), Doha, Qatar, 14–17 December 2012; pp. 1–4.

https://doi.org/10.3390/mi17010052

29. Crespo-Yepes, A.; Nasarre, C.; Garsot, N.; Martin-Martinez, J.; Rodriguez, R.; Barajas, E.; Aragones, X.; Mateo, D.; Nafria, M. CMOS inverter performance degradation and its correlation with BTI, HCI and OFF state MOSFETs aging. *Solid-State Electron.* **2022**, *191*, 108264. [CrossRef]
30. Berbel, N.; Fernandez, R.; Gil, I. Modelling and experimental verification of the impact of negative bias temperature instability on CMOS inverter. *Microelectron. Reliab.* **2009**, *49*, 1048–1051. [CrossRef]
31. Unutulmaz, A.; Helms, D.; Eilers, R.; Metzdorf, M.; Kaczer, B.; Nebel, W. Analysis of NBTI Effects on High Frequency Digital Circuits. In Proceedings of the 2016 Design, Automation & Test in Europe Conference & Exhibition (DATE), Dresden, Germany, 14–18 March 2016; pp. 223–228.
32. Veselinović, N.; Petrović, M.; Đorđević, D.; Tasić, L.; Mitrović, N.; Veljković, S.; Marjanović, M.; Živanović, E.; Davidović, V.; Danković, D. Impact of NBTS and Thermal Relaxation on Characteristics of CMOS inverter. In Proceedings of the IEEE 34th International Conference on Microelectron (MIEL), Niš, Serbia, 13–16 October 2025; pp. 83–87.
33. Prijić, A.; Danković, D.; Vračar, L.; Manić, I.; Prijić, Z.; Stojadinović, N. A method for negative bias temperature instability (NBTI) measurements on power VDMOS transistors. *Meas. Sci. Technol.* **2012**, *23*, 085003. [CrossRef]
34. Lange, A.; Velarde Gonzalez, F.A.; Giering, K.-U.; Vervantidis, A.; Hahne, L.; Heinig, A.; Jancke, R. A General Approach for Degradation Modeling to Enable a Widespread Use of Aging Simulations in IC Design. *Microelectron. Reliab.* **2022**, *137*, 114775. [CrossRef]
35. International Rectifier. *IRF510 Datasheet*; International Rectifier: El Segundo, CA, USA, 1998.
36. Kuznetsov, V.; Andreev, D.; Andreev, V.; Piskunov, S.; Popov, A.I. The Compact Model Synthesis for the RADFET Device. *Technologies* **2025**, *13*, 492. [CrossRef]
37. Li, Q.; Gao, B.; Deng, H.; Wang, L.; Yang, D.; Wang, L.; Luo, J.; Han, Z. Failure analysis of the VDMOS device with Vsd and Rds (on) exceeded limit based on reliability physics. In Proceedings of the Prognostics and System Health Management Conference (PHM-Chengdu), Chengdu, China, 19–21 October 2016; pp. 1–5.
38. Fischer Elektronik. *Fischer Elektronik Heatsink SK 104 50.8 STIS, Datasheet*; Fischer Elektronik: Ludenscheid, Germany, 2025.
39. Alutronic. *Alutronic Heatsink PR17/35II/SE, Datasheet*; Alutronic: Halver, Germany, 2025.

 https://doi.org/10.3390/mi17010052

micromachines

MDPI

Review

A Review of Tunnel Field-Effect Transistors: Materials, Structures, and Applications

Shupeng Chen *, Yourui An, Shulong Wang and Hongxia Liu

Key Laboratory of Wide Bandgap Semiconductor Materials, Faculty of Integrated Circuit, Ministry of Education, Xidian University, Xi'an 710071, China; 23111213718@stu.xidian.edu.cn (Y.A.); slwang@xidian.edu.cn (S.W.); hxliu@mail.xidian.edu.cn (H.L.)

* Correspondence: spchen@xidian.edu.cn

Abstract: The development of an integrated circuit faces the challenge of the physical limit of Moore's Law. One of the most important "Beyond Moore" challenges is the scaling down of Metal-Oxide-Semiconductor Field-Effect Transistors (MOSFETs) versus their increasing static power consumption. This is because, at room temperature, the thermal emission transportation mechanism will cause a physical limitation on subthreshold swing (*SS*), which is fundamentally limited to a minimum value of 60 mV/decade for MOSFETs, and accompanied by an increase in off-state leakage current with the process of scaling down. Moreover, the impacts of short-channel effects on device performance also become an increasingly severe problem with channel length scaling down. Due to the band-to-band tunneling mechanism, Tunnel Field-Effect Transistors (TFETs) can reach a far lower *SS* than MOSFETs. Recent research works indicated that TFETs are already becoming some of the promising candidates of conventional MOSFETs for ultra-low-power applications. This paper provides a review of some advances in materials and structures along the evolutionary process of TFETs. An in-depth discussion of both experimental works and simulation works is conducted. Furthermore, the performance of TFETs with different structures and materials is explored in detail as well, covering Si, Ge, III-V compounds and 2D materials, alongside different innovative device structures. Additionally, this work provides an outlook on the prospects of TFETs in future ultra-low-power electronics and biosensor applications.

Keywords: tunnel field-effect transistor (TFET); ultra-low-power electronics; subthreshold swing (*SS*); band-to-band tunneling (BTBT)

1. Introduction

With the continuous advancement of CMOS integrated circuit (IC) technology, key process dimensions in semiconductor manufacturing process have been continuous scaled down with the Moore's Law. However, traditional Metal-Oxide-Semiconductor Field-Effect Transistors (MOSFETs) face the 60 mV/decade (mV/dec) physical limitation of subthreshold swing (*SS*), which is governed by the thermal emission process and the Fermi–Dirac distribution of carriers. This constraint creates a significant trade-off between power consumption and performance, making it challenging to enhance performance without increasing power consumption. Furthermore, the scaling down of traditional MOSFETs causes the short-channel effects, resulting in degraded device performance [1]. The recent appearance of Tunnel Field-Effect Transistors (TFETs) shows a great potential to solve the 60 mV/dec limitation problem faced by CMOS devices. Unlike traditional MOSFETs, TFETs leverage a band-to-band tunneling carrier injection strategy as the primary transport mechanism, enabling carriers to traverse the channel barrier via quantum tunneling rather than the thermal emission process in MOSFETs. Consequently, TFETs bypass the Fermi–Dirac distribution limitation, delivering enhanced switching characteristics at reduced voltages and positioning them as promising candidates for ultra-low-power applications [2,3].

The concept of TFETs dates back to the 1970s, initially featuring a homojunction structure, where both the source and drain were processed by same semiconductor material [4]. The first tunnel field-effect transistor (TFET) was fabricated in 2004 by Joerg Appenzeller, featuring a carbon nanotube as the channel. It achieved a small *SS* of 63 mV/dec [5]. In 2007, Woo Young Choi et al. successfully fabricated a 70 nm n-channel silicon-based TFET, achieving a low *SS* drop to 52.8 mV/dec at room temperature; this result paved the way for the development of sub-60 mV/dec *SS* in silicon-based TFETs [6]. In 2008, Tejas Krishnamohan et al. proposed a double-gate TFET design, which enhanced electric field control and increased tunneling probability by incorporating a second gate, thereby improving device performance [7]. This structure demonstrated higher on-currents and reduced leakage currents, showing promise in initial experiments. In 2011, Wei Cao introduced the p-n-i-n structure with a pocket, further enhancing TFET performance through a simulation analysis method [8]. In 2012, a team from Seoul National University developed an L-shaped channel TFET, with an original intention to increase the tunneling area by extending the point tunneling mechanism to line tunneling mechanism [9]. In 2013, Munetaka Noguchi fabricated an InGaAs TFET with a switching ratio of 10^6; this result greatly promotes the application of TFET as a logic switch [10]. In the same year, M. J. Kumar proposed and simulated a dopingless tunnel field-effect transistor (DL-TFET) [11] and B. Ghosh proposed and simulated a Junctionless Tunnel Field Effect Transistor (JL-TFET) [12], which simplified the fabrication process of TFET by avoiding the doping process of abrupt tunneling junction. In 2015, Deblina Sarkar fabricated a TFET by using a two-dimensional material MoS_2 [13], paving the way for extensive exploration into 2D-material-based TFETs, which has sparked a research boom on 2D-material-based TFET structures. In 2017, Shupeng Chen et al. proposed a novel Symmetric U-shaped TFET with bidirectional current flow and proper performance [14], which can simplify the TFET digital design. In 2020, Seungho Kim et al. utilized bulk BP (black phosphorus) and monolayer BP to construct a homojunction 2D-TFET [15]. In 2021, Clarissa Convertino et al. fabricated an InGaAs/GaAsSb TFET by using nearly standard silicon processes, which provided a valuable reference for the fabrication process of TFETs [16]. In 2022, Shupeng Chen proposed a novel TFET (QB-TFET) based on quasi-broken gap energy band alignment [17], achieving an ultra-low *SS* and an exceptionally high I_{on}/I_{off} ratio.

In order to review the development history of TFETs more clearly, the remaining discussion of this review article is divided into three major parts. The first part mainly discusses the experimental research results of TFETs; this section shows the feasible ways to manufacture TFETs by using the current state-of-the-art technology and reveals the benchmark of the TFETs performance with experiment result. The second part focuses on simulation research results of TFETs; this section indicates the theoretical potential of TFET performance and shows the directions for TFET structure and material optimization in the future. The third part briefly introduces the potential applications and popular areas of TFETs in the future.

2. Experiment Research Works on TFETs

2.1. Si/Ge-Based TFETs

Silicon-based TFETs are among the earliest proposed and experimentally studied TFET devices. Woo Young Choi et al. developed a TFET using Silicon-On-Insulator

(SOI) technology [6]. This work represents the first experimental observation of a silicon-based TFET achieving sub-60 mV/dec SS, with a minimum SS of 52.8 mV/dec at room temperature. The on-state and off-state currents were measured as 12.1 μA/μm and 5.4 nA/μm with V_D = 1 V, respectively. Although the SS was lower than that of MOSFETs, the switching ratio (I_{on}/I_{off}) was inferior to that of MOSFETs. As illustrated in Figure 1a, the source and drain were located on the side and top of the "SOI channel", with the gate positioned above the "SOI channel", separated by an oxide layer. The study highlighted the influence of gate oxide thickness and SOI thickness on SS values, showing a reduction in SS with thinner oxide and SOI layers. Experimental data indicated that with a gate oxide thickness (t_{ox}) of 2 nm and an SOI thickness (t_{SOI}) of 70 nm, the SS remained below 60 mV/dec.

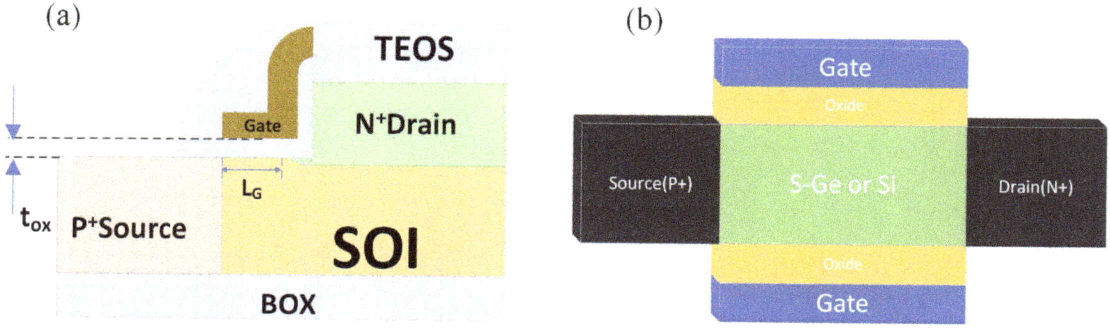

Figure 1. (**a**) Schematic structure of the silicon-based TFET. (**b**) Schematic of the double-gate TFET.

However, silicon faces some challenges as a TFET material due to its large and indirect bandgap. On the one hand, during band-to-band tunneling, electrons have to pass through a relatively thick triangular tunneling barrier. On the other hand, electrons must simultaneously abide by the conservation of energy and momentum. This will lead to a low tunneling rate and result in a low tunneling current. Additionally, traditional silicon-based TFET fabrication requires heavy doping, which is still difficult to achieve with the state-of-the-art processing techniques. Krishna Mohan et al. proposed a Si/Ge heterojunction structure combined with an innovative double-gate (DG) design, as illustrated in Figure 1b. Their experimental results demonstrated a DG, strained-Ge (s-Ge) TFET achieving a record on-state current (I_{on}) of approximately 300 μA/μm and an SS of around 50 mV/dec [7].

A novel silicon-based line tunneling TFET has recently emerged through the work of Weijun Cheng et al. [18]. This innovative structure, presented in Figure 2a,b, features an N_+ pocket layer integrated between source and gate terminals to facilitate carrier tunneling along the entire interface. The enlarged tunneling cross-section substantially enhances tunneling current generation. The experimental results revealed a minimum subthreshold swing (SS_{min}) of 69 mV/dec and an average subthreshold swing (SS_{avg}) of 80 mV/dec at room temperature. Additionally, the device achieved a high I_{on} of 40 μA/μm at V_{DS} = 1 V while maintaining an off-state current (I_{off}) below 1 pA/μm, as shown in Figure 2c.

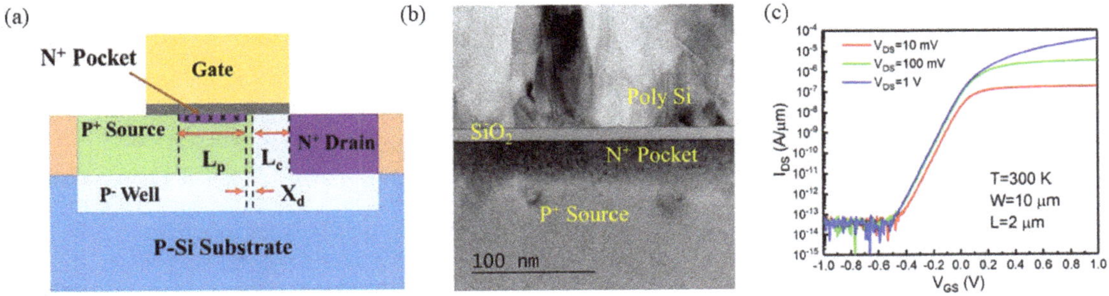

Figure 2. (**a**) Schematic structure of the Si line tunneling TFET. (**b**) The TEM image of the Si line tunneling TFET. (**c**) Transfer characteristic curves of the Si line tunneling TFET [18]. Reproduced with permission from ref. [18].

2.2. Heterojunction TFETs Based on III-V Compounds

III-V compound materials, owing to their good electrical properties, have also been one of the major topics of discuss in the fabrication of TFETs. Compared to Si/Ge, the direct and small bandgap of some III-V compounds can give a lower tunneling barrier for band-to-band tunneling of carriers. Additionally, their lower effective masses for electrons and holes, combined with extremely high electron mobility and steep density of states (DOS), significantly enhance tunneling efficiency. From the fabrication perspective, III-V compounds can be readily integrated into silicon-based processes through heteroepitaxy, offering greater flexibility for designing high-performance and low-power integrated circuits. Munetaka Noguchi et al. fabricated a planar heterojunction TFET by doping the source region of 100 nm $In_{0.53}Ga_{0.47}As$/S.I.InP wafers with Zn [19]. Figure 3a illustrates the structure of the device, and Figure 3b describes the specific fabrication process of the device. The solid-phase diffusion of Zn formed a steep-profile and defect-free p^+/n source junction due to Zn diffusion characteristics in InGaAs, significantly enhancing TFET performance. This device achieved a minimum *SS* of 64 mV/dec with I_{on} of 10^{-6} A/μm, I_{off} of 10^{-12} A/μm, and I_{on}/I_{off} of 10^6.

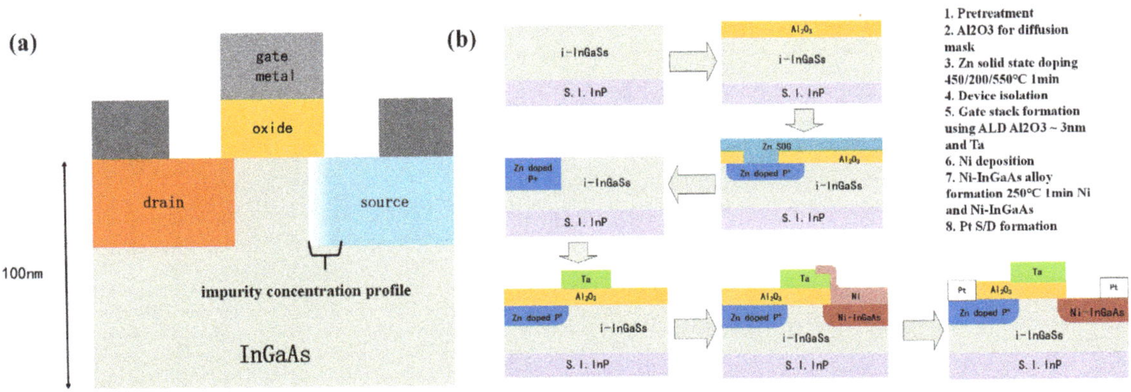

Figure 3. (**a**) Schematic structure of the $In_{0.53}Ga_{0.47}As$ TFET. (**b**) Schematic fabrication process of the $In_{0.53}Ga_{0.47}As$ TFET.

Additionally, nanowire TFETs are a popular research field, especially in III-V compounds TFETs. Due to the high surface-to-volume ratio of their nanowire structures, the gate can comprehensively envelop the channel and provide a strong control. This "Gate-All-Around" structure effectively suppresses short-channel effects and minimizes leakage currents, thereby enhancing device performance such as I_{on}/I_{off} [20]. Moreover, nanowire TFETs are well suited for compact nanoscale fabrication processes and exhibit strong compatibility with existing technologies [21]. This makes them particularly advantageous for achieving high-density integrated circuits applications. Anil W. Dey et al. developed a nanowire heterojunction TFET, which exhibited relatively higher I_{on} [22]. This device utilized a GaSb/InAs (Sb) heterojunction with a type II broken band alignment and was fabricated through metal–organic vapor phase epitaxy. As shown in

Figure 4a, the structure of the InAsSb/GaSb NW-TFET is presented. However, studies have revealed that a significant number of dislocations are generated at the heterojunction interface between InAs and GaSb, which adversely affects carrier tunneling [23,24].

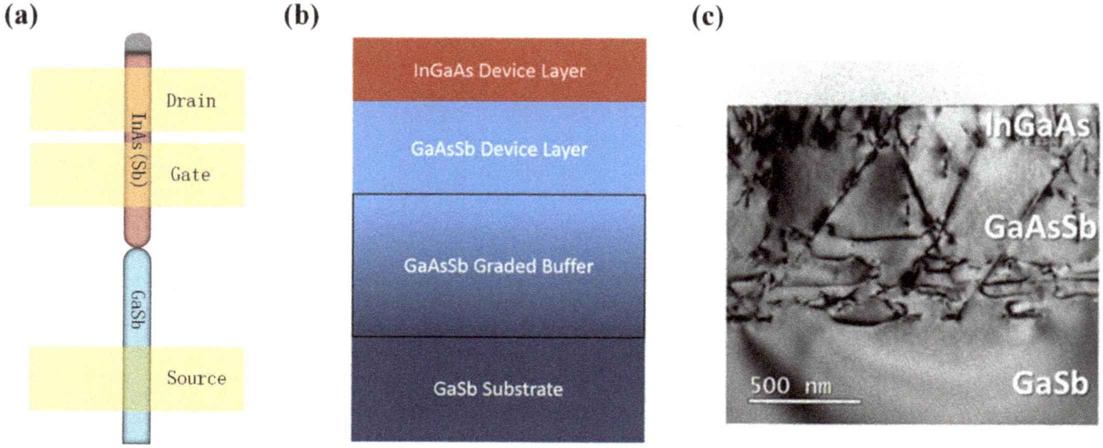

Figure 4. (**a**) Schematic of the GaSb/InAs(Sb) heterojunction TFET. (**b**) Schematic of the InGaAs/GaAsSb NW-TFET. (**c**) TEM images of the InGaAs/GaAsSb tunnel junctions [25]. (**b**,**c**) are reproduced with permission from ref. [25].

To address the issue of dislocations at the heterojunction interface, Ryan M. et al. fabricated an $In_{0.8}Ga_{0.2}As/GaAs_{0.3}Sb_{0.7}$ NW-TFET using metal-organic chemical vapor deposition (MOCVD) [25]. In Figure 4b, the growth structure of the tunnel diode is illustrated. The fabrication process involved first growing a GaAsSb buffer layer on a GaSb substrate, followed by the growth of an InGaAs/GaAsSb functional layer, effectively reducing dislocations caused by lattice mismatches. However, according to TEM images, as shown in Figure 4c, a significant number of threading and misfit dislocations were observed at the InAs/GaSb and InGaAs/GaAsSb interfaces despite the relatively small lattice mismatch. This phenomenon may be attributed to the MOCVD growth process, as the use of gaseous reactants in MOCVD can introduce impurities. Furthermore, during material growth, MOCVD may result in Ga carryover effects due to methyl exchange reactions between trimethylindium (TMIn) and trimethylgallium (TMGa), potentially causing Ga atoms to migrate to the upper layers and leading to dislocation formation [23,26–32] and elemental exchange [33,34]. Such issues can have a significant negative impact on achieving high-quality tunnel junction interfaces. In 2018, Jheng-Sin Liu et al. used solid-source molecular beam epitaxy (MBE) to achieve heterogeneous integration of an InAs/GaSb NW-TFET using a 200 nm strained $GaAs_{1-y}Sb_y$ dislocation-filter buffer layer [35]. Figure 5a shows the growth structure, while Figure 5b presents the *J-V* characteristics of the fabricated tunnel junction. At room temperature, the tunnel junction exhibited a tunneling current of 5×10^{-5} A/μm^2 at 2 V reverse bias under room temperature.

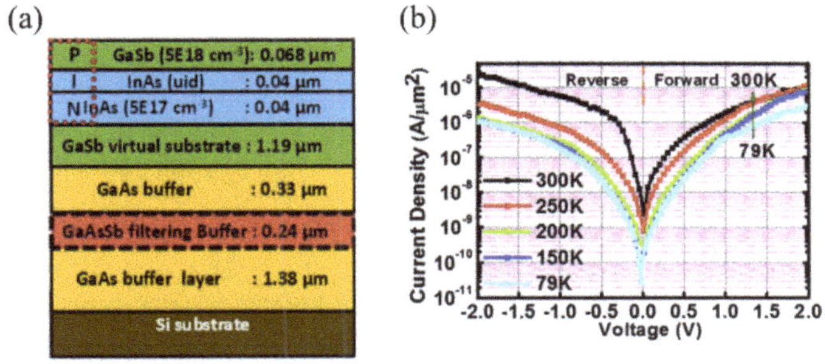

Figure 5. (**a**) Structure of the InAs/GaSb NW-TFET. (**b**) J-V characteristics of the InAs/GaSb NW-TFET [35]. Reproduced with permission from ref. [35].

Beyond InGaAs/GaSb, the InGaAs/InAs heterojunction has also been explored for TFET design. Xin Zhao et al. developed a nanowire TFET featuring a heterojunction between p_+-InGaAs and intrinsic InGaAs, with a 2 nm i-InAs/8 nm i-$In_{0.7}Ga_{0.3}As$ "notch" layer inserted to reduce the tunneling barrier and enhance drive current [36]. Process optimizations included improved ALD chamber conditioning to ensure uniform Al_2O_3 deposition, followed by five digital etch cycles post-RIE to refine nanowire diameter to 20 nm, improving surface quality. Rapid thermal annealing was also employed to passivate interface traps, thereby suppressing trap-assisted tunneling (TAT) effects. The device demonstrated a temperature-insensitive SS, achieving 53 mV/dec at V_{DS} = 0.3 V at room temperature and maintaining 37 mV/dec at V_{DS} = 0.05 V down to 77 K.

The adoption of a Gate-All-Around (GAA) structure can enhance the performance of Nanowire TFETs by further improving the surface-to-volume ratio. In 2019, T. Vasen et al. fabricated Vertical Gate-All-Around Nanowire GaSb-InAs Core-Shell n-Type Tunnel FETs (VGAA C-S TFET) [37]. The structure and corresponding SEM images are shown in Figure 6a–c. The testing results, illustrated in Figure 6d,e, demonstrate an SS_{min} of 40 mV/dec at V_D = 10mV. The device achieves a current drive of 39.4 µA/µm, normalized to a GaSb core diameter of 35 nm, at V_D = 0.3 V and V_G = 0.5 V. Recent research on nanowire TFETs continues to attract significant interest [38,39]. For instance, the newly proposed ferroelectric tunnel field-effect transistor utilizes a nanowire structure [40], which enables frequency doubling without generating additional harmonics, making it highly suitable for signal transmission applications.

In 2021, Clarissa Convertino et al. fabricated an InGaAs MOSFET and an InGaAs/GaAsSb TFET using near-standard silicon manufacture processes [16]. The structure and main fabrication process of the device are shown in Figure 7a–d. Unlike pure MOSFET processes, the epitaxial regrowth of electrode contact zones was conducted in two separate steps, forming n-type and p-type regions independently. MOSFET electrodes and TFET drain terminal comprised epitaxially grown n^+ Sn-doped $In_{53}Ga_{47}As$ (1×10^{19} cm^{-3}), while the TFET source was formed from p^+ Zn-doped $Ga_{0.5}As_{0.5}Sb$ (4×10^{19} cm^{-3}), lattice-matched to InGaAs. This design enabled clear differentiation between TFETs and MOSFETs at the device level, facilitating flexible hybrid logic block design. Measurements showed that at room temperature, with V_{DS} = 50 mV, the minimum SS for the GaAsSb TFET at L_G = 25 nm and W_{FIN} = 25nm was 43 mV/dec. The MOSFET counterpart measured a minimum SS of 62 mV/dec under similar conditions, with electrical stability demonstrated across varying temperatures (Figure 7e,f). This study confirmed the feasibility of manufacturing TFETs using standard silicon-CMOS-compatible processes, paving the way for designing ultra-low-power hybrid circuits.

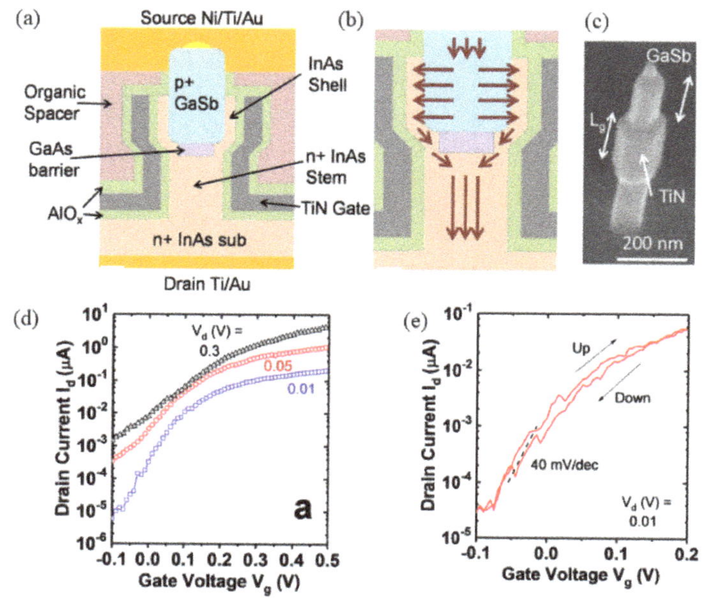

Figure 6. (**a**) Structural depiction of the VGAA C-S TFET. (**b**) Enhanced magnification highlighting electron trajectories within the device. (**c**) Scanning electron micrograph (SEM) of the nanowire (NW) segment positioned above the gate electrode following selective digital removal of the InAs shell layer. (**d**,**e**) I_D–V_G characteristics [37]. Reproduced with permission from ref. [37].

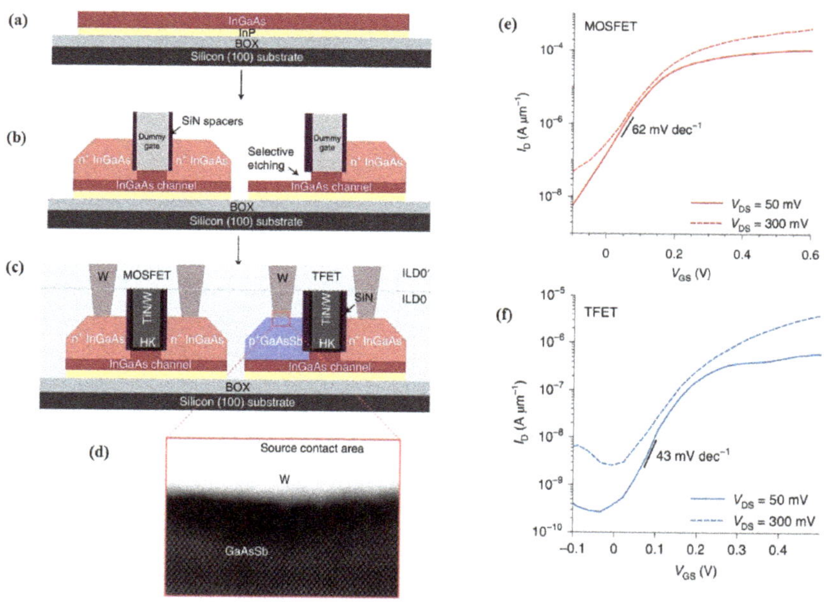

Figure 7. (**a**–**c**) Schematics for the fabrication process steps of the InGaAs MOSFET and the InGaAs/GaAsSb TFET. (**d**) High-resolution TEM image assessing the crystalline quality in the GaAsSb exposed region. (**e**) Transfer characteristic of the InGaAs MOSFET. (**f**) Subthreshold characteristic for a TFET with L_G = 25 nm and W_{FIN} = 25 nm [16]. Reproduced with permission from ref. [16].

2.3. TFETs Based on 2D Materials

In recent years, many TFETs have been fabricated using two-dimensional (2D) materials including Molybdenum disulfide (MoS_2), black phosphorus (BP), tungsten diselenide (WSe_2), etc. [41–44]. Two-dimensional semiconductors exhibit excellent gate controllability due to their ultra-thin layered structure. However, their bandgap can be influenced by the quantum confinement effect depending on the number of layers. The fabrication of 2D TFETs must therefore consider both material selection and the number of layers. TFETs based on 2D materials are categorized as homojunction or heterojunction TFETs, with heterojunctions further divided into 2D/2D and 2D/3D types.

Heterojunction TFETs leverage staggered or broken band alignments, generating sharper band profiles than those achievable through doping modulation alone. Therefore, they enhance significant enhancement in both on-current and subthreshold steepness. For example, vertical heterojunction TFETs using materials such as WSe_2 can exhibit band offset transitions from a staggered gap (type-II) to a broken gap (type-III) under high positive back-gate voltages, inducing tunneling at channel biases [45,46]. In 2015, Tania Roy et al. constructed a vertical 2D/2D heterojunction TFET using MoS_2 and WSe_2, mechanically transferred onto a Si/SiO_2 substrate, with top-gate stacking to enable dual-gate control (Figure 8a,b) [47]. In the same year, Rusen Yan et al. developed a BP (black phosphorus)/ $SnSe_2$ heterojunction TFET, further validating tunneling mechanisms (Figure 8c,d) [48]. In 2016, Tania Roy incorporated ZrO_2 as a gate dielectric to enhance gate control and reduce *SS* in $WSe_2/SnSe_2$ vertical TFETs, achieving a minimum *SS* of 100 mV/dec, though still above 60 mV/dec (Figure 9a–c) [49]. In 2017, Xiao Yan fabricated a $WSe_2/SnSe_2$ vertical TFET with an ultra-thin Al_2O_3 gate dielectric, achieving a minimum *SS* of 37 mV/dec and an average *SS* of approximately 80 mV/dec, with a maximum on-current of ~1.5 μA and an I_{on}/I_{off} exceeding 10^6 (Figure 9d–f) [50].

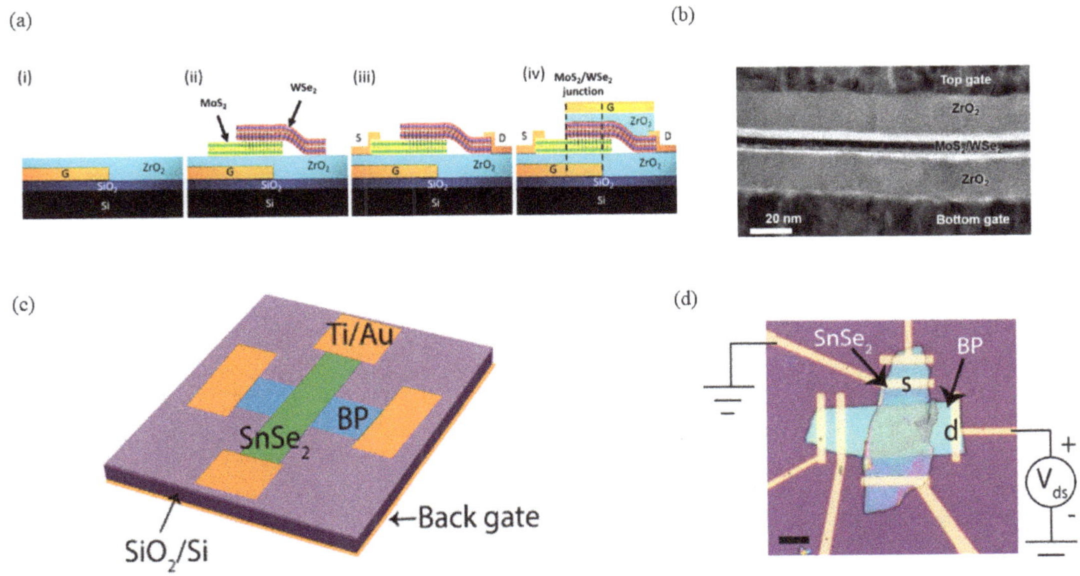

Figure 8. (**a**) Schematics for the fabrication process steps of the MoS_2/WSe_2 TFET. (**b**) Cross-sectional TEM image of the MoS_2/WSe_2 TFET [47]. (**c**,**d**) Schematic illustration and optical image of the $BP/SnSe_2$ heterojunction TFET [48]. (**a**,**b**) are reproduced with permission from ref. [47]. (**c**,**d**) are reproduced with permission from ref. [48].

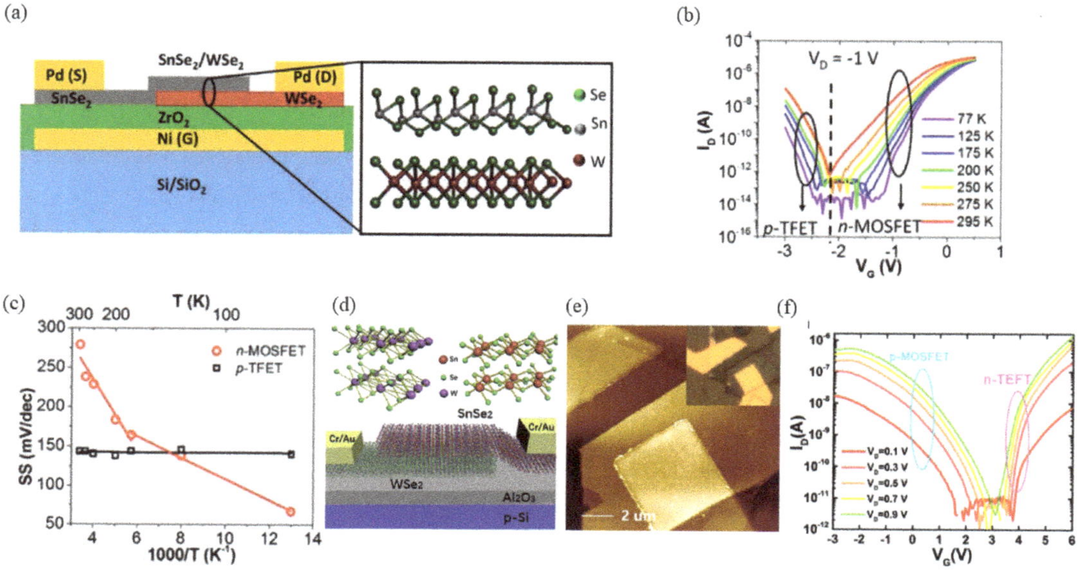

Figure 9. (**a**) Device structure of the $WSe_2/SnSe_2$ TFET. (**b**) Transfer characteristic of the $WSe_2/SnSe_2$ TFET. (**c**) The subthreshold swing (*SS*) of both devices varies with temperature [49]. (**d**) Schematic of the $WSe_2/SnSe_2$ TFET. (**e**) Atomic force microscopy (AFM) topography of the fabricated device. (**f**) Transfer characteristic of $WSe_2/SnSe_2$ heterostructure [50]. (**a–c**) are reproduced with permission from ref. [49]. (**d–f**) are reproduced with permission from ref. [50].

To further enhance the performance of 2D/2D heterojunction TFETs, Hyun Bae Jeon et al. introduced the use of an ion gel as a top-gate dielectric in MoS_2/WSe_2 heterojunction TFETs [51]. Figure 10a shows the structure of the device, and Figure 10b presents the microscope image of the device. The ion gel dielectric exhibited high capacitance, creating strong electric fields in small-channel regions, enhancing gate control and reducing *SS*. This approach also enabled modulation from type-II to type-III (broken gap) band alignment, facilitating band-to-band tunneling (BTBT) with steep *SS* and high I_{on}/I_{off}. This device achieved a minimum *SS* of 36 mV/dec, with an I_{on}/I_{off} of 10^6.

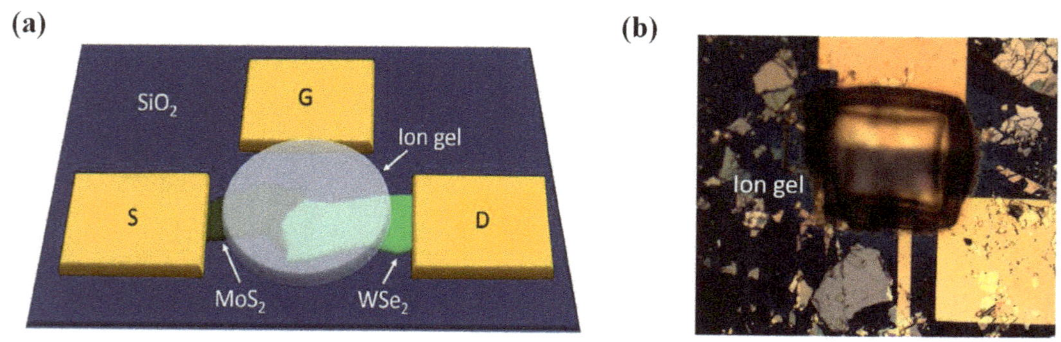

Figure 10. (**a**) WSe_2/MoS_2 van der Waals heterostructure TFET schematic. (**b**) Optical micrograph post ion-gel top-gate deposition. [51]. Reproduced with permission from ref. [51].

In addition to using 2D/2D heterojunctions, vertical tunneling field-effect transistors can also incorporate 2D/3D heterojunctions. Deblina Sarkar et al. transferred MoS_2 onto p-Ge to form a TFET structure, as illustrated in Figure 11a. This device utilized double-layer MoS_2 synthesized through chemical vapor deposition (CVD), chosen for its high electron affinity compared to other 2D materials. Meanwhile, Ge, as a 3D material, exhibited lower electron affinity than other Group III-V compounds and IV elements. The experimental results (Figure 11b) revealed an average *SS* of 46.4 mV/dec at room temperature. However, further studies indicated that the formation of a Ge oxide layer during the process limited carrier tunneling probability [13]. In 2018, Gwang Hyuk Shin et al. fabricated a vertical MoS_2/Si TFET using a trilayer MoS_2 and heavily doped Si (10^{19} cm^{-3}), with Al_2O_3 serving as the gate dielectric, as depicted in Figure 11c. Figure 11d displays the transfer characteristic curves of the device, and the device demonstrated a minimum *SS* of 23 mV/dec and an I_{on}/I_{off} exceeding 10^7 [52]. In 2021, Jinshui Miao and collaborators constructed a 2D/3D TFET using n-type 2D InSe and heavily p-doped 3D silicon. Their process involved etching square windows on the $SiO_2/p^{++}Si$ substrate to expose the underlying $p^{++}Si$ layer, followed by transferring mechanically exfoliated 2D InSe flakes onto the exposed surface. The device structure is shown in Figure 11e. Testing at room temperature with $V_{p\text{-}Si} = -1$ V yielded a minimum *SS* of 6.4 mV/dec and an average *SS* of 30 mV/dec, with a maximum on-current density of $I_{p\text{-}Si} = 0.3$ μA/μm. A comparison with InSe MOSFETs revealed a minimum *SS* of ~100 mV/dec (Figure 11f) [53].

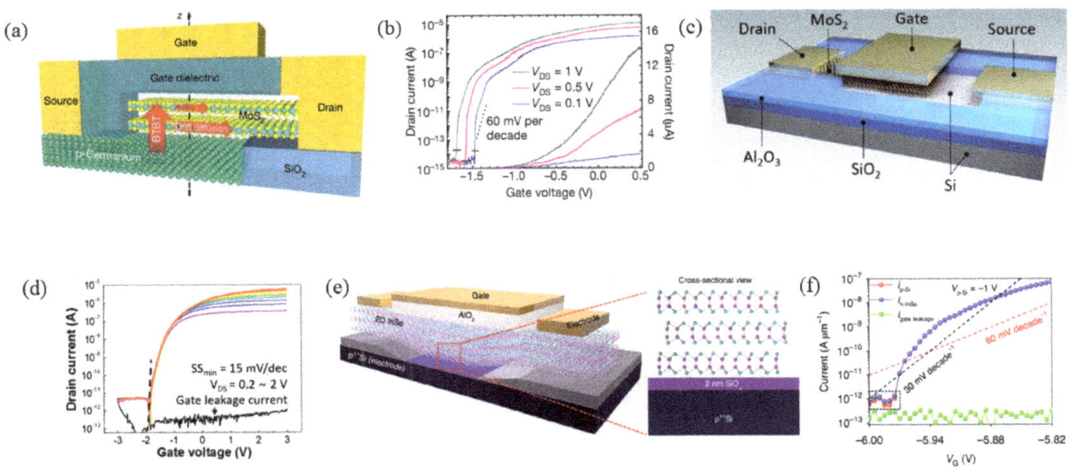

Figure 11. (**a**) Cross-sectional schematic of MoS_2/Ge heterojunction TFET. (**b**) Corresponding transfer characteristics (I_D-V_G) [13]. (**c**) Device structure of the MoS_2/Si TFET. (**d**) I_{DS}-V_{GS} transfer characteristics of the MoS_2/Si TFET [52]. (**e**) Schematic of an n-InSe/p^{++}Si 2D/3D HJ-TT. (**f**) Transfer characteristic of 2D/3D HJ-TT [53]. (**a**,**b**) are reproduced with permission from ref. [13]. (**c**,**d**) are reproduced with permission from ref. [52]. (**e**,**f**) are reproduced with permission from ref. [53].

Hexagonal boron nitride (h-BN), a wide-bandgap two-dimensional (2D) material, is frequently employed as a dielectric or tunneling barrier in 2D material-based memory devices. Ruiqing Cheng et al. proposed a novel tunneling field-effect transistor (TFET) structure by inserting a thin layer of h-BN between two tunnel-capable 2D materials, resulting in a Cr/hBN/MoS_2 TFET (Figure 12a) [54]. When the thickness of the h-BN layer is relatively small (~3.2 nm), the device exhibits direct tunneling characteristics. However, for thicker h-BN layers (~9.6 nm), the device demonstrates clear Fowler–Nordheim tunneling (FNT) behavior. A linear relationship in the $\ln(I/V^2) - 1/V$ plot is observed above 5.8 V (Figure 12b), confirming the triangular barrier-induced FNT mechanism, with an extracted barrier height of h-BN of approximately 3.43 eV. Under FNT conditions, the tunneling current can be effectively modulated via gate control of the Fermi level in the graphene electrode. The device achieves an on/off current ratio of 5×10^3. Furthermore, it demonstrates a high rectification ratio of 7×10^5 and excellent non-volatile memory performance, with a program/erase (P/E) ratio exceeding 10^5 and retention time over 400 s (Figure 12c). This work is a significant advancement, offering new insights into the development of 2D-TFETs and opening avenues toward high-performance 2D rectifiers and memory devices.

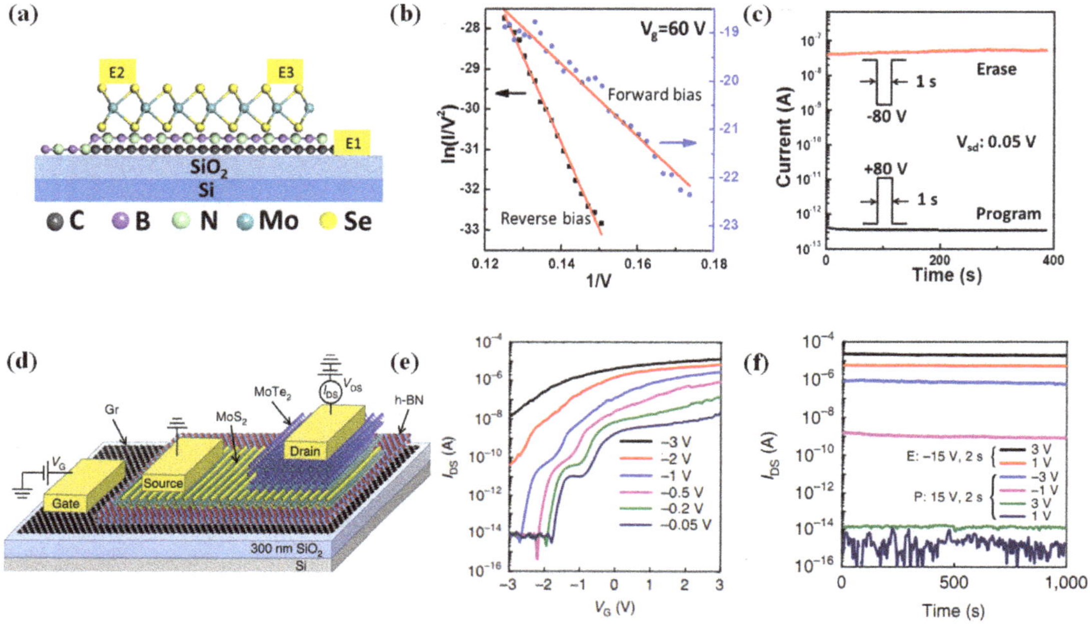

Figure 12. (**a**) Device structure of the Cr/hBN/MoS_2 TFET. (**b**) $\ln(I/V^2) - 1/V$ curves. (**c**) Retention characteristic of the memory [54]. (**d**) Structure schematic of the Cr/hBN/MoS_2/$MoTe_2$ TFET. (**e**) I_{DS}-V_G transfer curves. (**f**) Retention characteristics of programmed and erased states under varying read bias conditions [55]. (**a–c**) are reproduced with permission from ref. [54]. (**d–f**) are reproduced with permission from ref. [55].

Building upon this design, the team further improved the device by replacing the MoS_2 channel with a MoS_2/$MoTe_2$ heterostructure, forming an asymmetric Cr/hBN/MoS_2/$MoTe_2$ TFET (Figure 12d) [55]. Under positive drain bias ($V_{DS} > 0$), the device behaves like a conventional p–n junction dominated by thermionic emission, whereas under negative bias, it transitions into a TFET regime. In this mode, the gate voltage (V_G) effectively modulates the tunneling barrier width, enabling efficient control over the tunneling process. The experimental results reveal an impressive on/off current ratio up to 6×10^8 (Figure 12e), a five-order-of-magnitude improvement over the previous structure. Additionally, the rectification ratio is enhanced to 10^8, and the memory performance is significantly improved with a P/E ratio > 10^9 VD and retention time exceeding 1000 s (Figure 12f). The device also exhibits excellent photoresponse, suggesting strong potential for 2D material-based optoelectronic and memory-integrated systems.

Besides heterostructures, 2D materials can be employed to fabricate homojunction TFETs, avoiding trap-related issues arising from contact between different materials and thus enhancing device reliability. Unlike heterostructures that rely on van der Waals (vdWs) linkages and multiple stacking layers, homojunctions are less challenging for modern manufacturing processes. In 2019, Peng Wu et al. developed a homojunction TFET using black phosphorus (BP), exploiting electrostatic doping through different gates. The device structure is depicted in Figure 13a,b. Applying opposing bias polarities to gates (G1, G2) and the middle top gate (G) created tunnel p-i-n or n-i-p junctions; for instance, a positive V_{G1} doped the BP base n-type, while a negative V_{G1} induced p-type doping (Figure 13c,d). This approach enabled the creation of p-type or n-type 2D-MOSFETs and TFETs. Tests revealed a minimum SS of 170 mV/dec and an on-current (I_{on}) of 0.6 μA/μm (Figure 13e) [56].

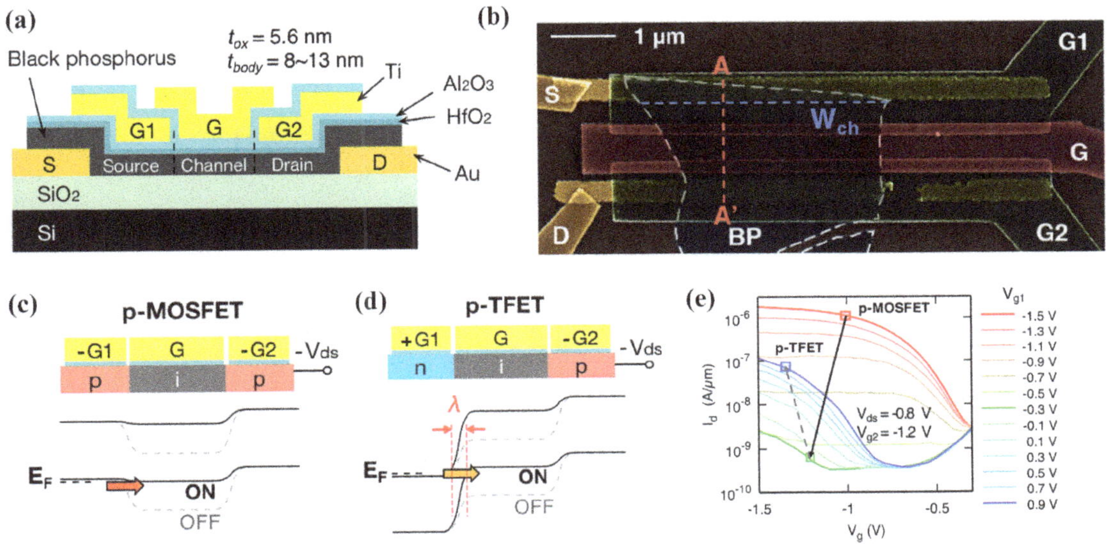

Figure 13. (**a**) BP RED-TFET architecture. (**b**) False-colored SEM image of the BP RED-TFET. (**c**) p-MOSFET schematic and band diagram. (**d**) p-TFET operational schematic and band diagram. (**e**) ransfer characteristics under p-configuration [56]. Reproduced with permission from ref. [56].

In addition to utilizing gate voltage for electrical doping, the unique thickness-dependent electrical properties of 2D materials can also be exploited to fabricate TFETs. In 2023, Tomohiro Fukui et al. utilized homojunctions formed by MoS_2 layers of varying thickness to create TFETs. Figure 14a shows the structure of the device, and Figure 14b presents the microscope image of the device. Due to changes in bandgap (E_G) with layer thickness, they grew Nb-doped p^+ MoS_2 with a hole concentration of 2×10^{19} cm^{-3} via chemical vapor transport (CVT). Mechanical exfoliation, however, caused S vacancies on the MoS_2 surface, leading to an n-type transition due to higher electron concentrations from vacancies than Nb-induced hole concentrations. Consequently, a homojunction was formed between MoS_2 layers of different thicknesses. Measurements showed a minimum SS of 140 mV/dec at room temperature (Figure 14c,d) [57].

In 2020, Seungho Kim et al. utilized bulk BP and monolayer BP to construct a homojunction TFET. The source region consisted of bulk BP, the channel region was formed by a monolayer of BP, and the drain was capped with ultra-thin hBN and graphene (Figure 15a). Representative transfer characteristics of two BP NHJ TFET devices are presented in Figure 15b,c. Device 1 incorporated a 285 nm SiO_2 bottom-gate dielectric and 10 nm hBN top-gate dielectric, with channel dimensions $L \approx 0.7$ μm and $W \approx 1$ μm. This device exhibited p-type TFET operation, achieving $SS_{ave,4dec} < 60$ mV/dec. At $V_D = -0.6$ V, Device 1 surpassed all previously reported TFETs (including n-type variants) in both $SS_{ave,4dec}$ and I_{60} metrics, with $I_{60} \approx 65$ μA/μm, aligning with the preferred operational range (1–10 μA/μm). The I_D-V_{TG} transfer curve (10 nm hBN gate dielectric) yielded $SS_{ave,4dec} \approx 23.7$ mV/dec and $I_{60} \approx 65$ μA/μm. Device 2 employed a 3 nm hBN bottom-gate dielectric and 5 nm hBN top-gate dielectric, with $L \approx 0.5$ µm and $W \approx 1$ µm. Under n-type operation at V_D = +0.7 V, this configuration achieved record-low $SS_{ave,4dec}$ along with the highest I_{60} reported for sub-thermionic TFETs. The measurement results (3 nm hBN dielectric) indicated $SS_{ave,4dec} \approx 24.0$ mV/dec and $I_{60} \approx 0.054$ µA/µm. Notably, both devices required substantially lower switching voltages (Device 1: ΔV_{TG} = 0.15 V; Device 2: ΔV_{BG} = 0.2 V) compared to state-of-the-art MOSFETs ($\Delta V_G \geq 0.7$ V), signifying significantly reduced power consumption in BP NHJ-TFETs [15]. However, BP's poor stability in air remains a challenge for practical applications [58].

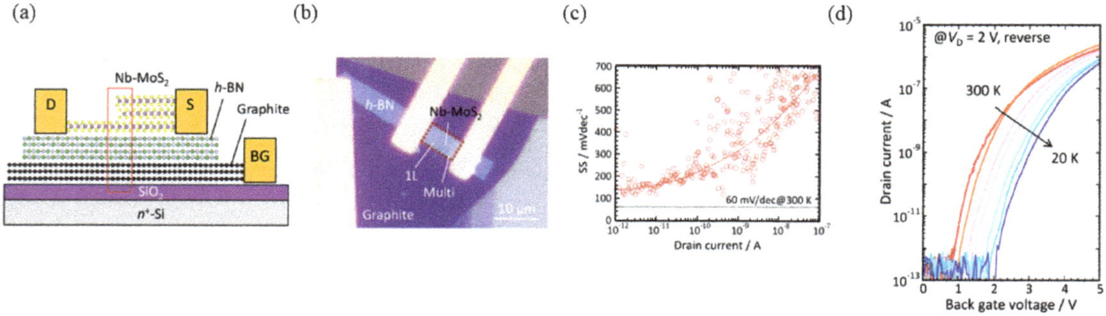

Figure 14. (**a**) Schematic the TM-homojunction Nb-doped MoS_2 TFET. (**b**) Optical micrograph of the TM-homojunction Nb-doped MoS_2 TFET. (**c**) *SS* of the Nb-doped MoS_2 TFET as a function of temperature (**d**) Temperature dependence of transfer characteristics of Nb-doped MoS_2 TFET [57]. Reproduced with permission from ref. [57].

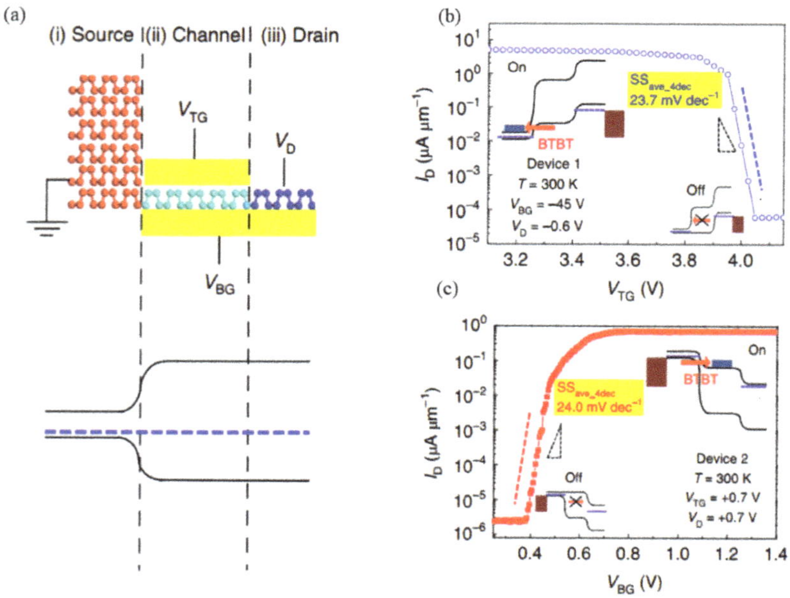

Figure 15. (**a**) Schematic structure and energy band diagrams of BP NHJ-TFET at source, channel, and drain regions. (**b**) p-type transfer characteristics for Device 1 with indicated subthreshold swing (*SS*). (**c**) n-type transfer characteristics for Device 2 with corresponding *SS* [15]. Reproduced with permission from ref. [15].

2.4. Carbon Nanotubes TFET

Carbon nanotubes (CNTs), with their unique cylindrical geometry, have also emerged as a material of choice for TFET fabrication [59,60]. Their natural cylindrical shape allows for partial or full gate wrapping, enhancing gate control [60]. The CNT-TFET was first proposed by J. Appenzeller et al., utilizing band-to-band tunneling induced by applying a negative back-gate voltage and a positive top-gate voltage, resulting in band bending and an *SS* around 40 mV/dec [5]. When a negative voltage is applied to the top gate, the device behaves like a traditional MOSFET with thermionic emission dominating and the *SS* exceeding 60 mV/dec.

In 2021, Chin-Sheng Pang and colleagues developed a CNT TFET using a multi-gate structure, as shown in Figure 16a. And Figure 16b shows an SEM image of the device. Similar to the aforementioned BP-based TFET [57], the device utilized V_{TGS} control to adjust transistor states (Figure 16d). As V_{TGS} increased, the material beneath gradually transitioned from p-type to n-type, forming an n-i-p tunnel junction. Tests showed a minimum *SS* of 41 mV/dec at room temperature, with a maximum tunneling current of 100 nA (Figure 16c) [61]. While CNT TFETs hold promise, further improvements are necessary to fully realize their potential.

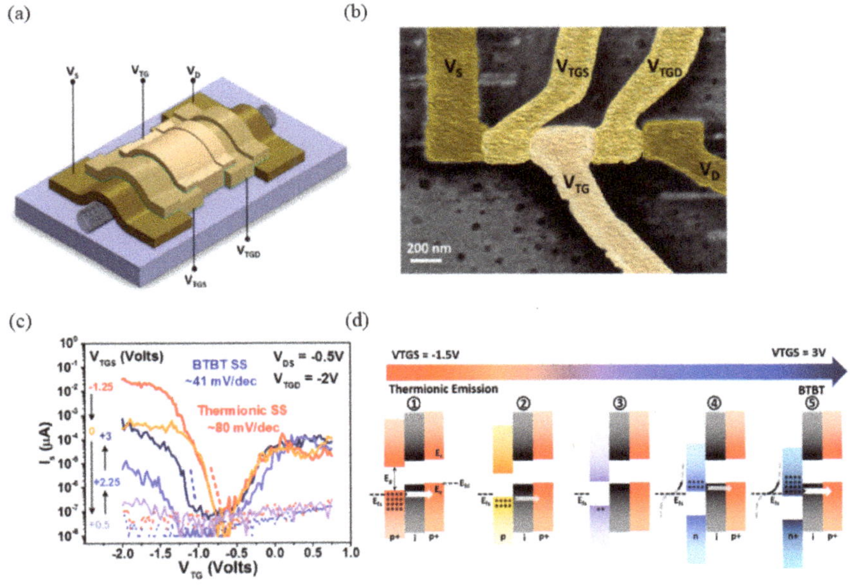

Figure 16. (**a**) Electrode configuration with voltage bias scheme. (**b**) Scanning electron micrograph of fabricated CNT triple-gate device (scale bar: 200 nm). (**c**) Transfer characteristics under varying top-gate voltage conditions. (**d**) Energy band diagrams at different gate biases [61]. Reproduced with permission from ref. [61].

Enhancing tunneling efficiency with suppressed leakage necessitates band-alignment engineering via controlled doping in CNTs. Chongwu Zhou et al. explored surface doping by adsorbing potassium atoms via chemical vapor deposition. However, this approach suffered from poor long-term stability [62]. To address this, Maguang Zhu et al. proposed an internal cavity doping method using encapsulated perovskites instead of surface modifiers (Figure 17a,b) [63]. This strategy greatly improved device reliability, with HAADF-STEM confirming no degradation of the perovskite crystal structure even after 30 days. Electrical characteristics remained stable, achieving *SS* = 35.2 mV/dec, an I_{on} of

4.94 μA per tube, and an I_{on}/I_{off} ratio exceeding 10^5 at room temperature and $V_D = -0.1$ V (Figure 17c).

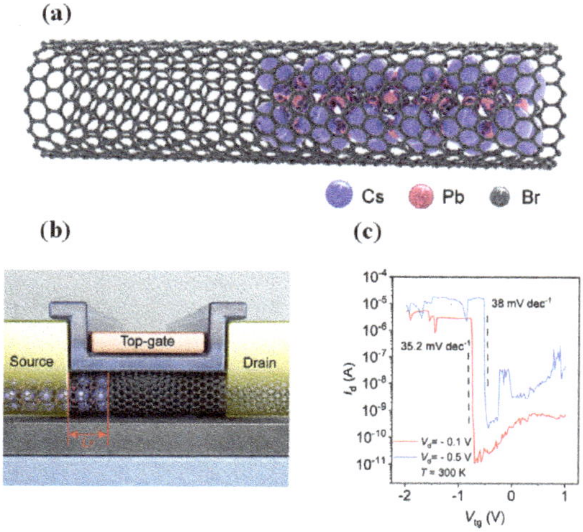

Figure 17. (**a**) Schematic of the coaxial $CsPbBr_3$/CNT. (**b**) Schematic of the $CsPbBr_3$/CNT heterojunction FET. (**c**) Transfer characteristics of the $CsPbBr_3$/CNT heterojunction FET [63]. Reproduced with permission from ref. [63].

Table 1 presents an experimental comparison of key electrical parameters for the aforementioned TFET devices. Overall, 2D materials exhibit a promising trend of surpassing 3D materials in the field of TFETs. It is worth noting that 2D materials are still in the early stages of development. Therefore, further simulations and experimental research are necessary to comprehensively explore and establish the design-level correlations between 2D material systems, device design strategies, and synthesis/manufacturing technologies. This approach will ultimately enable CMOS-level performance at the nanoscale.

Table 1. Experimental comparison of electrical parameters for fabricated TFETs.

Device Name	*SS* (mV/dec)	I_{on} (μA/μm)	I_{on}/I_{off}	V_d (V)	Ref.
SOI Si-TFET	52.8	12.1	$\sim 2.2 \times 10^3$	1.0	[6]
Si/Ge DG TFET	~50	~300	-	-	[7]
Si Line Tunneling TFET	69	40	$>10^7$	1.0	[18]
InGaAs TFET	64	1.0	10^6	-	[19]
InGaAs/GaAsSb TFET	43	-	-	0.05	[16]
InAs/GaSb NW-TFET	-	5×10^{-5} (A/μm^2)	-	2.0	[35]
WSe_2/MoS_2 TFET	36	100	10^6	−1.0	[51]
Cr/hBN/MoS_2 TFET	-	-	5×10^3	-	[54]
Cr/hBN/$MoS_2/MoTe_2$ TFET	-	-	6×10^8	−1.0	[55]
$WSe_2/SnSe_2$ TFET	37	1.5	$>10^6$	-	[50]
MoS_2/Ge TFET	46.4	-	-	-	[13]
MoS_2/Si TFET	23	-	$>10^7$	-	[52]
InSe/Si TFET	6.4	0.3	-	−1.0	[53]
BP RED-TFET	170	0.6	-	0.8	[56]
Nb-doped MoS_2 TFET	140	-	-	2.0	[57]
BP Homojunction TFET	23.7–24.0	-	-	0.6/0.7	[15]
CNT Triple Gate TFET	41	0.1	-	-	[61]
$CsPbBr_3$ CNT TFET	35.2	4.94	$>10^5$	−0.1	[63]

3. Simulation Research Works on TFETs

Given that TFET has become one of the international hot research topics in the post-Moore era, performance optimization of TFET in terms of structures, materials, working mechanisms, etc., needs to be maximized excavation and research. This results in a large number of researchers using simulation methods to analyze and study the performance of TFETs and has produced a large number of research results till now. This section will focus on the simulation research results on TFETs; the research findings in this part indicate the theoretical potential of TFET performance and show the directions for TFETs structure and material optimization in the future. Table 2 summarizes the key performance parameters of the TFETs to be discussed below.

Table 2. Simulated performance metrics of advanced TFET architectures.

Device Name	SS (mV/dec)	I_{on} (μA/μm)	I_{on}/I_{off}	V_d (V)	Ref.
NC-LTFET	18.3	0.24	-	-	[64]
S-TFET	45	-	10^5	0.5	[65]
SUTFET	15.2	13.5	4.4×10^6	0.5	[14]
HTG-TFET	36.59	7.02	-	0.2	[66]
OGDL-TFET	20.6	75.5	9×10^{13}	0.5	[67]
DF-TFET	18.2	58.8	-	-	[68]
SNPJL-TFET	20	54.3	1.45×10^{14}	0.5	[69]
QB-TFET	4.9	921	-	0.5	[17]

3.1. p-n-i-n TFET

To improve the tunneling probability of the traditional p-i-n structure, Wei Cao et al. proposed the p-n-i-n configuration [8]. This modified architecture, shown in Figure 18a, inserts a thin n-doped interlayer at the p-i tunneling junction to enhance the tunneling electric field (E_y). A progressive increase in the doping density in the interlayer was experimentally demonstrated to enhance overall device conductance. Simulations quantifying gate oxide influence indicated significantly suppressed threshold voltage fluctuations induced by oxide variations, thereby establishing a foundational framework for further TFET optimization.

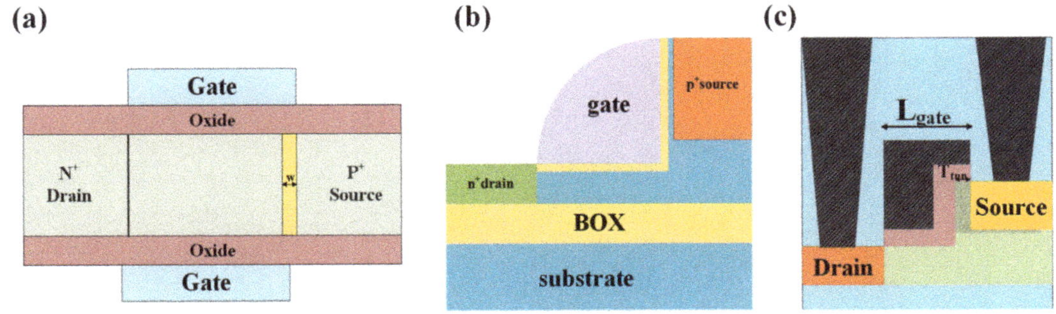

Figure 18. (**a**) p-n-i-n TFET schematic showing layer configuration and contacts. (**b**) Schematic of L-shaped TFET configuration. (**c**) Fabrication process flow.

3.2. L-Shaped TFET

The L-shaped TFET was pioneered by Sang Wan Kim et al. [9], with its structural configuration depicted in Figure 18b. To enhance band-to-band tunneling (BTBT) efficiency, minimizing the tunneling barrier width (W_t) between E_V and E_C is essential, given the exponential relationship between tunneling probability and W_t [70]. Conventional TFETs exhibit W_t variation modulated by gate voltage (V_G), as it is governed by junction depletion dynamics. At low V_G values, the significant W_t dimension results in diminished tunneling probability, yielding gradual switching transitions and elevated subthreshold swing (SS) values. Channel length reduction further degrades performance due to increased off-state current (I_{off}) from thermal emission, manifesting punch-through currents analogous to MOSFET behavior [71]. In contrast, L-shaped TFETs maintain a fixed W_t determined by the L_i dimension. This configuration achieves substantially steeper SS values compared to conventional designs. Crucially, whereas conventional TFETs restrict effective tunneling to nanometer-scale regions beneath the gate electrode, thereby limiting on-state current (I_{on}) [72], L-shaped TFETs establish tunneling orthogonally to the channel direction (Figure 18b). This geometric advantage significantly expands the effective tunneling cross-section, enabling simultaneous improvement in both SS steepness and I_{on} magnitude.

Sang Wan Kim et al. continued to explore the fabrication process of an L-shaped TFET and successfully produced devices with improved performance [73], as shown in Figure 18c. Experimental characterization reveals that the L-shaped TFETs have a less ideal I_{off} compared to simulation results. But even though this is a less-ideal result, an I_{on}/I_{off} ratio of 5 orders of magnitude is achieved by the experimental results at V_d = 0.05V. Thus, the L-shaped TFET still has great potential in applications of ultra-low-power circuits [74].

In 2022, Zhang, H et al. optimized the original L-shaped TFET design by proposing the Negative Capacitance LTFET (NC-LTFET) (Figure 19a) [64]. Inspired by the p-n-i-n structure [8], an n^+ pocket layer was introduced at the p^+ source. Additionally, drawing on Masaharu Kobayashi et al.'s research on NC-TFETs [75] (Figure 19b), a ferroelectric gate oxide layer was incorporated between the gate and the channel. According to Kobayashi's findings, using ferroelectric materials in this configuration creates negative capacitance, generating an electric field that facilitates carrier tunneling. Si (SiO_2-doped hafnium oxide) was employed as the gate oxide, with the manufacturing process for the ferroelectric materials (Si) detailed in reference [76]. The simulations yielded the transfer characteristic curves of the NC-LTFET (Figure 19c). The NC-LTFET demonstrated an SS of 18.3 mV/dec over a drain current range from 4×10^{-17} to 1×10^{-9} A/µm. Additionally, its large I_{on} of 2.4×10^{-7} A/µm at V_{GS} = 1 V suggests strong driving capability in digital circuits. Its small I_{off} of 4×10^{-17} A/µm at V_{GS} = 0 V also indicates minimal static power consumption in the off state. Consequently, the NC-LTFET is a highly promising candidate for low-power applications and merits further investigation.

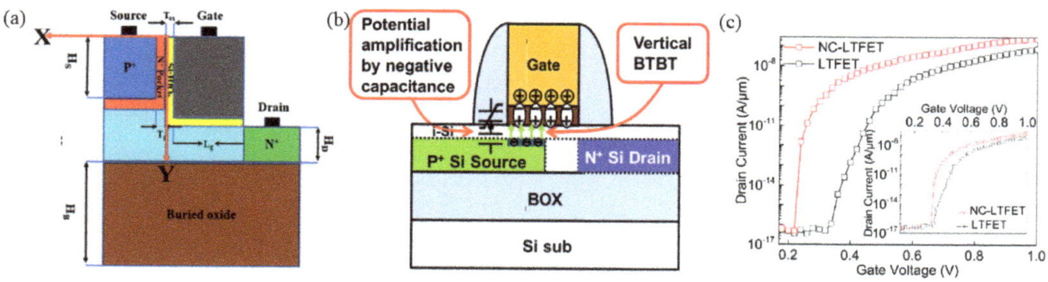

Figure 19. (**a**) The cross-section view of NC-LTFET. [64] (**b**) Schematic of NCTFET [75]. (**c**) The transfer characteristic curve of NC-LTFET and LTFET [64]. (**a**,**c**) are reproduced with permission from ref. [64]. (**b**) is reproduced with permission from ref. [75].

3.3. U-Shaped TFET

In 2014, Wei Wang et al. proposed the U-shaped TFET (UTFET), with its structure illustrated in Figure 20a [77]. Unlike planar TFETs, the UTFET channel is recessed into the substrate, transforming the lateral p-i-n configuration into a vertical direction, significantly enhancing tunneling efficiency and improving device performance [78]. Further optimizations were made, such as adding an additional heavily doped n-type auxiliary tunneling barrier layer at the source tunneling junction to increase tunneling efficiency [8]. The use of a heavily doped p-type germanium source and a silicon/germanium heterojunction as the tunneling junction further increased the tunneling efficiency compared to LTFETs [9]. This work compared five TFET configurations—Ge-UTFET, SiGe-UTFET, Si-UTFET, planar Si-TFET, and planar SiGe-TFET—by evaluating their transfer characteristics. The Ge-UTFET demonstrated significantly superior current drive capability relative to both Si-UTFET and planar Si-TFET under V_G = 0.7V bias. Substituting the Ge source with $Si_{0.6}Ge_{0.4}$ substantially reduced I_{on} and I_{off}—an effect principally attributed to SiGe's wider bandgap (0.84 eV vs. Ge's 0.67 eV). Crucially, all devices achieved optimized switching at V_D = 0.7V, exhibiting an average subthreshold swing of 60 mV/dec with drain current modulation spanning over six orders of magnitude.

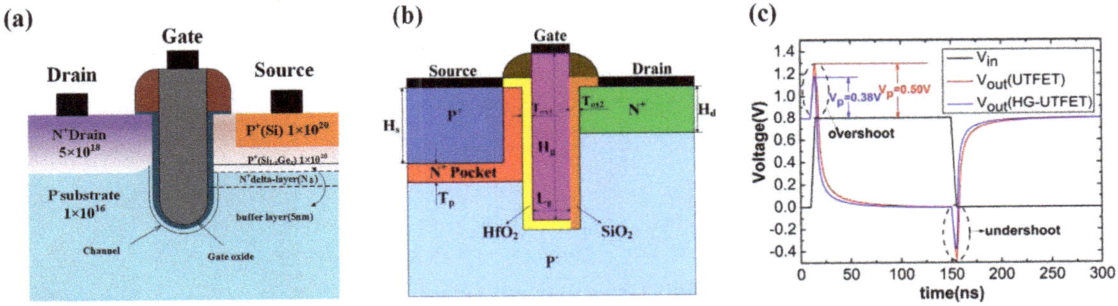

Figure 20. (**a**) Comparative SiGe-UTFET configurations with/without n_+-delta layer. (**b**) Schematic of the HG -UTFET. (**c**) Transient response of TFET-based inverters demonstrating signal overshoot/undershoot [79]. (**b**,**c**) are reproduced with permission from ref. [79].

In 2017, Wei Li et al. introduced an improvement to the U-shaped TFET by proposing the heterogeneous gate UTFET (HG-UTFET) [79], with its structure shown in Figure 20b. This design aimed to address the issue of large Miller capacitance in TFETs. The Miller capacitance significantly impacts the frequency response of analog circuits and the delay characteristics of digital circuits [80–82]. When present, the Miller capacitance amplifies the effective capacitance transferred from the input to the output by a factor of (1 + A_v), where A_v is the voltage gain. In digital circuits, delay is proportional to capacitance, while in analog circuits, frequency is inversely proportional to capacitance. Thus, the presence of Miller capacitance negatively affects both digital and analog circuit performance. The large Miller capacitance in TFETs arises from their band-to-band tunneling conduction mechanism [83–85]. During gate voltage transitions from off- to on-state, the inversion layer of the TFET expands from the channel surface toward the source region, eventually covering the entire channel surface. Elevated gate–drain capacitance (C_{gd}) results in the Miller capacitance being dominated by the gate–drain capacitance, with gate–source capacitance (C_{gs}) contributing minimally. This behavior differs from MOSFETs, where the gate–source and gate–drain capacitances are similar before the formation of the inversion layer [86]. Once the inversion layer forms, the gate capacitance is mainly determined by the C_{gs}.

To address the issue of large C_{gd} in TFETs, researchers have proposed several solutions. Primarily, these include the use of heterogeneous gate dielectrics and gate–drain exposure methods [87–90]. The heterogeneous gate dielectric method involves dividing the TFET gate oxide layer into two sections: a high-k material is used near the source region to maintain strong gate control, while a low-k material is applied near the drain region to reduce gate-drain overlap capacitance [91]. Consequently, the left side of the HG-UTFET employs a high-k dielectric (HfO_2), while the right side uses low-k SiO_2. To demonstrate that the HG-UTFET offers a smaller Miller capacitance (*CM*) compared to the UTFET, Figure 20c shows that both inverters experience output signal overshoot/undershoot. HG-UTFET—demonstrates a ∼45.6% reduction in Miller capacitance compared to conventional UTFET. Furthermore, *CM* significantly affects TFET inverter delay characteristics, with HG-UTFET inverters exhibiting 30.7% shorter falling delay than UTFET implementations.

3.4. Symmetric Tunnel Field-Effect Transistor

Hyohyun Nam et al. pioneered the Symmetric TFET (S-TFET) featuring a unique p-i-p architecture [65]. As depicted in Figure 21a, this design employs n-type-doped silicon channels (N_{Si} = 10^{16} cm^{-3}) with 5 nm thick lightly p-doped silicon pads ($N_{Si\text{-}pad}$ = 10^{15} cm^{-3}). Critical innovation lies in the 15 nm thick germanium source/drain regions (N_{Ge} = 2 × 10^{19} cm^{-3}), where narrow-bandgap materials enhance I_{on} performance. Contrasting conventional p-i-n configurations, the p-i-p topology facilitates bidirectional current flow while achieving sub-60 mV/decade subthreshold swing. The electrical characterization presented in Figure 21b,c establishes that under a bias voltage of V_{DD} = 0.5 V, S-TFET devices with germanium source/drain doping concentrations exceeding 10^{19} cm^{-3} exhibit significantly enhanced electrical properties. The drive current consistently surpasses 5 μA/μm, while the I_{on}/I_{off} ratios exceed 10^5, accompanied by minimal subthreshold swing values approximating 45 mV/decade. Performance optimization reaches a plateau at the critical doping concentration $N_{Ge,S/D}$ = 2 × 10^{19} cm^{-3}, beyond which further increases in doping density yield no substantial improvement in either current magnitude or switching characteristics.

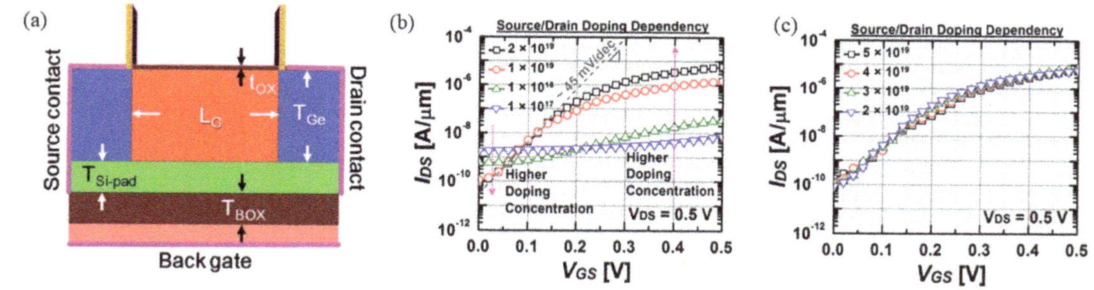

Figure 21. (**a**) Cross-sectional schematic of symmetric tunnel FET (S-TFET) highlighting structural layers. (**b**,**c**) I_{DS}-V_{GS} characteristics at V_{DS} = 0.5 V, showing source/drain doping effects: Source/drain doping concentration (**b**) lower than 2 × 10^{19} cm^3 and (**c**) higher than 2 × 10^{19} cm^3 [65]. Reproduced with permission from ref. [65].

In 2017, Shupeng Chen et al. proposed a novel Symmetric U-shaped Gate TFET (SUTFET) [14]. The structure, as shown in Figure 22a, features a gate resembling the shape of the letter "U" with p$^+$ Ge source/drain regions flanking both gate sides. Two n$^+$ pockets adjoining the gate enhance tunneling efficiency. The n-Si channel and p-Si pad are situated beneath the gate to create an I_{on} channel for the device. Unlike conventional TFETs, SUTFET enables planar MOSFET-like bidirectional current flow, rendering it suitable for ultra-low-power ICs. Figure 22b presents simulation results comparing the transfer characteristic curves of the symmetric TFET (S-TFET) [65], L-TFET [73], U-TFET [77], and SUTFET [14], all derived using identical key device parameters, including the same n$^+$ Si pocket height and width. The simulation results indicate that at room temperature with V_{DD} = 0.5V, the min *SS* can reach 15.2 mV/dec, while the I_{on} is 13.5 μA/μm, and the I_{on}/I_{off} is approximately 4.4 × 10^6.

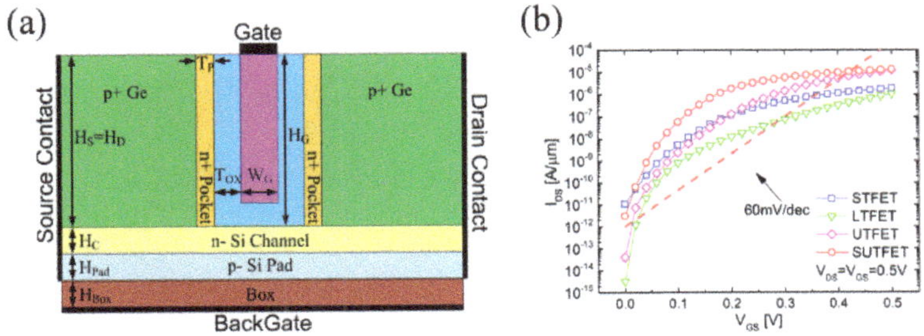

Figure 22. (**a**) Schematic of the SUTFET. (**b**) Transfer characteristic curves of S-TFET, L-TFET, U-TFET, and SUTFET [14]. Reproduced with permission from ref. [14].

3.5. HTG-TFET

Wei Li et al. innovatively integrated the L-shaped TFET [9] and SUTFET [14] architectures, proposing a T-gate heterojunction TFET (HTG-TFET), illustrated in Figure 23a [66]. This configuration expands the effective tunneling cross-section, significantly enhancing drive current. Under optimal simulation parameters, the HTG-TFET demonstrated superior performance, achieving an I_{on} of 7.02 μA/μm and a minimum SS of 36.59 mV/dec at V_G = 0.2 V (Figure 23b).

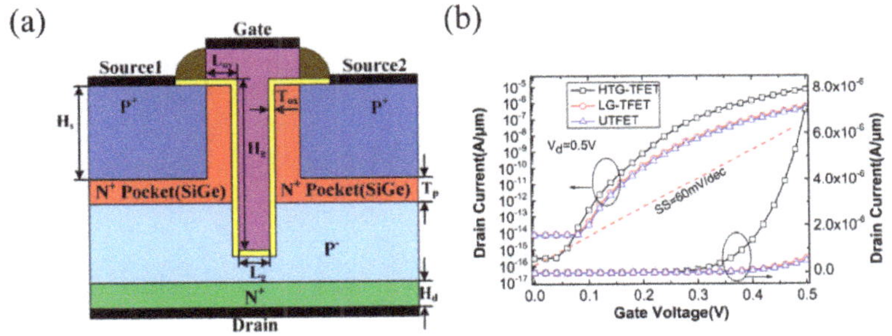

Figure 23. (**a**) Schematic structures of HTG-TFET. (**b**) Transfer characteristics of different devices [66]. Reproduced with permission from ref. [66].

However, in practical TFET fabrication, heavy doping of the source and drain regions can lead to diffusion into the channel region, thereby affecting the steepness of the PN junction and altering the tunneling barrier width at the junction. Additionally, achieving uniform heavy doping is challenging due to the random fluctuations introduced by the ion implantation process. Researchers have studied the impact of doping variability on TFET performance, and generally, doping fluctuations can cause variations in the I_{on} by up to two orders of magnitude [92–94], significantly affecting TFET performance. High-temperature annealing, often required during fabrication, further adds to the process cost.

3.6. DL-TFET and JL-TFET

To address these issues, researchers proposed a dopingless TFET (DL-TFET) [11,95–97] and a junctionless TFET (JL-TFET) [12,98–100]. In dopingless TFETs, the source, channel, and drain regions are intrinsic materials, with the tunneling PN junction modulated by the work functions of the gate, source, and drain. Unlike conventional devices, the source and drain regions of dopingless TFETs form Schottky contacts instead of Ohmic contacts. By appropriately adjusting the work functions of the source and drain regions, holes accumulate in the source region, while electrons accumulate in the channel, forming a "pseudo-PN junction" through charge plasma effects. In contrast, all regions of the junctionless TFET are uniformly doped. The source region achieves inversion, and by tuning the metal work function of the source, holes accumulate to form a p-type region. Figure 24 illustrates the structural schematics of both dopingless and junctionless TFET devices. Existing studies have shown that these devices exhibit ultra-low *SS* and minimal I_{off}. However, their primary limitation is the inability to achieve line tunneling, resulting in lower I_{on} compared to conventional TFETs. Therefore, enabling line tunneling in DL-TFETs and JL-TFETs through effective strategies not only addresses fabrication challenges but also enhances the overall performance of TFETs.

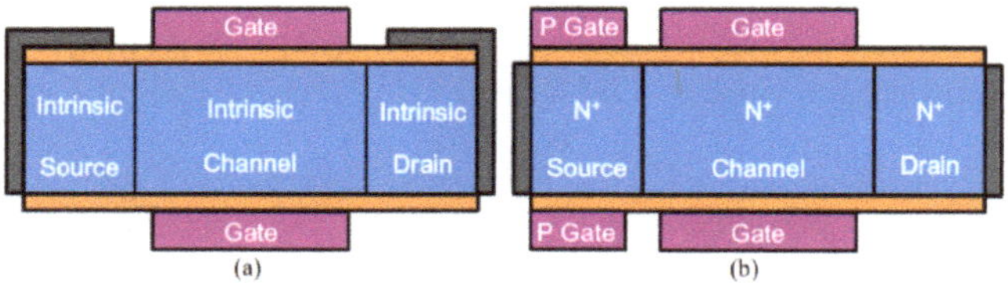

Figure 24. Schematic structures of (**a**) DL-TFET and (**b**) JL-TFET.

In 2019, Shupeng Chen improved the DL-TFET by proposing a germanium-based overlapping gate dopingless tunnel FET (OGDL-TFET) [67]. As illustrated in Figure 25a, this architecture positions gate and backgate electrodes on opposing channel surfaces. Differential work-function engineering induces energy band bending within their overlap region, establishing a line-tunneling junction via charge plasma principles. This line tunneling junction enables significantly increased I_{on}/I_{off} switching ratios through simultaneous I_{on} amplification and I_{off} suppression. The simulation results (Figure 25b) demonstrate that, at V_{DS} = 0.5 V, the OGDL-TFET delivers I_{on} = 75.5 μA/μm, $I_{on}/I_{off} = 9 \times 10^{13}$, SS_{min} = 1.9 mV/dec, and SS_{avg} = 20.6 mV/dec. Subsequently, in 2020, Shupeng Chen proposed a dopingless fin-shaped SiGe channel TFET (DF-TFET) [68]. This device also enables line tunneling for carriers, with its structure depicted in Figure 25c. The use of a fin structure in the DF-TFET significantly reduces the footprint compared to planar line tunneling TFETs. Consequently, the DF-TFET achieves an I_{on} of 58.8 μA/μm, an SS_{min} of 2.8 mV/dec, and an SS_{avg} of 18.2 mV/dec (Figure 25d).

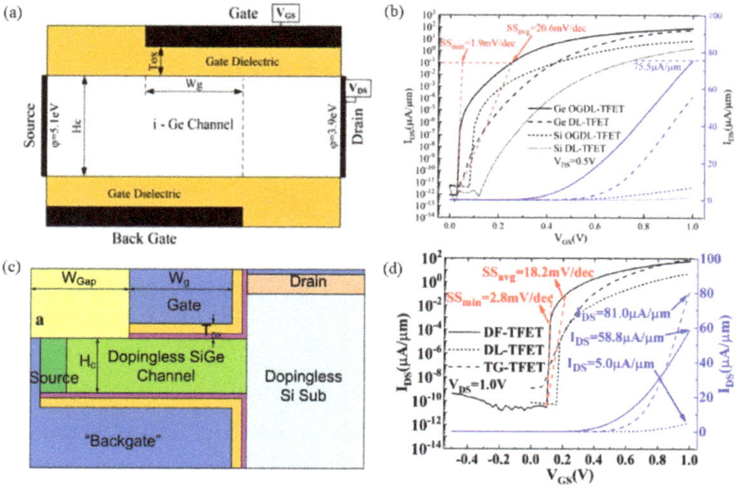

Figure 25. (**a**) Schematic of the OGDL-TFET. (**b**) Transfer characteristic curves of OGDL-TFET [67]. (**c**) Schematic of the DF-TFET. (**d**) Transfer characteristic curves of DF-TFET [68]. (**a**,**b**) are reproduced with permission from ref. [67]. (**c**,**d**) are reproduced with permission from ref. [68].

In 2021, Sazzad Hussain et al. introduced a modified JL-TFET design termed SNPJL-TFET, which incorporated a pocket gate into the original structure, forming a p–n–i–n configuration to enhance tunneling probability (Figure 26a) [69]. Additionally, a heterogeneous gate dielectric stack was used, balancing gate control and parasitic capacitance. Compared to all-SiO_2 or all-HfO_2 gate stacks, the hybrid stack enhanced C_g control while minimizing C_{gd}, making the device better suited for high-frequency applications. Under V_D = 0.5 V, the device achieved I_{on} = 5.43 × 10^{-5} A, I_{on}/I_{off} = 1.45 × 10^{14}, and SS = 20 mV/dec. It also exhibited a peak transconductance of 210 μS/μm, a cut-off frequency (F_T) of 65.4 GHz, and a gain-bandwidth product (GBP) of 14.9 GHz (Figure 26b,c).

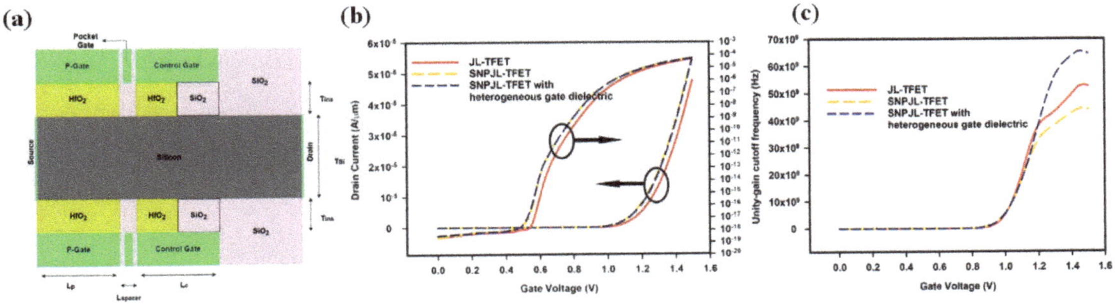

Figure 26. (**a**) Cross-section view of SNPJL-TFET with heterogeneous gate dielectric. (**b**) Transfer curve. (**c**) Unity-gain cutoff frequency (F_T) of JL-TFET, SNPJL-TFET and SNPJL-TFET with heterogeneous gate dielectric [69]. Reproduced with permission from ref. [69].

3.7. QB-TFET

In 2022, Shupeng Chen et al. developed a quasi-broken gap aligned tunneling FET (QB-TFET) to optimize subthreshold swing and switching efficiency [17]. As illustrated in Figure 27a, to achieve high I_{on}, the QB-TFET employs an InGaAs/GaAsSb heterojunction with quasi-broken gap energy band alignment, which improves the band-to-band tunneling rate. An intrinsic InGaAs spacer is incorporated between the n_+ InGaAs drain and p_+ GaAsSb source to minimize off-state leakage resulting from source-drain tunneling pathways. Additionally, TiO_2 is used as the gate dielectric to improve the gate voltage control over the channel. Consequently, under a supply voltage of 0.5 V, the QB-TFET achieves a high I_{on} of 921 μA/μm and an SS_{avg} as low as 4.9 mV/dec (Figure 27b).

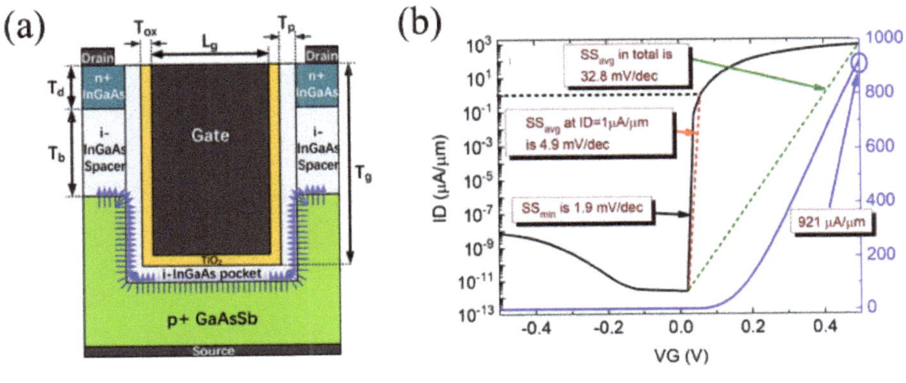

Figure 27. (**a**) Schematic of the QB-TFET. (**b**) Transfer characteristic curves of QB-TFET [17]. Reproduced with permission from ref. [17].

4. Applications of TFETs

TFETs, especially 2D-TFETs, offer significant advantages in circuit design due to their low power consumption. Arnab Pal et al. used Hspice simulations to analyze the performance, robustness, and power consumption of circuits like inverters, ring oscillators, and SRAMs designed with WTe_2/MoS_2-HJ-TFETs, comparing them to those based on 7 nm FinFET technology, as shown in Figure 28 [101]. Among these, the 2D-TFET inverter outperformed its 7 nm FinFET-based counterpart. Although both transistors demonstrate comparable transfer curves (Figure 28a), the 2D-TFET inverter achieves 48% higher voltage gain owing to enhanced drain current saturation (Figure 28b). For TFET-based ring oscillators, the oscillation frequencies reached 10 GHz and 57 MHz, with single-stage delays of 10 ps and 1.6 ns, respectively, indicating weaker performance compared to FinFET-based oscillators (Figure 28c,d). The designed 2D-TFET SRAM (Figure 28e) demonstrated relatively ideal performance, as shown in Figure 28f. The all-2D-TFET SRAM cell exhibits read, write, and hold margins of 133 mV, 304 mV, and 296 mV, respectively, at V_{DD} = 0.7 V. Overall, these circuit modules performed well and can be readily integrated into more complex circuits.

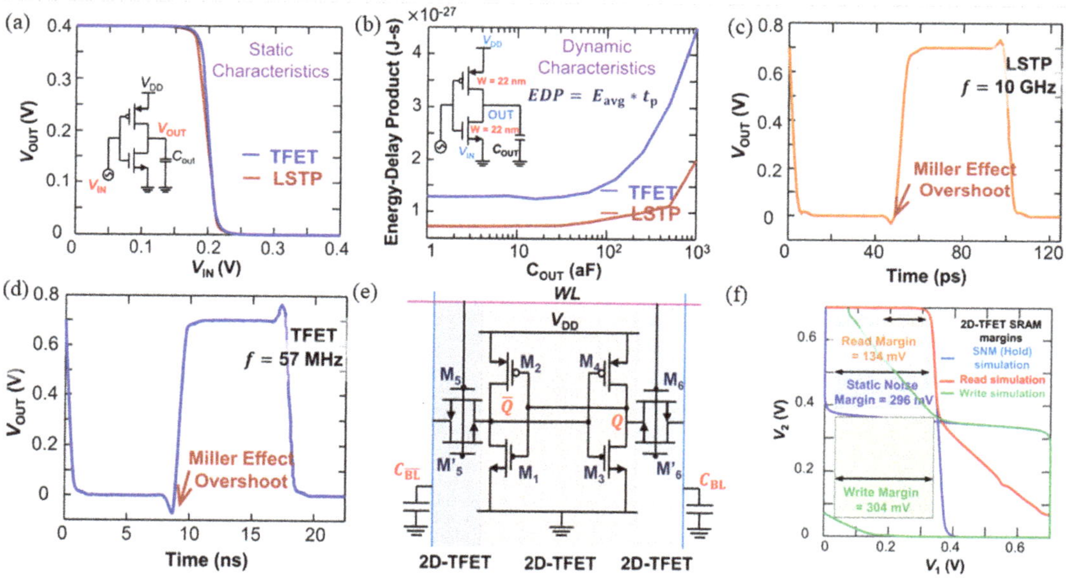

Figure 28. (**a**) Static characteristics of minimum-size 2D-TFET/LSTP inverters (V_{DD} = 0.4 V). (**b**) Energy-delay product (EDP) comparison across capacitive loads. (**c**) LSTP and (**d**) 2D-TFET oscillatory waveforms (10 GHz vs. 57 MHz), noting Miller overshoot in (**d**). (**e**) Unidirectional current-addressed all-2D-TFET SRAM circuit. (**f**) Static noise margin analysis for hold/read/write operations (V_{DD} = 0.7 V) [101]. Reproduced with permission from ref. [101].

Recent progress has also been made in hybrid TFET-MOSFET circuit designs. Wenjuan Lu et al. proposed an 11-transistor (11T) SRAM architecture that integrates TFETs and MOSFETs (Figure 29a) [102]. In this design, the read path is implemented using MOSFETs, and an additional write-assist transistor is included, both contributing to significantly reduced read and write delays compared to traditional SRAM architectures (Figure 29b,c). This translates to much lower dynamic power consumption. The use of TFETs in the storage nodes lowers the minimum operating voltage, which, in turn, enhances read/write speed advantages over MOSFET-based SRAMs at reduced voltages. While all-TFET SRAMs suffer from increased static power due to forward p-i-n leakage at higher V_{DD}, the 11T design mitigates this by using MOSFETs for write-access transistors, eliminating the forward p-i-n current and thereby lowering static power. This approach presents a viable route to low-power SRAM design.

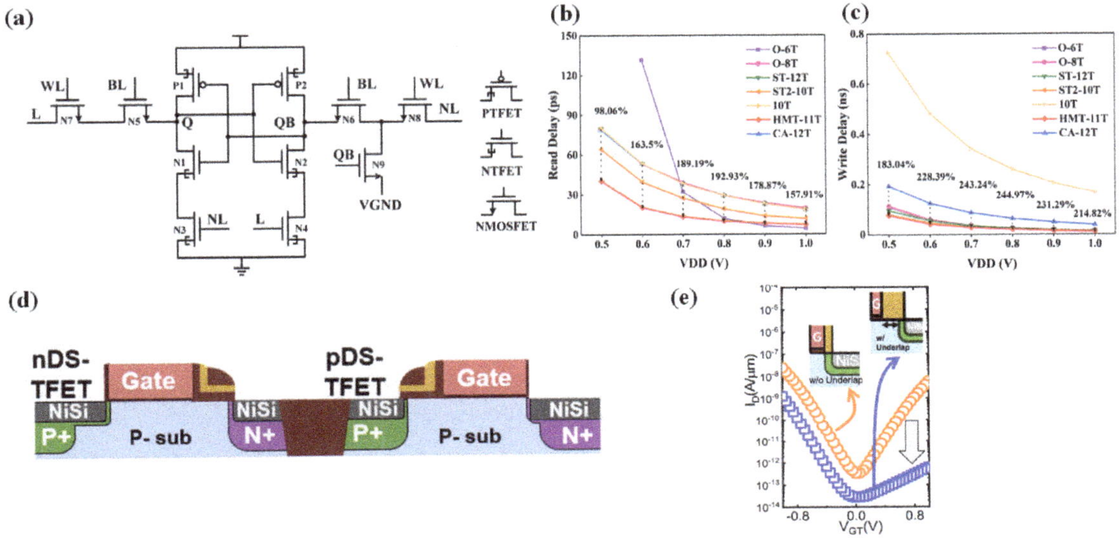

Figure 29. (**a**) Schematic of the HMT-11T SRAM cell. (**b**,**c**) Comparative read and write access latencies across SRAM cell architectures [102]. (**d**) Schematic structures of C-DS-TFETs. (**e**) Transfer curves of pDS-TFETs [103]. (**a**–**c**) are reproduced with permission from ref. [102]. (**d**,**e**) are reproduced with permission from ref. [103].

In another breakthrough, Kaifeng Wang et al. achieved monolithic integration of a dopant-segregated TFET (DS-TFET) (Figure 29d) into a 300 mm CMOS baseline process [103]. The core innovation lies in leveraging silicide-induced dopant segregation effects. And the standard NiSi process forms a sharp doping gradient at the Si interface (SIMS reveals a gradient > 5×10^{20} cm^{-3}/nm), improving junction abruptness by three orders of magnitude over conventional ion-implanted junctions. This effect, combined with a self-aligned gate architecture, enables high electric field tunneling at the gate edge, resulting in a 1000× increase in p-type TFET drive current. To suppress ambipolar leakage, a thick CMOS spacer was used as a hard mask to define a non-overlapped drain region, reducing the off-current to ~10^{-13} A (Figure 29e). This technology offers a practical path toward mass production of TFET-CMOS hybrid low-power circuits.

As previously discussed, 2D-TFETs exhibit considerable promise for non-volatile memory (NVM) applications. In 2024, Guangdi Feng et al. introduced a vertical 2D-TFET utilizing a ferroelectric layer, with a MoS_2/hBN/metal tunneling junction (Figure 30a) [104]. The ferroelectric polarization effectively modulates the Fermi level of MoS_2, thereby tuning the tunneling current across the MoS_2/metal interface. Inserting the hBN barrier not only suppresses the off-state current but also enhances the on-state current. This device achieved an exceptional on/off current ratio of 10^9, with an on-current exceeding 20 μA and an off-current as low as $10{-}^{14}$ A. Moreover, the write energy was reduced to 0.16 fJ, demonstrating outstanding non-volatile memory characteristics and ultra-high multi-level storage capacity with 22 distinct stable states (Figure 30b). This work overcomes conventional contact injection limitations and provides a novel design paradigm for low-power, memory-in-computer systems.

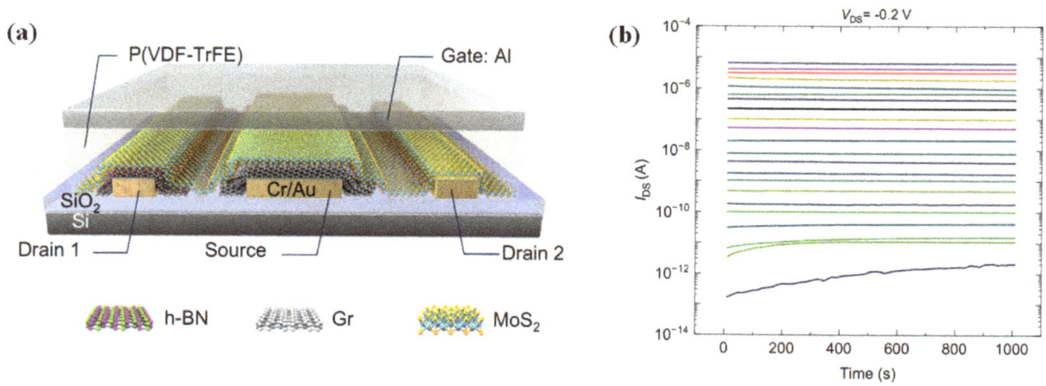

Figure 30. (**a**) Schematic of the MoS_2-FeFET. (**b**) Retention characteristic of the memory [104]. Reproduced with permission from ref. [104].

TFETs have extensive applications in sensors [105–115], with biosensors being a prominent example [105–109]. Compared to MOSFETs, TFETs offer better suitability for biosensing due to their shorter response time and lower leakage. In 2023, Iman Chahardah Cherik and Saeed Mohammadi designed a novel TFET-based biosensor with dual dopingless tunneling junctions (DMDS-TFET) [105]. The device functions similarly to DL-TFETs [11], as illustrated in Figure 31a. The sensor detects biomolecules by measuring changes in the dielectric constant within a cavity when biomolecules enter, thereby affecting device performance. Higher dielectric constants for biomolecules increase the probability of carrier tunneling and raise the I_{on}. They also analyzed the device's fabrication process (Figure 31b), which aligns with traditional CMOS technologies. Beyond biosensors, TFETs also find great potential in temperature sensors [112], pH sensors [113] and gas sensors [114,115] applications.

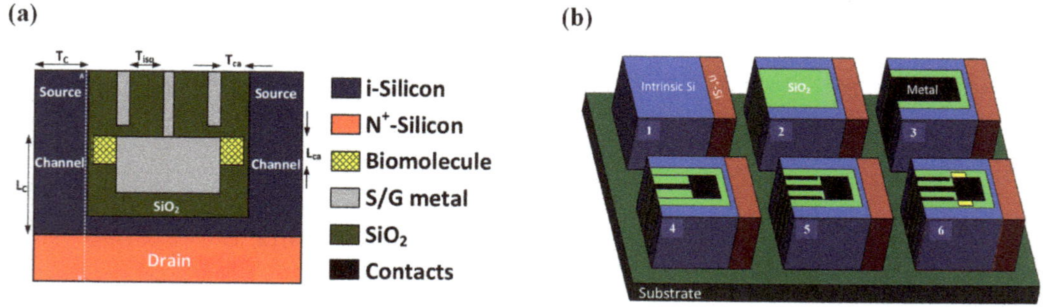

Figure 31. (**a**) A schematic cross-sectional view of the proposed DMDS-TFET biosensor structure. (**b**) Fabrication process steps (1–6) for realizing DMDS-TFET structure [105]. Reproduced with permission from ref. [105].

5. Conclusions

Experimental investigations have demonstrated that Si and Ge possess relatively wide bandgaps and high carrier effective masses, which limit the probability of band-to-band tunneling and consequently result in a low I_{on}. This makes it challenging to achieve high-performance switching at low supply voltages [116]. Moreover, the fabrication of TFETs based on these materials typically requires heavy doping concentrations, leading to increased manufacturing costs. To overcome these limitations, research efforts have shifted towards alternative material systems, including 2D materials, III-V compound semiconductors, and carbon nanotubes (CNTs), which represent promising directions for TFET development. Among these, 2D-material-based TFETs have emerged as the most active research branch due to their atomically flat surfaces free of dangling bonds and excellent gate electrostatic control. The experimental results suggest that the use of heterojunction architectures can significantly reduce *SS* and enhance the on/off current ratio. Furthermore, integrating wide bandgap materials such as hexagonal boron nitride (hBN) with FNT properties has enabled the development of TFET-based memory devices, offering great potential for non-volatile applications. In particular, these memory-capable 2D TFETs are considered promising building blocks for future in-memory computing circuits based on non-von Neumann architectures. Materials such as monolayer MoS_2, which is a direct-bandgap semiconductor, exhibit strong light absorption in the visible range, making them particularly suitable for optoelectronic and photodetection applications. On the fabrication side, significant progress has been made in the wafer-scale growth of single-crystal 2D semiconductors, laying a solid foundation for the future large-scale integration of 2D-TFETs [117]. III-V semiconductors, on the other hand, are compatible with current fabrication technologies and exhibit strong potential for TFET applications. However, further efforts are needed to address interface trap issues in order to reduce off-state leakage currents and improve the on/off current ratio. As for CNT-TFETs, advances in fabrication techniques and reliability design are essential to mitigate issues arising from impurities, oxidation, and associated degradation mechanisms [118].

Simulation-based studies primarily focus on optimizing TFET device architectures and selecting appropriate material systems. Techniques such as band engineering (QB-TFET) and alternative structural configurations like p-i-n-i TFETs can enhance tunneling rates. Dopingless (DL) or junctionless (JL) structures have also been proposed to eliminate the need for high doping concentrations, simplifying fabrication. With respect to the gate dielectric, the use of high-k dielectrics or negative capacitance effects can improve gate control and reduce SS. Nevertheless, it is critical to ensure the practical feasibility of proposed structures during simulation, as designs that significantly improve performance but are unrealizable in practice should be avoided.

In terms of practical applications, TFETs exhibit great promise across various domains. In the area of biosensing, most TFET-based applications remain at the simulation stage. When conducting such simulations, realistic considerations must be made—for instance, recessed structures should not be too narrow, as this may prevent biomolecules from entering the sensing region. Alternative device structures, such as ISFETs, may also be adopted to enhance practicality. At the circuit level, hybrid TFET-CMOS design has become one of the most actively explored directions. By leveraging the low subthreshold swing and low-power characteristics of TFETs while using MOSFETs to compensate for their shortcomings, hybrid designs—such as the 11T-SRAM architecture—have shown promising potential. Research by Kaifeng Wang et al. also provides valuable insights into process integration. However, compared to advanced FinFET/CMOS technologies, TFETs still face multiple challenges. For instance, circuit performance is highly sensitive to variations in TFET device parameters; process variability can significantly affect Ion, SS, delay, and power consumption, posing stringent requirements for circuit uniformity and yield control [119]. Additionally, in low-voltage design regimes, gate leakage through ultra-thin dielectrics and trap-assisted tunneling can lead to increased Ioff and degraded SS, thereby diminishing TFET effectiveness in ultra-low-power scenarios [120]. If these challenges can be adequately addressed, TFETs are poised for widespread adoption in the future.

In conclusion, the development of TFETs marks a transformative shift in addressing the physical limitations of traditional CMOS technology, particularly in achieving low power consumption, low operating voltage and better switching performance. Through extensive advancements in materials, including Si, Ge, III-V compounds and 2D materials, and structural innovations such as heterojunction and line tunneling, TFETs demonstrate the ability to surpass conventional limits of *SS* and power efficiency. The studies on novel TFET structures further exemplify the potential for performance enhancement of TFETs. Despite the challenges in uniform doping, interface stability, and process variability, ongoing research and simulation studies continue to optimize TFET performance and widen their application scope. TFETs hold substantial promise in areas such as digital circuits, sensors, and hybrid systems, paving the way for their role in next-generation electronic technologies. On the basis of the tremendous existing research results, coupled with continuous innovation in material science and fabrication processes, it is believed that, in the near future, TFETs will be essential for realizing their full potential in ultra-low-power, high-performance switching and sensing applications soon.

Funding: This research is financially supported by the National Natural Science Foundation of China (Grant No. U2241221) and the Opening Project of Science and Technology on Reliability Physics and Application Technology of Electronic Component Laboratory (ZHD202301).

Conflicts of Interest: The authors declare no conflict of interest.

References

1. Tajalli, A.; Leblebici, Y. Design trade-offs in ultra-low-power digital nanoscale cmos. *IEEE Trans. Circuits Syst. I Regul. Pap.* **2011**, *58*, 2189–2200. [CrossRef]
2. Franco, J.; Alian, A.; Vandooren, A.; Verhulst, A.S.; Linten, D.; Collaert, N.; Thean, A. Intrinsic robustness of tfet subthreshold swing to interface and oxide traps: A comparative pbti study of ingaas tfets and mosfets. *IEEE Electron Device Lett.* **2016**, *37*, 1055–1058. [CrossRef]
3. Núñez, J.; Avedillo, M.J. Comparison of tfets and cmos using optimal design points for power–speed tradeoffs. *IEEE Trans. Nanotechnol.* **2016**, *16*, 83–89.
4. Seabaugh, A.C.; Zhang, Q. Low-voltage tunnel transistors for beyond cmos logic. *Proc. IEEE* **2010**, *98*, 2095–2110. [CrossRef]
5. Lin, Y.M.; Appenzeller, J.; Avouris, P. Novel carbon nanotube fet design with tunable polarity. In Proceedings of the IEDM Technical Digest. IEEE International Electron Devices Meeting, 2004, San Francisco, CA, USA, 13–15 December 2004.
6. Choi, W.Y.; Park, B.G.; Lee, J.D.; Liu, T.J.K. Tunneling field-effect transistors (tfets) with subthreshold swing (ss) less than 60 mv/dec. *IEEE Electron Device Lett.* **2007**, *28*, 743–745. [CrossRef]
7. Krishnamohan, T.; Kim, D.; Raghunathan, S.; Saraswat, K. Double-Gate Strained-Ge Heterostructure Tunneling FET (TFET) With record high drive currents and 60 mV/dec subthreshold slope. In Proceedings of the 2008 IEEE International Electron Devices Meeting, San Francisco, CA, USA, 15–17 December 2008.
8. Cao, W.; Yao, C.J.; Jiao, G.F.; Huang, D.; Yu, H.Y.; Li, M.F. Improvement in reliability of tunneling field-effect transistor with p-n-i-n structure. *IEEE Trans. Electron Devices* **2011**, *58*, 2122–2126. [CrossRef]
9. Kim, S.W.; Choi, W.Y.; Sun, M.C.; Kim, H.W.; Park, B.G. Design guideline of si-based l-shaped tunneling field-effect transistors. *Jpn. J. Appl. Phys.* **2012**, *51*, 501–504.
10. Noguchi, M.; Kim, S.H.; Yokoyama, M.; Ji, S.M.; Ichikawa, O.; Osada, T. High ion/ioff and low subthreshold slope planar-type ingaas tunnel fets with zn-diffused source junctions. In Proceedings of the 2013 IEEE International Electron Devices Meeting, Washington, DC, USA, 9–11 December 2013.
11. Kumar, M.J.; Janardhanan, S. Doping-less tunnel field effect transistor: Design and investigation. *IEEE Trans. Electron Devices* **2013**, *60*, 3285–3290. [CrossRef]
12. Ghosh, B.; Akram, M.W. Junctionless tunnel field effect transistor. *IEEE Electron Device Lett.* **2013**, *34*, 584–586. [CrossRef]
13. Sarkar, D.; Xie, X.; Liu, W.; Cao, W.; Kang, J.; Gong, Y.; Kraemer, S.; Ajayan, P.M.; Banerjee, K. A subthermionic tunnel field-effect transistor with an atomically thin channel. *Nature* **2015**, *526*, 91–95. [CrossRef]
14. Chen, S.; Wang, S.; Liu, H.; Li, W.; Wang, Q.; Wang, X. Symmetric u-shaped gate tunnel field-effect transistor. *IEEE Trans. Electron Devices* **2017**, *64*, 1343–1349. [CrossRef]
15. Kim, S.; Myeong, G.; Shin, W.; Lim, H.; Kim, B.; Jin, T.; Chang, S.; Watanabe, K.; Taniguchi, T.; Cho, S. Thickness-controlled black phosphorus tunnel field-effect transistor for low-power switches. *Nat. Nanotechnol.* **2020**, *15*, 203–206. [CrossRef] [PubMed]
16. Convertino, C.; Zota, C.B.; Schmid, H.; Caimi, D.; Moselund, K.E. A hybrid iii–v tunnel fet and mosfet technology platform integrated on silicon. *Nat. Electron.* **2021**, *4*, 162–170. [CrossRef]

17. Chen, S.; Wang, S.; Liu, H.; Han, T.; Zhang, H. A high performance trench gate tunneling field effect transistor based on quasi-broken gap energy band alignment heterojunction. *Nanotechnology* **2022**, *33*, 225205. [CrossRef]
18. Cheng, W.; Liang, R.; Xu, G.; Yu, G.; Xu, J. Fabrication and characterization of a novel si line tunneling tfet with high drive current. *IEEE J. Electron Devices Soc.* **2020**, *8*, 336–340. [CrossRef]
19. Noguchi, M.; Kim, S.; Yokoyama, M.; Ichikawa, O.; Osada, T.; Hata, M.; Takenaka, M.; Takagi, S. High ion/ioff and low subthreshold slope planar-type ingaas tunnel fets with zn-diffused source junctions. *J. Appl. Phys.* **2015**, *118*, 045712. [CrossRef]
20. Singh, S.; Solay, L.R.; Anand, S.; Kumar, N.; Ranjan, R.; Singh, A. Implementation of Gate-All-Around Gate-Engineered Charge Plasma Nanowire FET-Based Common Source Amplifier. *Micromachines* **2023**, *14*, 1357. [CrossRef]
21. Vishnupriyan, J.; Chayadevi, S.K.; Megala, V.; Karpagam, R. Design and qualitative analysis of 5-nm nanowire tfet with spacer engineering. *J. Electron. Mater.* **2023**, *52*, 2094–2099. [CrossRef]
22. Dey, A.W.; Borg, B.M.; Ganjipour, B.; Ek, M.; Dick, K.A.; Lind, E.; Thelander, C.; Wernersson, L.E. High-current gasb/inas(sb) nanowire tunnel field-effect transistors. *IEEE Electron Device Lett.* **2013**, *34*, 211–213. [CrossRef]
23. Iutzi, R.M.; Fitzgerald, E.A. Microstructure and conductance-slope of inas/gasb tunnel diodes. *J. Appl. Phys.* **2014**, *115*, 2095–2096. [CrossRef]
24. Iutzi, R.M.; Fitzgerald, E.A. Defect and temperature dependence of tunneling in inas/gasb heterojunctions. *Appl. Phys. Lett.* **2015**, *107*, 133504. [CrossRef]
25. Iutzi, R.M.; Fitzgerald, E.A. Conductance slope and curvature coefficient of ingaas/gaassb heterojunctions at varying band alignments and its implication on digital and analog applications. *J. Appl. Phys.* **2015**, *118*, 235702. [CrossRef]
26. Zhang, X.B.; Ryou, J.H.; Dupuis, R.D.; Xu, C.; Mou, S.; Petschke, A.; Hsieh, K.C.; Chuang, S.L. Improved surface and structural properties of in as/ga sb superlattices on (001) gasb substrate by introducing an inassb layer at interfaces. *Appl. Phys. Lett.* **2007**, *90*, 1116.
27. Booker, G.R.; Klipstein, P.C.; Lakrimi, M.; Lyapin, S.; Walker, P.J. Growth of inas/gasb strained layer superlattices. I. *J. Cryst. Growth* **1994**, *145*, 778–785. [CrossRef]
28. Wang, S.; Wu, P.C.; Su, V.C.; Lai, Y.C.; Chen, M.K.; Kuo, H.Y.; Chen, B.H.; Chen, Y.H.; Huang, T.T.; Wang, J.H.; et al. A broadband achromatic metalens in the visible. *Nat. Nanotechnol.* **2018**, *13*, 227–232. [CrossRef] [PubMed]
29. Huang, Y.; Ryou, J.; Dupuis, R.D.; Petschke, A.; Mandl, M.; Chuang, S. Inas/gasb type-ii superlattice structures and photodiodes grown by metalorganic chemical vapor deposition. *Appl. Phys. Lett.* **2010**, *96*, 2545. [CrossRef]
30. Lackner, D.; Pitts, O.J.; Najmi, S.; Sandhu, P.; Kavanagh, K.L.; Yang, A.; Steger, M.; Thewalt, M.L.; Wang, Y.; McComb, D.W.; et al. Growth of inassb/inas mqws on gasb for mid-ir photodetector applications. *J. Cryst. Growth* **2009**, *311*, 3563–3567. [CrossRef]
31. Zhu, Y.; Jain, N.; Vijayaraghavan, S.; Mohata, D.K.; Datta, S.; Lubyshev, D.; Fastenau, J.M.; Liu, W.K.; Monsegue, N.; Hudait, M.K. Role of inas and gaas terminated heterointerfaces at source/channel on the mixed as-sb staggered gap tunnel field effect transistor structures grown by molecular beam epitaxy. *J. Appl. Phys.* **2012**, *112*, 2568. [CrossRef]
32. Zhu, Y.; Mohata, D.K.; Datta, S.; Hudait, M.K. Reliability studies on high-temperature operation of mixed as/sb staggered gap tunnel fet material and devices. *IEEE Trans. Device Mater. Reliab.* **2014**, *14*, 245–254. [CrossRef]
33. Xie, Q.; Van Nostrand, J.E.; Brown, J.L.; Stutz, C.E. Arsenic for antimony exchange on gasb, its impacts on surface morphology, and interface structure. *J. Appl. Phys.* **1999**, *86*, 329–337. [CrossRef]
34. Kaspi, R. Compositional abruptness at the inas-on-gasb interface: Optimizing growth by using the sb desorption signature. *J. Cryst. Growth* **1999**, *201–202*, 864–867. [CrossRef]
35. Liu, J.S.; Clavel, M.; Pandey, R.; Datta, S.; Hudait, M.K. Heterogeneous integration of inas/gasb tunnel diode structure on silicon using 200 nm gaassb dislocation filtering buffer. *AIP Adv.* **2018**, *8*, 105108. [CrossRef]
36. Zhao, X.; Vardi, A.; del Alamo, J.A. Sub-Thermal Subthreshold Characteristics in Top–Down InGaAs/InAs Heterojunction Vertical Nanowire Tunnel FETs. *IEEE Electron Device Lett.* **2017**, *38*, 855–858. [CrossRef]
37. Vasen, T.; Ramvall, P.; Afzalian, A.; Doornbos, G.; Holland, M.; Thelander, C.; Dick, K.A.; Wernersson, L.E.; Passlack, M. Vertical gate-all-around nanowire gasb-inas core-shell n-type tunnel fets. *Sci. Rep.* **2019**, *9*, 202. [CrossRef]
38. Solay, L.R.; Kumar, N.; Amin, S.I.; Kumar, P.; Anand, S. Design and performance analysis of gate-all-around negative capacitance dopingless nanowire tunnel field effect transistor. *Semicond. Sci. Technol.* **2022**, *37*, 115001. [CrossRef]
39. Arun, A.V.; Sreelekshmi, P.S.; Jacob, J. Implementation of an Efficient Charge Pump using Gate All Around Nanowire TFET for Energy Harvesting Applications. In Proceedings of the 2024 28th International Symposium on VLSI Design and Test (VDAT), Vellore, India, 1–3 September 2024.
40. Zhu, Z.; Persson, A.E.O.; Wernersson, L. Reconfigurable signal modulation in a ferroelectric tunnel field-effect transistor. *Nat. Commun.* **2023**, *14*, 2530. [CrossRef] [PubMed]
41. Kaniselvan, M.; Yoon, Y. Strain-tuning ptse$_2$ for high on-current lateral tunnel field-effect transistors. *Appl. Phys. Lett.* **2021**, *119*, 073102. [CrossRef]
42. Nakamura, K.; Nagamura, N.; Ueno, K.; Taniguchi, T.; Nagashio, K. All 2D heterostructure tunnel field effect transistors: Impact of band alignment and heterointerface quality. *ACS Appl. Mater. Interfaces* **2020**, *12*, 51598–51606. [CrossRef]

43. Iordanidou, K.; Mitra, R.; Shetty, N.; Lara-Avila, S.; Dash, S.; Kubatkin, S.; Wiktor, J. Electric field and strain tuning of 2D semiconductor van der waals heterostructures for tunnel field-effect transistors. *ACS Appl. Mater. Interfaces* **2022**, *15*, 1762–1771. [CrossRef]
44. Lemme, M.C.; Akinwande, D.; Huyghebaert, C.; Stampfer, C. 2D materials for future heterogeneous electronics. *Nat. Commun.* **2022**, *13*, 1392. [CrossRef]
45. Pande, G.; Siao, J.Y.; Chen, W.L.; Lee, C.J.; Sankar, R.; Chang, Y.M.; Chen, C.D.; Chang, W.H.; Chou, F.C.; Lin, M.T. Ultralow schottky barriers in hexagonal boron nitride-encapsulated monolayer WSe_2 tunnel field-effect transistors. *ACS Appl. Mater. Interfaces* **2020**, *12*, 18667–18673. [CrossRef]
46. Zhou, W.X.; Chen, K.Q. First-principles determination of ultralow thermal conductivity of monolayer wse2. *Sci. Rep.* **2015**, *5*, 15070. [CrossRef]
47. Roy, T.; Tosun, M.; Cao, X.; Fang, H.; Javey, A. Dual-gated MoS_2/WSe_2 van der waals tunnel diodes and transistors. *ACS Nano* **2015**, *9*, 2071–2079. [CrossRef]
48. Yan, R.; Fathipour, S.; Han, Y.; Song, B.; Xiao, S.; Li, M.; Ma, N.; Protasenko, V.; Muller, D.A.; Jena, D.; et al. Esaki diodes in van der waals heterojunctions with broken-gap energy band alignment. *Nano Lett.* **2015**, *15*, 5791–5798. [CrossRef]
49. Roy, T.; Tosun, M.; Hettick, M.; Ahn, G.H.; Hu, C.; Javey, A. 2D-2D tunneling field-effect transistors using $WSe_2/SnSe_2$ heterostructures. *Appl. Phys. Lett.* **2016**, *108*, 083111. [CrossRef]
50. Yan, X.; Liu, C.; Li, C.; Bao, W.; Zhou, P. Tunable $SnSe_2/WSe_2$ heterostructure tunneling field effect transistor. *Small* **2017**, *13*, 68. [CrossRef] [PubMed]
51. Jeon, H.B.; Shin, G.H.; Lee, K.J.; Choi, S.Y. Vertical-tunneling field-effect transistor based on WSe_2-MoS_2 heterostructure with ion gel dielectric. *Adv. Electron. Mater.* **2020**, *6*, 2000091. [CrossRef]
52. Shin, G.; Koo, B.; Park, H.; Woo, Y.; Lee, J.; Choi, S. Vertical-tunnel field-effect transistor based on a Silicon-Mos_2 three-dimensional-two-dimensional heterostructure. *ACS Appl. Mater. Interfaces* **2018**, *10*, 40212–40218. [CrossRef] [PubMed]
53. Miao, J.; Leblanc, C.; Wang, J.; Gu, Y.; Liu, X.; Song, B.; Zhang, H.; Krylyuk, S.; Hu, W.; Davydov, A.V.; et al. Heterojunction tunnel triodes based on two-dimensional metal selenide and three-dimensional silicon. *Nat. Electron.* **2022**, *5*, 744–751. [CrossRef]
54. Cheng, R.; Wang, F.; Yin, L.; Xu, K.; Ahmed Shifa, T.; Wen, Y.; Zhan, X.; Li, J.; Jiang, C.; Wang, Z.; et al. Multifunctional tunneling devices based on graphene/*h*-BN/$MoSe_2$ van der Waals heterostructures. *Appl. Phys. Lett.* **2017**, *110*, 173507. [CrossRef]
55. Cheng, R.; Wang, F.; Yin, L.; Wang, Z.; Wen, Y.; Shifa, T.A.; He, J. High-performance, multifunctional devices based on asymmetric van der Waals heterostructures. *Nat. Electron.* **2018**, *1*, 356–361. [CrossRef]
56. Wu, P.; Ameen, T.; Zhang, H.; Bendersky, L.A.; Ilatikhameneh, H.; Klimeck, G.; Rahman, R.; Davydov, A.V.; Appenzeller, J. Complementary black phosphorus tunneling field-effect transistors. *ACS Nano* **2019**, *13*, 377–385. [CrossRef]
57. Fukui, T.; Nishimura, T.; Miyata, Y.; Ueno, K.; Taniguchi, T.; Watanabe, K.; Nagashio, K. Single-Gate MoS_2 Tunnel FET with a Thickness-Modulated Homojunction. *ACS Appl. Mater. Interfaces* **2024**, *16*, 8993–9001. [CrossRef]
58. Tran, V.; Soklaski, R.; Liang, Y.; Yang, L. Layer-controlled band gap and anisotropic excitons in few-layer black phosphorus. *Phys. Rev. B* **2014**, *89*, 817–824. [CrossRef]
59. Ahmadchally, A.A.; Gholipour, M. Investigation of 6-armchair graphene nanoribbon tunnel fets. *J. Comput. Electron.* **2021**, *20*, 1114–1124. [CrossRef]
60. Appenzeller, J.; Lin, Y.M.; Knoch, J.; Avouris, P. Band-to-band tunneling in carbon nanotube field-effect transistors. *Phys. Rev. Lett.* **2004**, *93*, 196805. [CrossRef] [PubMed]
61. Pang, C.-S.; Han, S.-J.; Chen, Z. Steep slope carbon nanotube tunneling field-effect transistor. *Carbon* **2021**, *180*, 237–243. [CrossRef]
62. Zhou, C.; Kong, J.; Yenilmez, E.; Dai, H. Modulated Chemical Doping of Individual Carbon Nanotubes. *Science* **2000**, *290*, 1552–1555. [CrossRef]
63. Zhu, M.; Yin, H.; Cao, J.; Xu, L.; Lu, P.; Liu, Y.; Ding, L.; Fan, C.; Liu, H.; Zhang, Y.; et al. Inner Doping of Carbon Nanotubes with Perovskites for Ultralow Power Transistors. *Adv. Mater.* **2024**, *36*, 2403743. [CrossRef] [PubMed]
64. Zhang, H.; Chen, S.; Liu, H.; Wang, S.; Wang, D.; Fan, X.; Chong, C.; Yin, C.; Gao, T. Polarization Gradient Effect of Negative Capacitance LTFET. *Micromachines* **2022**, *13*, 344. [CrossRef] [PubMed]
65. Nam, H.; Cho, M.H.; Shin, C. Symmetric tunnel field-effect transistor (s-tfet). *Curr. Appl. Phys.* **2015**, *15*, 71–77. [CrossRef]
66. Li, W.; Liu, H.; Wang, S.; Chen, S.; Yang, Z. Design of high performance si/sige heterojunction tunneling fets with a t-shaped gate. *Nanoscale Res. Lett.* **2017**, *12*, 198. [CrossRef]
67. Chen, S.; Liu, H.; Wang, S.; Han, T.; Li, W.; Wang, X. A novel Ge based overlapping gate dopingless tunnel FET with high performance. *Jpn. J. Appl. Phys.* **2019**, *58*, 100902. [CrossRef]
68. Chen, S.; Wang, S.; Liu, H.; Han, T.; Chong, C. A novel dopingless fin-shaped sige channel tfet with improved performance. *Nanoscale Res. Lett.* **2020**, *15*, 202. [CrossRef] [PubMed]
69. Hussain, S.; Mustakim, N.; Hasan, M.; Saha, J.K. Performance enhancement of charge plasma-based junctionless TFET (JL-TFET) using stimulated n-pocket and heterogeneous gate dielectric. *Nanotechnology* **2021**, *32*, 335206. [CrossRef] [PubMed]
70. Sze, S.M. *Physics of Semiconductor Devices*, 3rd ed.; John Wiley & Sons: Hoboken, NJ, USA, 2006; p. 422.

71. Yu, T. Massachusetts Institute of Technology. InGaAs/GaAsSb Quantum-Well Tunnel-FETs for Ultra-Low Power Applications. Ph.D. Thesis, Massachusetts Institute of Technology, Cambridge, MA, USA, 2016.
72. Kim, S.H.; Agarwal, S.; Jacobson, Z.A.; Matheu, P.; Hu, C.; Liu, T.J.K. Tunnel field effect transistor with raised germanium source. *IEEE Electron Device Lett.* **2010**, *31*, 1107–1109. [CrossRef]
73. Kim, S.W.; Kim, J.H.; Liu, T.J.K.; Choi, W.Y.; Park, B.G. Demonstration of l-shaped tunnel field-effect transistors. *IEEE Trans. Electron Devices* **2015**, *63*, 1774–1778. [CrossRef]
74. Chander, S.; Sinha, S.K.; Chaudhary, R.; Singh, A. Ge-source based l-shaped tunnel field effect transistor for low power switching application. *Silicon* **2022**, *14*, 7435–7448. [CrossRef]
75. Kobayashi, M.; Jang, K.; Ueyama, N.; Hiramoto, T. Negative capacitance for boosting tunnel fet performance. *IEEE Trans. Nanotechnol.* **2017**, *16*, 253–258. [CrossRef]
76. Bscke, T.S.; Miiller, J.; Brauhaus, D.; Schroder, U.; Bottger, U. Ferroelectricity in hafnium oxide thin films. *Appl. Phys. Lett.* **2011**, *99*, 5397. [CrossRef]
77. Wang, W.; Wang, P.F.; Zhang, C.M.; Lin, X. Design of u-shape channel tunnel fets with sige source regions. *IEEE Trans. Electron Devices* **2013**, *61*, 193–197. [CrossRef]
78. Wang, W.; Zang, S.G.; Lin, X.; Liu, X.Y.; Zhang, D.W. A novel recessed-channel tu nneling fet design with boosted drive current and suppressed leakage current. In Proceedings of the 2012 IEEE 11th International Conference on Solid-State and Integrated Circuit Technology, Xi'an, China, 29 October–1 November 2012.
79. Li, W.; Liu, H.; Wang, S.; Chen, S. Reduced miller capacitance in u-shaped channel tunneling fet by introducing heterogeneous gate dielectric. *IEEE Electron Device Lett.* **2017**, *38*, 403–406. [CrossRef]
80. Krenik, W.R.; Izzi, L.J. Output Buffer Circuits with Controlled Miller Effect Capacitance. United. States Patent US 5,274,284, 28 December 1993.
81. Israelsohn, J. Miller on edge: The role of miller capacitance in nonlinear circuits. EDN 2007. Available online: https://www.edn.com/miller-on-edge-the-role-of-miller-capacitance-in-nonlinear-circuits/ (accessed on 25 June 2025).
82. Jiang, C.; Wu, D.D. Asymmetric S/D Structure to Improve Transistor Performance by Reducing Miller Capacitance. U.S. Patent 5,925,914, 20 July 1999.
83. Mookerjea, S.; Krishnan, R.; Datta, S.; Narayanan, V. On enhanced miller capacitance effect in interband tunnel transistors. *IEEE Electron Device Lett.* **2009**, *30*, 1102–1104. [CrossRef]
84. Mookerjea, S.; Krishnan, R.; Datta, S.; Narayanan, V. Effective capacitance and drive current for tunnel fet (tfet) cv/i estimation. *IEEE Trans. Electron Devices* **2009**, *56*, 2092–2098. [CrossRef]
85. Yang, Y.; Tong, X.; Yang, L.T.; Guo, P.F.; Fan, L.; Yeo, Y.C. Tunneling field-effect transistor: Capacitance components and modeling. *IEEE Electron Device Lett.* **2010**, *31*, 752–754. [CrossRef]
86. Elmasry, M.I. Capacitance calculations in mosfet vlsi. *Electron Device Lett. IEEE* **1982**, *3*, 6–7. [CrossRef]
87. Upasana Narang, R.; Gupta, M.; Saxena, M. Simulation study for dual material gate hetero-dielectric tfet: Static performance analysis for analog applications. In Proceedings of the 2013 Annual IEEE India Conference (INDICON), Mumbai, India, 13–15 December 2013.
88. Goswami, R.; Bhowmick, B. Hetero-gate-dielectric gate-drain underlap nanoscale tfet with a δp^+ Si1−xGex layer at source-channel tunnel junction. In Proceedings of the 2014 International Conference on Green Computing Communication and Electrical Engineering (ICGCCEE), Coimbatore, India, 6–8 March 2014.
89. Paul, D.J.; Abdullah-Al-Kaiser, M.; Islam, M.S.; Khosru, Q.D.M. Fringing-field-based 2-D analytical model for a gate-underlap double-gate TFET. *J. Comput. Electron.* **2018**, *17*, 1567–1577. [CrossRef]
90. Ankur, B.; Nandakishor, Y.; Prasad, S.A.; Kumar, V.S. Analog/rf characteristics of a 3D-cyl underlap gaa-tfet based on a ge source using fringing-field engineering for low-power applications. *J. Comput. Electron.* **2018**, *17*, 1650–1657.
91. Singh, G.; Amin, S.I.; Anand, S.; Sarin, R.K. Design of si0.5ge0.5 based tunnel field effect transistor and its performance evaluation. *Superlattices Microstruct.* **2016**, *92*, 143–156. [CrossRef]
92. Damrongplasit, N.; Kim, S.H.; Liu, T.J.K. Study of random dopant fluctuation induced variability in the raised-ge-source TFET. *Electron Device Lett.* **2013**, *34*, 184–186. [CrossRef]
93. Damrongplasit, N.; Kim, S.H.; Shin, C.; Liu, T.J. Impact of gate line-edge roughness (ler) versus random dopant fluctuations (rdf) on germanium-source tunnel fet performance. *IEEE Trans. Nanotechnol.* **2013**, *12*, 1061–1067. [CrossRef]
94. Damrongplasit, N.; Shin, C.; Kim, S.H.; Vega, R.A.; King Liu, T.J. Study of random dopant fluctuation effects in germanium-source tunnel fets. *IEEE Trans. Electron Devices* **2011**, *58*, 3541–3548. [CrossRef]
95. Anand, S.; Sarin, R.K. Hetero-material gate doping-less tunnel fet and its misalignment effects on analog/rf parameters. *J. Electron. Mater.* **2018**, *47*, 2988–2996. [CrossRef]
96. Yadav, D.S.; Verma, A.; Sharma, D.; Tirkey, S.; Raad, B.R. Comparative investigation of novel hetero gate dielectric and drain engineered charge plasma tfet for improved dc and rf performance. *Superlattices Microstruct.* **2017**, *111*, 123–133. [CrossRef]

97. Sharma, S.; Basu, R.; Kaur, B. Temperature Analysis of a Dopingless TFET Considering Interface Trap Charges for Enhanced Reliability. *IEEE Trans. Electron Devices* **2022**, *69*, 2692–2697. [CrossRef]
98. Aghandeh, H.; Ziabari, S.A. Gate engineered heterostructure junctionless tfet with gaussian doping profile for ambipolar suppression and electrical performance improvement. *Superlattices Microstruct.* **2017**, *111*, 103–114. [CrossRef]
99. Singh, D.; Pandey, S.; Nigam, K.; Sharma, D.; Yadav, D.S.; Kondekar, P. A charge-plasma-based dielectric-modulated junctionless tfet for biosensor label-free detection. *IEEE Trans. Electron Devices* **2016**, *64*, 271–278. [CrossRef]
100. Gupta, S.; Nigam, K.; Pandey, S.; Sharma, D.; Kondekar, P.N. Effect of interface trap charges on performance variation of heterogeneous gate dielectric junctionless-tfet. *IEEE Trans. Electron Devices* **2017**, *64*, 4731–4737. [CrossRef]
101. Pal, A.; Chai, Z.; Jiang, J.; Cao, W.; Davies, M.; De, V.; Banerjee, K. An ultra energy-efficient hardware platform for neuromorphic computing enabled by 2D-tmd tunnel-fets. *Nat. Commun.* **2024**, *15*, 3392. [CrossRef] [PubMed]
102. Lu, W.; Wang, C.; Hu, W.; Dai, C.; Peng, C.; Lin, Z.; Wu, X. Hybrid MOSFET-TFET 11T SRAM cell with high write speed and free half-selected disturbance. *Microelectron. J.* **2025**, *156*, 106498. [CrossRef]
103. Wang, K.; Huang, Q.; Wu, Y.; Ren, Y.; Wei, R.; Wang, Z.; Yang, L.; Zhang, F.; Geng, K.; Li, Y.; et al. A Novel Energy-Efficient Salicide-Enhanced Tunnel Device Technology Based on 300mm Foundry Platform Towards AIoT Applications. *arXiv* **2024**, arXiv:2410.12390.
104. Feng, G.; Liu, Y.; Zhu, Q.; Feng, Z.; Luo, S.; Qin, C.; Chen, L.; Xu, Y.; Wang, H.; Zubair, M.; et al. Giant tunnel electroresistance through a Van der Waals junction by external ferroelectric polarization. *Nat. Commun.* **2024**, *15*, 9701. [CrossRef]
105. Cherik, I.C.; Mohammadi, S. Impact of trap-related non-idealities on the performance of a novel tfet-based biosensor with dual doping-less tunneling junction. *Sci. Rep.* **2023**, *13*, 11495. [CrossRef]
106. Krsihna, B.V.; Chowdary, G.A.; Ravi, S.; Reddy, K.V.; Kavitha, K.R.; Panigrahy, A.K.; Prakash, M.D. Tunnel field effect transistor design and analysis for biosensing applications. *Silicon* **2022**, *14*, 10893–10899. [CrossRef]
107. Anam, A.; Anand, S.; Amin, S.I. Design and Performance Analysis of Tunnel Field Effect Transistor With Buried Strained Si1−xGex Source Structure Based Biosensor for Sensitivity Enhancement. *IEEE Sens. J.* **2020**, *20*, 13178–13185. [CrossRef]
108. Dwivedi, P.; Singh, R.; Sengar, B.S.; Kumar, A.; Garg, V. A New Simulation Approach of Transient Response to Enhance the Selectivity and Sensitivity in Tunneling Field Effect Transistor-Based Biosensor. *IEEE Sens. J.* **2021**, *21*, 3201–3209. [CrossRef]
109. Kumar, P.; Sharma, S.K.; BalwinderRaj. Comparative analysis of nanowire tunnel field effect transistor for biosensor applications. *Silicon* **2021**, *13*, 4067–4074. [CrossRef]
110. Gayduchenko, I.; Xu, S.G.; Alymov, G.; Moskotin, M.; Bandurin, D.A. Tunnel field-effect transistors for sensitive terahertz detection. *Nat. Commun.* **2021**, *12*, 543. [CrossRef]
111. Singh, P.; Raman, A.; Yadav, D.S.; Kumar, N.; Dixit, A.; Ansari, H.R. Ultra thin finger-like source region-based tfet: Temperature sensor. *IEEE Sens. Lett.* **2024**, *8*, 2501104. [CrossRef]
112. Nigam, K.; Pandey, S.; Kondekar, P.N.; Sharma, D. Temperature sensitivity analysis of polarity controlled electrically doped hetero-tfet. In Proceedings of the 2016 12th Conference on Ph. D. Research in Microelectronics and Electronics (PRIME) 2016, Lisbon, Portugal, 27–30 June 2016.
113. Yousf, N.; Anam, A.; Rasool, Z.; Amin, S.I. Ultralow-Power DST-TFET pH Sensor Exceeding the Nernst Limit with Influence of Temperature on Sensitivity. *ACS Appl. Bio Mater.* **2024**, *7*, 4562–4572. [CrossRef]
114. Singh, S.; Verma, A.; Singh, J.; Wadhwa, G. Investigation of n + SiGe gate stacked VTFET based on dopingless charge plasma for gas sensing application. *Silicon* **2022**, *14*, 6205–6218. [CrossRef]
115. Ghosh, S.; Rajan, L.; Varghese, A. Junctionfree Gate Stacked Vertical TFET Hydrogen Sensor at Room Temperature. *IEEE Trans. Nanotechnol.* **2022**, *21*, 655–662. [CrossRef]
116. Esseni, D.; Pala, M.; Palestri, P.; Alper, C.; Rollo, T. A review of selected topics in physics based modeling for tunnel field-effect transistors. *Semicond. Sci. Technol.* **2017**, *32*, 083005. [CrossRef]
117. Yin, L.; Cheng, R.; Ding, J.; Jiang, J.; Hou, Y.; Feng, X.; Wen, Y.; He, J. Two-Dimensional Semiconductors and Transistors for Future Integrated Circuits. *ACS Nano* **2024**, *18*, 7739–7768. [CrossRef]
118. Noyce, S.G.; Doherty, J.L.; Cheng, Z.; Han, H.; Bowen, S.; Franklin, A.D. Electronic Stability of Carbon Nanotube Transistors Under Long-Term Bias Stress. *Nano Lett.* **2019**, *19*, 1460–1466. [CrossRef]
119. Rendón, M.; Cao, C.; Landázuri, K.; Garzón, E.; Prócel, L.M.; Taco, R. Performance Benchmarking of TFET and FinFET Digital Circuits from a Synthesis-Based Perspective. *Electronics* **2022**, *11*, 632. [CrossRef]
120. Majumdar, K.; Bhat, N.; Majhi, P.; Jammy, R. Effects of Parasitics and Interface Traps on Ballistic Nanowire FET in the Ultimate Quantum Capacitance Limit. *IEEE Trans. Electron Devices* **2010**, *57*, 2264–2273. [CrossRef]

MDPI AG
Grosspeteranlage 5
4052 Basel
Switzerland
Tel.: +41 61 683 77 34

Micromachines Editorial Office
E-mail: micromachines@mdpi.com
www.mdpi.com/journal/micromachines

www.ingramcontent.com/pod-product-compliance
Lightning Source LLC
Chambersburg PA
CBHW040826250926
51850CB00007BA/1270
* 9 7 8 3 7 2 5 8 8 0 8 1 2 *